Tables and figures useful for reference

D0301781

Semiconductor and dopant properties

Properties of silicon dioxide

Photolithographic properties

Properties of thin-film materials

Properties of gases

Ion implantation

Typical fabrication sequences

Troubleshooting guides

MICROELECTRONIC PROCESSING
An Introduction to the Manufacture of Integrated Circuits

McGraw-Hill Series in Electrical Engineering

Consulting Editor

Stephen W. Director, *Carnegie-Mellon University*

CIRCUITS AND SYSTEMS
COMMUNICATIONS AND SIGNAL PROCESSING
CONTROL THEORY
ELECTRONICS AND ELECTRONIC CIRCUITS
POWER AND ENERGY
ELECTROMAGNETICS
COMPUTER ENGINEERING
INTRODUCTORY
RADAR AND ANTENNAS
VLSI

Previous Consulting Editors

Ronald N. Bracewell, Colin Cherry, James F. Gibbons, Willis W. Harman, Hubert Heffner, Edward W. Herold, John G. Linvill, Simon Ramo, Ronald A. Rohrer, Anthony E. Siegman, Charles Susskind, Frederick E. Terman, John G. Truxal, Ernst Weber, and John R. Whinnery

VLSI

Consulting Editor
Stephen W. Director, *Carnegie-Mellon University*

Elliott: *Microlithography Process Technology for IC Fabrication*
Offen: *VLSI Image Processing*
Ruska: *Microelectronic Processing: An Introduction to the Manufacture of Integrated Circuits*
Sze: *VLSI Technology*
Tsividis: *Operation and Modeling of the MOS Transistor*
Walsh: *Choosing and Using CMOS*

MICROELECTRONIC PROCESSING

An Introduction to the Manufacture of Integrated Circuits

W. Scot Ruska, Ph.D.

Intel Corporation
Livermore, California

McGraw-Hill Book Company

New York St. Louis San Francisco Auckland Bogotá Hamburg
London Madrid Mexico Milan Montreal New Delhi
Panama Paris São Paulo Singapore Sydney Tokyo Toronto

To My Wife
Paula

Many women have done excellently,
but you surpass them all.
Proverbs 31:29

This book was set in Times Roman by Kachina Typesetting, Inc.
The editor was Sanjeev Rao;
the cover was designed by Infield + D'Astolfo Associates;
cover photo by Dan McCoy, Rainbow;
the production supervisors were Leroy A. Young and Fred Schulte.
Project supervision was done by Cobb/Dunlop Publisher Services, Incorporated.
R. R. Donnelley & Sons Company was printer and binder.

MICROELECTRONIC PROCESSING
An Introduction to the Manufacture of Integrated Circuits

2 3 4 5 6 7 8 9 0   DOCDOC   8 9 4 3 2 1 0 9 8 7

ISBN 0-07-054280-5

Library of Congress Cataloging-in-Publication Data

Ruska, Walter E. W.
  Microelectronic processing.

  (McGraw-Hill series in electrical engineering.
Computer engineering)
    Includes index.
    1. Integrated circuits—Design and construction.
  I. Title. II. Series.
  TK7874.R87 1987   621.381'73   86–15362
  ISBN 0–07–054280–5 (text)
  ISBN 0–07–054281–3 (solutions manual)

CONTENTS

ABOUT THE AUTHOR

W. SCOT RUSKA has ten years' practical experience in the field of microelectronics manufacture, both in engineering and management. He received his bachelor's degree, majoring in chemistry, from Rice University and his master's degree and Ph.D. in chemistry from the University of California at Berkeley. As an engineer, his assignments have included diffusion, photolithography, and thin-film deposition responsibilities. He has also held managerial positions, both in engineering and manufacturing, with major integrated circuit companies. His experience includes bipolar and advanced MOS technologies as well as thin-film recording devices. He is a member of several professional societies, including Sigma Xi. Currently, Dr. Ruska is a production manager with Intel Corporation.

PREFACE

Thirty years ago, the transistor was a laboratory curiosity. As late as the mid-1960s integrated circuits contained only handfuls of transistors and were intended mainly for exotic applications in defense and aerospace. And yet today the transistor and its microelectronic offspring are commonplace in everyday life, appearing in applications as innovative as the personal computer and as traditional as the alarm clock. Without doubt, no new technology has ever had a more profound effect more quickly on a society.

The explosive growth of microelectronics has had a number of consequences. One of them has been the emergence of a new engineering specialist, skilled in the fabrication of microelectronic components. This is the microelectronic process engineer. As the microelectronics industry has matured, processing has emerged from the shadow of design and assumed a position as an independent and important area of study. As a result, microelectronic processing is increasingly finding a place in the academic curriculum.

The purpose of this book is to give a comprehensive, fundamentally sound, and practical introduction to microelectronic processing. The aim is to prepare the reader to enter a laboratory or fabrication area ready to become a competent and informed contributor. To accomplish that aim, a book must have several characteristics.

First, a processing text should be organized around processing methods rather than device applications. Therefore, the first three major sections of this text present the three major types of fabrication methods: diffusion, lithography, and thin-film deposition. A fourth section demonstrates how these methods are integrated into processes for the construction of various microelectronic devices. Information on device physics and device design is presented when necessary to understand the purpose or limitations of various fabrication strategies. This is the reverse of the approach used in many texts, which are written more from a design viewpoint.

Second, an introductory text must confront the interdisciplinary nature of processing. Processing professionals are drawn from both the chemical and electrical engineering communities as well as other scientific fields, and an introductory text must take account of their varying backgrounds. Consequently, this book requires a minimum of specialized prerequisites. I assume only that knowledge common to the science-engineering community: mathematics through calculus, basic mechanics and electromagnetism, and some exposure to quantum mechanics. Introductory chemistry would be quite helpful, but is not required. A background in electronic device physics is specifically not assumed.

From this assumption, it follows that many readers will already be familiar with some of the subject matter of this book. Specifically, the reader with prior exposure to semiconductor device behavior may want to treat the coverage of semiconductor physics in Sections 2-1 through 2-4, and the presentation of bipolar and MOS transistors in Sections 10-1 and 11-1, as review. Students with a chemical or materials background will be well acquainted with chemical concentrations, molecular weight, and acid-base reactions as reviewed in Chapter 6, with reaction kinetics as presented in Chapter 8, and perhaps with the crystallographic information of Section 2-5. Conversely, the reader new to any of these subjects may well wish for a fuller presentation of them. I have attempted to strike a reasonable balance, informing the novice while not overly boring the veteran. Instructors may wish to fine-tune their coverage of these sections to fit the background of their classes.

A third obligation assumed by any technical textbook is to deal soundly with fundamentals: the underlying scientific principles that make processing techniques practical, desirable, or necessary. Processing, in particular, has been a field prone to undue empiricism, probably because the processing frontier was first settled by self-taught pioneers emigrating from other disciplines. With the growing availability of academic courses in the field, I have tried to combat this empiricist tendency by especially emphasizing, for each fabrication technique, the pertinent physical and chemical principles. In general, the first section of each chapter is given over to a presentation of principles that explicitly links a processing method with the central body of physical and chemical knowledge.

Fourth, a processing text should also recognize that the processing professional exists in order to produce working microelectronic devices. This puts an emphasis on practical results, requiring a knowledge not only of principles, but also of equipment, rules-of-thumb, tricks of the trade, and even of slang and jargon. Such material seems especially fitting in a book written by an industrial rather than an academic practitioner. Therefore, most chapters contain at least one section on equipment, and practical details and terminology are introduced whenever possible.

A few additional points seem worthy of mention. I have given some attention to the problem set included with each chapter. By the time a student reaches this book, he has presumably won his mathematical spurs by surviving whatever course constitutes the traditional computational ordeal in his discipline. Therefore, I have tried to avoid algebraic complexity in problems, instead designing them to increase the student's understanding. Some are meant to emphasize underlying principles. Others illustrate relationships between phenomena or relative magnitudes of effects. Still others give practice in day-to-day computations required in practical processing. Where appropriate, problems are used to supplement the principal text with additional processing details.

A word is also in order on notation and units. Defining notation in a interdisciplinary field presents complications. Each basic scientific discipline has a traditional
notation that is reasonably unambiguous. However, in treating matter from different
disciplines, conflicts in notation quickly appear. Perhaps the worst example is the
versatile k, which is commonly used for Boltzmann's constant, for rate constants, the
wave number, a momentum vector, and frequently for other constants and integers.
The author has a choice of abandoning well-established usages, thus complicating the
"translation" of equations between this book and other works, or of accepting multiple
uses of the same characters. I have, in general, opted for the latter, with some
modifications (for example, I have cut k down to only five applications). Within each
chapter, notation is usually unique, and I believe all usages are clear in context. I have
also provided a comprehensive list of notation as an aid in preventing confusion. As to
units, SI units are used except where practice overwhelmingly mandates some other
usage. For example, the counting of particles in air is an application internationally
dominated by U.S. government specifications, which are expressed in counts per cubic
foot. The semi-disreputable Ångstrom is suppressed except in Chapter 6, where it is
the unquestioned unit of choice in dealing with etching.

One final comment concerns timeliness. Processing is a young, but maturing,
body of knowledge. Both microelectronics in general and the processing field in
particular are expanding and changing rapidly. Fabrication equipment still becomes
obsolescent on about a three-year cycle—about the gestation period for a book! This
complicates the task of publishing information before it is outdated. On the other hand,
while capabilities continually improve, the field is well enough established that basic
methods appear fixed. Although two or three new generations of devices may appear
each decade, they will probably all be built using the same basic methods: diffusion
and ion implantation, photolithography and etching, deposition of thin films. The
relative importance of techniques may change, but at this point the emergence of
entirely revolutionary new methods is unlikely. To reflect this reality, I have empha-
sized principles, which change slowly, over the latest details of application.

Before ending, I would like to acknowledge the assistance of many in the
preparation of this book. I am deeply indebted to my wife, Paula, for her patience and
support; to my colleagues Bill Young and particularly Josh Walden for reviewing
material and offering thoughtful suggestions; to others too numerous to name who
provided material and advice; and to Mr. San Rao, my editor at McGraw-Hill, for his
guidance and encouragement. In addition, I should express my gratitude to Intel
Corporation: to Intel management for permitting this project and providing an environment in which it could succeed; to Laura McHolm of the Legal Department for her
helpfulness, and especially to Jim Bradley, who reviewed the entire text on behalf of
Intel as well as providing invaluable advice.

In summary, this text attempts to offer a reasonably complete introduction to the
science and art of fabricating microelectronic devices. The reader who enters into this
field has ahead of him a fascinating experience of scientific discovery, practical
achievement, economic challenge, and above all change. Surely there can be no more
exciting endeavor with which to occupy the closing years of the twentieth century.

W. Scot Ruska

NOTATION

a	Lattice parameter	Crystals
a	Percent of wafer area free of area defects	Yields
A	Constant in oxidation law [Eq. (3-58)]	Oxidation
A	A (cross-sectional) area	General
A	Area of one die	Yields
A	Cross-sectional area of junction	Device physics
A	Alignment tolerance	Alignment
B	Constant in oxidation law [Eq. (3-58)]	Oxidation
B	Magnetic field	General
BV	Breakdown voltage (often followed by subscripts denoting applied voltages)	Device physics
c	Speed of light	General
c	A constant [Eq. (4-4)]	Lithography
c	Dopant gradient in linearly graded junction	Device physics
C	Capacitance per unit area	Device physics
C_o	Oxide capacitance per unit area	Device physics
C_{inv}	Capacitance per unit area, inversion	Device physics
d	Wafer diameter	Yield, etc.
d	Source-to-substrate distance	Evaporation
d	Density (grams per cubic centimeter)	General
d_R	Rayleigh depth (measurement of focus)	Optics
D	A dopant (chemical equations only)	Diffusion
$D\ (D')$	A distance along the optical axis, in object (image) space	Optics
D	Diffusion coefficient	Diffusion
D_{EB}	Diffusivity of carriers, emitter-base junction	Device physics
D	Defects density (defects per square centimeter)	Yield
DY	Die yield	Yield
$(DY)_o$	Maximum possible die yield	Yield

E	Energy	General
E_A	Activation energy	
E_c (E_v)	Energy of edge of conduction (valence) band	Device physics
E_F	Energy of Fermi level	
E_{Fn} (E_{Fp})	Fermi energy for n (p)-doped side	Device physics
E_G	Energy of gap	Device physics
E_i	Intrinsic Fermi energy	Device physics
E_0	Energy required for complete development	Lithography
E_1	A particular energy	
$\mathscr{E}$	Electric field	General
$\mathscr{E}_{crit}$	Critical electric field for breakdown	Device physics
$f(E)$	Probability of a site being occupied	Device physics
F	Force	General
F_{acc}	Accelerating force	General
F_r	Restraining force	General
F	Flow	Vacuum systems
FY	Final yield	Yield
f	Focal length	Optics
g	Acceleration of gravity	General
g	Gap (in proximity printing)	Optics
g_0	Conductance	MOS devices
g_m	Transconductance	MOS devices
G	Generation rate	Plasma
Gr	Grashof number	Fluid dynamics
h	An integer (with k and l)	General
h	"Henry's constant" term	Oxidation
h	Planck's constant	General
$\hbar$	Planck's constant divided by 2π	General
h_{FB}	Common-base gain	Bipolar device
h_{FE}	Common-emitter gain	Bipolar device
i	An integer	General
i	Either 1 or 2, in Eq. (10-5)	Bipolar device
I	Current	Electricity
I_E (I_B, I_C)	Current into emitter (base, collector) terminal	Bipolar device
I (I')	Light intensity in object (image) space	Optics
I_0	Light intensity at aperture center	Optics
I_{max} (I_{min})	Maximum (minimum) intensity	Optics
j	Flux	Diffusion
j_d	Flux due to drift	Diffusion
j_{total}	Total flux	Diffusion
j	An integer, with i	General
$\mathscr{J}$	Current density	Electricity
k	An integer (with j and l)	General
k	Boltzmann's constant	General
k	Reaction rate constant	Chemistry
k'	A particular reaction rate constant	Chemistry
k_s	Surface reaction rate constant	Chemistry

k	Segregation coefficient	Crystal growth
k	Momentum vector	Device physics
K	Factor dependent on equipment, etc.	Various
K	Chemical equilibrium constant	Chemistry
l	An integer (with j and k)	General
l	A length	General
L	Avogadro's number (molecules per mole)	General
L	Latent heat of fusion	Crystal growth
L	Channel length	MOS Devices
LY	Line yield	Yield
m	Mass	General
m_0	Electron rest mass	General
m^*	Effective mass	General
m	Magnification	Optics
$m_x\ (m_\alpha)$	Linear (angular) magnification	Optics
M	Molecular weight	Chemistry
M	Modulation	Optics
MFP	Mean free path	Vacuum
MTF	Modulation transfer function	Optics
n	Number of moles	Chemistry
$n\ (n')$	Refractive index (in image space)	Optics
n	Electron number density	Device physics
n_F	Frenkel defect density	Crystals
n_i	Intrinsic carrier density	Device physics
n_S	Schottky defects density	Crystals
$n_1,\ n_2$	Number of diffusing species	Diffusion
$n_+\ (n_-)$	Number of species diffusing right (left)	Diffusion
$N,\ N'$	Number density of vacancies	Crystals
N_0	Number density of reactant at surface	CVD
$N(x)$	Number density of oxidant at x	Oxidation
$N_A\ (N_D)$	Number density of acceptor (donor)	Device physics
$N_c\ (N_v)$	Number of states at conductance (valence) band edge	Device physics
$N_E\ (N_B)$	Dopant number density in emitter (base)	Bipolar device
N_G	Number density of reactant in gas bulk	Oxidation, CVD
$N_l\ (N_s)$	Number density of dopant in liquid (solid)	Crystal growth
N_1	Dopant number density, lightly doped side of junction	Device physics
N	Stepper magnification in "$N:1$"	Optics
N.A.	Numerical aperture ($2r_0/d$)	Optics
p	Number density of holes	Device physics
$P\ (P')$	A point in space (image space)	Optics, etc.
P	Percent solids in photoresist	Lithography
P	Pitch	Lithography
P	Power	Device physics
P	Pressure	General
q	Charge on the electron	General
Q	Charge per unit area	Device physics
Q_{inv}	Charge per unit area due to inversion	Device physics
Q_{ss}	Charge per unit area due to surface states	Device physics

r (r_0)	Radius of aperture, slit, etc	Optics
r	Etch, deposition of reaction rate	Various
r_{evap}	Evaporation rate (grams per second)	Evaporation
R	Radius	General
R	Gas constant	Chemistry
R	Resistance	Electricity
R_{sat}	Resistance in saturation	Bipolar device
R_p	Range	Ion implant
ΔR_p $(\Delta R_\perp)$	Straggle (transverse straggle)	Ion implant
Re	Reynolds number [Eq. (8-2)]	Fluid dynamics
s	Thickness of insulator in capacitor	Device physics
s	Spacing between probes	Ion implant
s	Element along a line integral	Various
S	Degree of coherence	Optics
S, S'	Paths	Etching
S	Pumping speed	Vacuum systems
S	Scaling factor	MOS devices
t	Time	General
t_r	Residence time	Vacuum, plasma
t_1	A particular time	General
T	Temperature	General
u	A dummy variable	General
v	Velocity	General
v_d	Drift velocity	Various
V	Voltage	General
V'	Effective voltage	Device physics
V_{app}	Applied voltage	Device physics
V_b	Bias voltage	Device physics
V_B	Voltage applied to semiconductor bulk	MOS device
V_D	Drain voltage	MOS device
V_{EB}, etc.	With two of E, B, and C; voltage between emitter, base, collector terminals	Bipolar device
V_f	Forward voltage	Device physics
V_{fb}	Flat-band voltage	MOS device
V_G	Gate voltage	MOS device
V_0	Voltage at zero ion velocity	Ion implant
V_{offset}	Offset or "turn-on" voltage	Bipolar device
V_S	Source voltage	MOS device
V	Volume	General
w	Width, especially of depletion region	General
w_{EB}	Width of emitter-base depletion region	General
W	Work function of metal	Device physics
x	Distance, x-coordinate, or a variable	General
x'	A variable	General
x (x')	Distance perpendicular to optical axis in object (image) space	Optics
x^*	A particular perpendicular distance	Optics

x_A (x_D)	Width of acceptor-(donor-)doped portion of depletion region	Device physics
x_d	Width of depletion region	Device physics
$x_{d,max}$	Maximum width of depletion region	Device physics
x_i	Initial oxide thickness [Eq. (3-58d)]	Oxidation
x_j	Junction depth	General
x_L (x_S)	Width of line (space)	Lithography
x_o	Thickness of oxide film	Oxidation, etc.
X	Proportion of melt solidified	Crystal growth
X	Property of material [Eq. (10-11)]	Bipolar device
y	Distance, y-coordinate	General
Y	Property of material [Eq. (10-11)]	Bipolar device
z	Distance, z coordinate	General
z	Resist film thickness	General
z	Wafer thickness	Various
z_0	Original resist film thickness	Lithography
z'	A particular film thickness	Various
Z	Electrical charge	General
Z_1, Z_2	Atomic numbers (nuclear charge)	Various
Z	Property of material [Eq. (10-11)]	Bipolar device
Z	Channel width	MOS device
α	Parameter in equation for plane (with β, γ)	Crystals
α	Parameter for restoring force, related to collision frequency [Eq. (3-20)]	Diffusion
α	Absorbance	Lithography
α	Common-base gain	Bipolar device
α, α'	Angles	General
β	Parameter in equation for plane (with α, γ)	Crystals
β	Thermal expansion coefficient (gas)	Fluid dynamics
β	Common emitter gain	Bipolar device
γ	Parameter in equation for plane (with α, β)	Crystals
γ	Number of oxidant molecules required to form a unit volume of oxide	Oxidation
γ	Contrast of a photoresist	Lithography
ϵ	Permittivity	Device physics
ϵ_0	Permittivity of free space	Device physics
ϵ_{ox}	Permittivity of silicon dioxide	Device physics
ϵ_S	Permittivity of semiconductor	Device physics
θ	An angle	General
θ_r (θ_1, θ_2)	Angles of reflection, incidence, and refraction	Optics
κ	Thermal conductivity	General
λ	Wavelength of light	General
λ	A length [Eq. (3-1)]	Diffusion
λ	Boundary layer thickness	Fluid dynamics

μ	Mobility	Device physics
μ_n	Electron mobility	Device physics
μ_p	Hole mobility	Device physics
μ	Fluid viscosity	Fluid dynamics
μm	Symbol for one micron (10^{-6} m)	General
ν	Frequency	General
ν_{int}	Interstitial "jump" frequency	Diffusion
ν_{subst}	Substitutional "jump" frequency	Diffusion
ν_0	Crystal or atomic vibration frequency	General
ρ	Charge distribution	Device physics
ρ	Resistivity	General
ρ_s	Sheet resistivity	General
τ	A time, usually a specific time	Various
τ	Carrier recombination lifetime	Bipolar device
τ	Time delay for switching, etc.	MOS device
ϕ	Electric potential	General
ϕ_{bi}	Built-in potential of a junction	Device physics
ϕ_F	Fermi potential	Device physics
Φ_{MS}	Metal-silicon work function	Device physics
χ	Electron affinity	Device physics
ψ (ψ')	Angles	General
ω	Rotational speed (RPM)	General
Ω	Solid angle	Various
Ω	Symbol for ohms	General

A PERSPECTIVE ON MICROELECTRONIC PROCESSING

This book is an introduction to microelectronic processing, which is the art and science of fabricating microelectronic devices. The processing professional is an expert in accomplishing precise changes in the small-scale properties of matter, so as to create microelectronic devices visualized by a designer. The first nine chapters of this book are devoted to individual techniques for modifying materials; the last three give examples of applications to various devices.

Processing techniques are traditionally divided into the three categories of diffusion, patterning, and thin-film deposition. A typical processing facility will have a department devoted to each category, and each is given a segment in this book. In Chapters 2 and 3, we will study semiconductor properties and the use of diffusion methods to modify them. In Chapters 4–6, we look at the creation of very small patterns by photolithography and etching. Chapters 7–9 present methods for depositing thin films of various materials as well as ion implantation. In Chapters 10–12, we turn to the application of these techniques to the construction of bipolar and MOS silicon integrated circuits, and to nonsilicon devices.

In this chapter, we introduce the subject of microelectronic processing and put it in perspective. In Section 1-1 we define the field more precisely and present the criteria used in selecting subject material for this text. Since processing constitutes only one phase in the manufacture of electronic devices, the next three sections give a temporal perspective on processing. In Section 1-2 we describe how silicon substrates are manufactured from raw materials, in Section 1-3 we give an overview of how processing techniques are combined in device manufacture, and in Section 1-4 we describe the test and packaging steps that follow completion of processing.

The last two sections of this chapter look at the economic and physical features of the processing environment. In Section 1-5 we look at *yield,* the ratio of functioning devices to substrates processed. This number is a key determinant of economic viability for a device and of professional accomplishment for the process engineer. In Section 1-6 we discuss the physical environment required for successful processing.

1-1 MICROELECTRONIC PROCESSING: A DEFINITION

Microelectronic processing is a field of endeavor easier to recognize than to define precisely. Certainly, the core of the subject is the fabrication of state-of-the-art silicon integrated circuits. But the nature of "mainstream" application changes with time: from germanium and silicon discrete transistors in the past to perhaps gallium arsenide in the future. And processing professionals are also involved in many alternate applications, from semiconductor lasers to bubble memories, all of which are valid parts of the field. In addition, attention must be given to possible new methods currently in the research phase. Therefore, it seems inappropriate to define processing solely in terms of the silicon integrated circuit.

However, the bundle of techniques used in processing have little logical connection beyond their efficiency in making small-scale changes in matter. This makes difficult a definition in terms of the methods themselves. So in the end an operational definition must be used: microelectronic processing is what microelectronic processing professionals do.

The result of this definition is some subjectivity in choice of subject matter, especially in an introductory text of limited length. I have made those decisions as follows. First, I have limited myself to that portion of the total manufacturing sequence normally performed by processing specialists: from the first step performed on the bare substrate to the last step prior to cutting the substrate into dice. Within these boundaries, I have given in-depth coverage to the principal techniques in current widespread commercial use in the production of silicon integrated circuits. For each method, the underlying scientific principles and the equipment required for successful application are also described. I have mentioned, frequently in less detail, those additional methods that appear likely to have significant commercial application in the next five years, either due to the introduction of new methods or a shift in the nature of the "mainstream." With these choices, the vast majority of processing professionals will find coverage in this text of most of his or her daily endeavors.

1-2 PREREQUISITES FOR PROCESSING: SUBSTRATE MANUFACTURE

Microelectronic processing is a matter of surfaces. Processing techniques are concerned with modifying properties no more than a few microns below or above the surface of a working material. The flat, macroscopic object on which processing takes place is called the *substrate*. In semiconductor processing, modification of substrate chemistry is an important operation; in other applications, the substrate may serve mainly as a mechanical support with the device constructed solely above the surface.

With semiconductors, especially silicon, the substrate is a single-crystal slice cut from a larger crystal and is called a *wafer*. Since the silicon wafer is by far the most common substrate, we will use it as an example in this section. In the rest of the book, we will use the term *substrate* rather than *wafer* to preserve generality.

In the early years of microelectronics, processing professionals were frequently concerned with the production of wafers, since high-quality commercial sources were not available. Now, however, finished, high-quality wafers are supplied to microelectronics producers by raw material vendors who are either specialists in semiconductor crystals or more broadly based chemical suppliers. A similar evolution from laboratory to commercial supply has occurred with other microelectronic substrates as technologies have matured. Thus we will not consider substrate production in the depth we devote to processing techniques, despite some of the very interesting science involved in crystal growth. However, we do want to acquaint the processing professional with some consequences of the growth process that can affect the behavior of substrates.

Figure 1-1 shows the sequence of events by which raw materials are converted into single-crystal silicon wafers. The electronic properties of semiconductors are sensitive to impurities in amounts undetectable by normal chemical tests; therefore, the first three steps are devoted to obtaining very pure silicon for crystal growth. Raw materials containing silicon (in the form of quartzite) and carbon (coal, coke, wood chips) are first combined in a submerged-electrode arc furnace. In this furnace, the raw materials are loaded into a crucible containing an electrode. The electrode arcs to the crucible, melting the materials and becoming submerged in the melt. As material sinks toward the bottom of the crucible, various reactions take place, producing silicon carbide (SiC) to react with the silicon dioxide (SiO_2) comprising the quartzite. These compounds then react to give gaseous SiO and CO according to

$$SiC + SiO_2 \rightarrow Si + CO + SiO \tag{1-1}$$

The gases escape, and the liquid silicon sinks to form a layer at the bottom of the crucible. This is drawn off through a vent to form metallurgical silicon, about 98% pure.

The metallurgical silicon is pulverized and reacted with hydrogen chloride in a fluidized bed at about 300°C, forming a liquid chloride of silicon:

$$Si + 3HCl \rightarrow SiHCl_3 + H_2 \tag{1-2}$$

Various impurities also form chlorides, and the material is purified by fractional distillation of the chlorides. The silicon chloride is then used to make *electronic grade silicon (EGS)* by chemical vapor deposition. Chemical vapor deposition, which we shall study in depth in Chapter 8, is the reaction of vapors or gases to produce a solid. In this case, the reaction is

$$SiHCl_3 + H_2 \rightarrow Si + 3HCl \tag{1-3}$$

The resulting silicon, in the form of polycrystalline rods, is "100%" pure by traditional chemical methods. Further analysis must be electrical, after single-crystal growth.[1]

The key step in making silicon wafers for microelectronic use is the growth of a large single crystal. The method used is the *Czochralski* or "CZ" method, named for

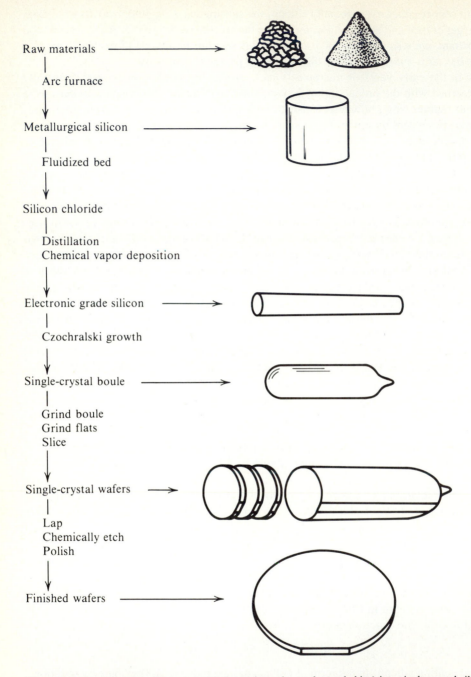

Raw materials

    Arc furnace

Metallurgical silicon

    Fluidized bed

Silicon chloride

    Distillation
    Chemical vapor deposition

Electronic grade silicon

    Czochralski growth

Single-crystal boule

    Grind boule
    Grind flats
    Slice

Single-crystal wafers

    Lap
    Chemically etch
    Polish

Finished wafers

Figure 1-1 The conversion of raw materials (quartzite, coke, coal, wood chips) into single-crystal silicon wafers ready for processing into microelectronic devices.

the inventor. The process begins with a large crucible into which electronic-grade silicon rods are broken and melted. A *seed* crystal is then brought into contact with the melt. Some of the melt solidifies onto the seed crystal, forming a larger crystal. The seed crystal is slowly "pulled" upward as the melt continues to deposit. Eventually, most of the melt has condensed and a very large bologna-shaped *boule* of single-crystal silicon, up to 6 in in diameter and several feet long, has been formed. Figure 1-2 shows the concepts of Czochralski crystal growth.

The electrical properties of silicon crystals depend both on chemical purity and on crystal quality. We shall discuss in Chapter 2 how impurities affect the electrical resistivity of the crystal and the way in which *crystal defects* disturb the order of the crystal. For good results, both of these factors must be well controlled during the crystal growth process.

One source of impurities is from the crystal grower. The crucible material is of quartz (SiO_2) and introduces some amount of oxygen into the silicon. This impurity can affect both electrical properties and defect distribution, and the results can be beneficial or adverse. Use of vacuum or an inert gas like argon as the crystal-growing ambient prevents additional oxygen contamination from the atmosphere. Carbon is

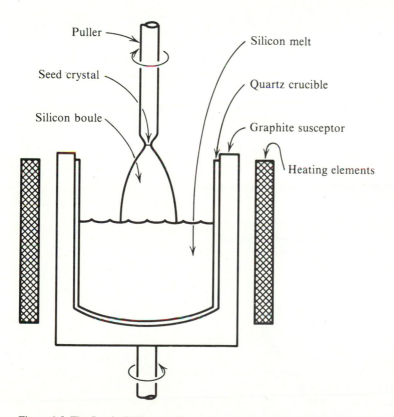

Figure 1-2 The Czochralski method for growing single crystals. A seed crystal is immersed in a melt, and a large bologna-shaped boule is pulled from the melt.

another common impurity, transmitted from graphite components such as the susceptor used to support the crucible; it is electrically inactive, but can influence defect formation.

Additional "impurities" are intentionally added during growth. These are *dopants:* electrically active elements added to lower the resistivity of the crystal and to make it either "*n*-type" (electron-rich) or "*p*-type" (electron-poor). We discuss the role of dopants at length in Chapter 2. Common dopants are boron (*p*-type) and phosphorus, arsenic, or antimony (*n*-type). The type and resistivity of the material are key properties that the vendor must guarantee to the buyer, so the distribution of these dopants must be well controlled, even though exceedingly small concentrations (parts per 10^9 or much less) are required. Dopants are added to the melt in the form of already doped pieces of polysilicon. Rotation of the crucible and the crystal helps assure uniform mixing so that the radial concentration gradient of dopants is small.

However, there is an unavoidable *axial* distribution gradient of dopants, related to the *segregation coefficient* of the dopant. Dopants—and other impurities—have a different equilibrium concentration N_s in the solid from the equilibrium concentration N_l in the liquid. Their ratio k is the segregation coefficient, i.e.,

$$k = \frac{N_s}{N_l} \tag{1-4}$$

Table 1-1 gives the segregation coefficients of a number of impurities. Since these values are generally much less than one, dopants tend to accumulate in the melt as the crystal grows. As a result, the "butt" end of the boule, which is grown last, will be considerably enriched in impurities. The impurity distribution as a function of fraction of melt consumed, X, is

$$N_s = kN_{l0}(1 - X)^{k-1} \tag{1-5}$$

Table 1-1 Segregation coefficients for common impurities in silicon

Impurity		Segregation coefficient k†
Aluminum	(Al)	0.002
Antimony	(Sb)	0.3
Arsenic	(As)	0.3
Boron	(B)	0.72–0.8
Carbon	(C)	0.07
Copper	(Cu)	4×10^{-4}
Gallium	(Ga)	0.007–0.008
Gold	(Au)	2.2×10^{-5}
Indium	(In)	4×10^{-4}
Iron	(Fe)	8×10^{-6}
Oxygen	(O)	1.25
Phosphorus	(P)	0.35

†Where multiple values are given, literature values vary.

where N_{l0} is the initial concentration in the melt.[2] (If most of the melt is consumed, X will approximate the percentage of boule length towards the butt end.) As a result, different portions of the boule will have different dopant concentrations and resistivities, and may be sold under different specifications.

An important determinant of crystal quality and homogeneity is *pull speed*, the speed with which the seed end of the boule is withdrawn from the melt. The maximum pull rate is set by the need to transfer heat fast enough to allow continued crystallization at the growing end. This limit can be found to be

$$v = \frac{\kappa}{Ld} \frac{dT}{dx} \tag{1-6}$$

where v is the pull rate in centimeters per second, κ is the thermal conductivity of the crystal, L is the latent heat of fusion, d the density of the solid, and dT/dx the temperature gradient along the crystal. In practice, pull rates rarely reach 50% of the theoretical maximum and vary inversely with crystal diameter.[3]

Higher pull rates, besides being more economical, have advantages for crystal quality. Crystal defects can agglomerate and grow in a hot crystal: faster cooling, which is associated with faster pulling, reduces the growth time. Also, a slowly pulled crystal is prone to remelting at the interface, which leads to *striations* in doping density. However, the higher the pull rate, the harder it is to maintain single-crystal growth at a uniform diameter, especially for large crystal diameters.

The challenge of the crystal grower is thus to achieve satisfactory pull rate and crystal quality with the largest possible diameter. Large diameters are important because the finished wafers are of the diameter of the boule, and so the larger the boule, the more circuits can be fabricated on each wafer. Each increment in crystal diameter leads to reduced costs and marks another milestone in the economic history of microelectronics. Starting with subinch diameters, crystal growers have achieved 2-, 3-, and 4-in standards; now 5- and 6-in wafers are widely available and 8-in boules have been displayed.

After the boule has been pulled, it is checked for defects by selective etching and for dopant concentration by resistivity measurement (methods we shall see in Chapters 2 and 3). Defective parts of the boule are discarded. The remainder is then ground on a lathelike machine, which converts the somewhat irregular boule into a cylinder. One or more flats are ground along the cylinder to aid in determining crystal planes for later use, and the boule is then sliced into round wafers, which are then lapped to give a smooth surface. Since sawing and lapping does crystal damage to the wafer surface, wafers are chemically etched to remove the surface layer and then given a final polishing step to produce a mirror finish on the working surface of the wafer. The wafers are then packaged and delivered to a microelectronic fabrication facility, where the true processing activity begins.

There are microelectronic applications using substrates other than Czochralski-grown silicon. Silicon solar cells and some other specialized applications use a method known as float-zoning. Some methods for materials other than silicon are mentioned in Chapter 12. Further information on substrate production can be found in Refs. 2–5.

1-3 MICROELECTRONIC PROCESSING: THE SEQUENTIAL APPLICATION OF TECHNIQUES

On the surface of the wafer or substrate, the processing professional creates a microelectronic device by the successive application of various fabrication techniques. As an illustration of how these methods are combined, we will look at a processing sequence for creation of a simple *diode*. A diode is a rectifying device (it passes electric current in one direction only) composed of a layer of *n*-type silicon in contact with a layer of *p*-type silicon. We will need to create this layer structure and provide the diode with electrical connections through which current can flow.

Figure 1-3 illustrates the sequence of operations we will use. The left-hand column of illustrations shows a cross-section of the process, showing the effect of operations below or above the substrate surface. Such a view would be possible if the substrate were sliced apart through the device. The right-hand column shows a top view, looking down on the substrate as one would with a microscope. While the top view corresponds both to what can actually be seen during manufacture, and to what is drawn by the designer, the cross-section reveals more about what is actually happening to the substrate.

We will begin with a *p*-type silicon wafer or substrate, doped with boron during crystal growth. The substrate forms one of our two layers, and thus we need only to form a *n*-type layer next to it. We can form this layer by creating an electron-rich zone in the silicon, which in turn is done by *diffusing* an electron-donating impurity or dopant, such as phosphorus, into the substrate. Sufficient phosphorus can overpower the boron doping in the substrate and convert the diffused region from *p*-type to *n*-type.

However, to simplify electrical connection later, we want both *n*- and *p*-type layers present at the surface, so we will need to perform selective diffusion of phosphorus into just a part of the surface. We thus proceed as follows. First, we expose the substrate to an oxidant (oxygen or steam) in a hot furnace. This grows a layer of *silicon dioxide* on the wafer. We will study oxidation methods in Chapter 3. Then, using a photographic method covered in Chapters 4, 5, and 6, we selectively etch a pattern of openings in the oxide. We now have a substrate with a patterned oxide film atop it.

The substrate is now put in a hot furnace containing a phosphorus-rich ambient (Chapter 3). The phosphorus diffuses through the holes in the oxide, doping the underlying silicon. However, in the regions where the oxide is intact, no doping takes place. The oxide pattern is now converted into a pattern of *p*- and *n*-doped silicon near the surface. At the bottom of each *n*-doped region, a diode has been formed.

Electrical connection of the diodes to form useful devices requires additional operations. We again oxidize the substrate, growing an oxide all over the wafer, and use the patterning process to form a new pattern of holes in the oxide. We now use a thin-film deposition method such as evaporation (Chapter 7) to cover the whole wafer with a layer of metallic aluminum. The aluminum touches the silicon where holes were cut in the oxide, making electrical contact. The oxide is an insulator and prevents contact elsewhere.

We again call upon a patterning operation, removing most of the aluminum and forming the remainder into a pattern of *leads* and *pads*. The thin leads carry electricity between pads and the contact openings, one each for the *p*- and the *n*-type regions. The big square pads are used to attach wires of macroscopic size, which allow the diode to be connected to the outside, human-sized world. As a result, we have an electrical path from the pad to the *p*-type region, through the *pn* boundary we formed by diffusion and thence from the *n*-region to the other pad. This electrical path through a *pn* junction or boundary is a diode.

This process sequence is simplified, but serves to illustrate several points. One is that processing involves a sequence of different kinds of operations, with the same kind of operation (such as oxidation or patterning) appearing repeatedly in the sequence. Another is that patterning steps alternate with deposition or diffusion steps, so that each deposition or diffusion is given a characteristic pattern. Another is the way that successive layers (here a diffusion, an oxidation, and a metal layer) are laid down in sequence to form the device. In Chapter 3, we will begin studying each process technique in isolation, not returning to the integration of process sequences until Chapter 10. So, in the meantime, it is important to keep in mind how the techniques interact to produce a finished device.

We note that, in our example, we created a multitude of diodes on the single substrate. It is this ability to create simultaneously a large number of identical devices that gives microelectronics its economic leverage. After processing is complete, the substrate can be cut up and each of these elemental devices can be separately packaged and sold. It is this final part of the manufacturing process we consider next.

1-4 SUBSEQUENT TO PROCESSING: TESTING AND PACKAGING

Microelectronic processing produces on each substrate a large number of identical devices. Usually, these are arranged in an array of rectangular elements called *dice* (the singular is *die*), with edge dimensions measured in millimeters. To obtain useful, salable devices, the substrate must be sliced up into individual dice or chips, and each die must be put into a package sturdy enough to protect it and large enough to make contact with the macroscopic world. Along the way, devices are tested to ensure functionality. This is the province of the test and assembly operation, which is usually separate from the processing organization. Figure 1-4 presents the steps performed during assembly.

The manufacture of microelectronic devices is extremely demanding. Therefore, of the many devices on each substrate, some will not be functional. To avoid the expense of packaging these useless dice, the wafer is *tested* or *sorted* immediately after processing is completed. Each device is subjected to a test sequence, which can be quite elaborate and contain hundreds of thousands of tests. In a typical sequence, the device is first checked for shorts or opens between each pair of pads. This identifies gross failures and prevents damage of the tester during subsequent tests. Devices are then tested in various ways for functionality and finally for compliance with all

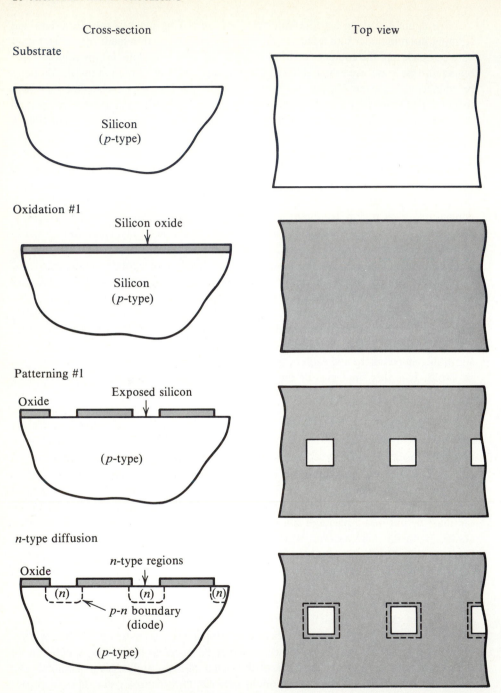

Cross-section   Top view

Substrate

Silicon
(*p*-type)

Oxidation #1

Silicon oxide

Silicon
(*p*-type)

Patterning #1

Oxide   Exposed silicon

(*p*-type)

n-type diffusion

Oxide   *n*-type regions

(*n*)   (*n*)   (*n*)

p-n boundary
(diode)

(*p*-type)

Figure 1-3 A process sequence for the fabrication of a set of diodes. The left-hand sketch in each set represents a cross-section through the substrate; the right-hand sketch illustrates a top view onto the substrate surface. Note the alternation of deposition and patterning steps, the repeated appearance of one type of operation (e.g., patterning) in the sequence, and the simultaneous manufacture of a large number of identical devices.

Cross-section   Top view

Oxidation #2

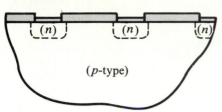

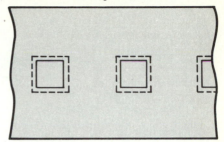

Patterning #2

Exposed *n*-type Si   Exposed *p*-type Si

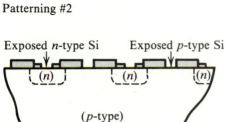

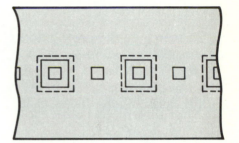

Aluminum deposition

Aluminum film

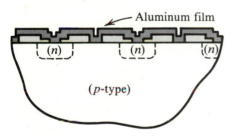

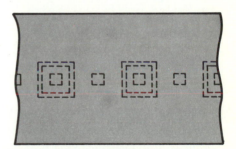

Patterning #3

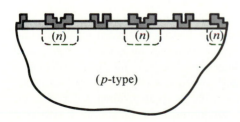

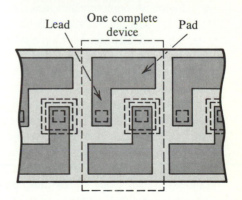

Lead   One complete device   Pad

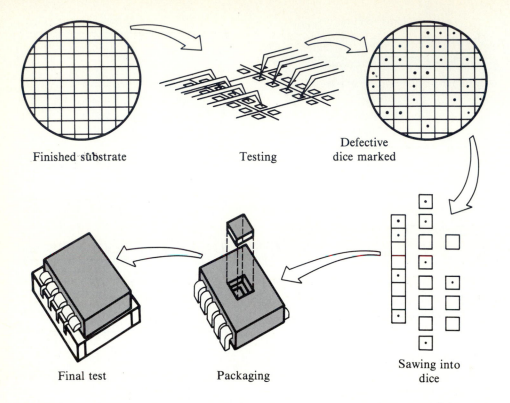

Finished substrate Testing Defective dice marked

Sawing into dice

Final test Packaging

Figure 1-4 Steps following completion of the process sequence: testing to identify nonfunctional devices, dicing of the substrate and discarding of bad devices, packaging, and final testing.

parameters of the product data sheet. To perform this sequence quickly and accurately, testing is automated and computerized. A set of needlelike *probes,* positioned on a *probe card* to match the device, is automatically positioned in turn over the pads on each device. A computerized tester exercises the test program for each die. Devices that fail any test are marked as defective by squirting a drop of ink onto the die. The tester automatically compiles a record of the yield (number of good dice) on each wafer and notes which tests were failed by each defective device.

The wafer is now scribed and broken or sawed into individual dice. The ink-marked defective dice are discarded, and the good devices forwarded for packaging.

The individual die or chip is small (some millimeters on a side), fragile, and vulnerable; it is far too tiny for convenient handling. The pads for electrical connection, though "large" by microelectronic standards (many tens of microns square) are far too small to receive wiring of macroscopic size. The chip is also vulnerable to physical damage, contamination, or corrosion by the environment. In operation, it will generate significant heat, which must be carried away. The role of the package is to provide mechanical support, electrical connection, protection, and heat removal for the chip.

Figure 1-5 shows the three principal kinds of packages: the dual in-line package (DIP), the chip carrier, and the pin array. A dotted line shows the position of the chip within the package. The three alternatives involve different ways of arranging the electrical connections on the outside of the package. The DIP is the oldest and most common package, but the higher density of interconnection suits the chip carrier and pin array for more demanding applications. Each of these types can be realized in several different materials. *Ceramics* afford hermetic sealing and great durability at a high cost. *Plastic* is inexpensive, but excludes moisture less efficiently. Various *refractory* technologies offer a compromise in both cost and hermeticity. The processing professional is rarely involved in packaging concerns, with one exception: moisture resistance. Cheaper, nonhermetic plastic packaging requires higher moisture tolerance and corrosion resistance, which often necessitates alterations in the process sequence.

Figure 1-6 shows a cross-section of a generalized package, showing how three principal operations are performed. The first is *die bonding:* the bonding of the die to the package. This can be done by metallizing the back of the die and then making a

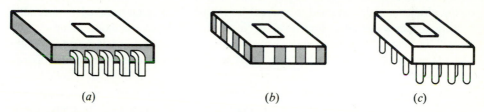

(a) (b) (c)

Figure 1-5 Different kinds of packages for integrated circuits. (*a*) The dual in-line package (DIP): electrical contact is made through two sets of leads running along the long sides of the package. (*b*) Chip carrier: contact is made to contacts along all sides of the package. (*c*) Pin array: contact is made to an array of pins on the bottom of the package.

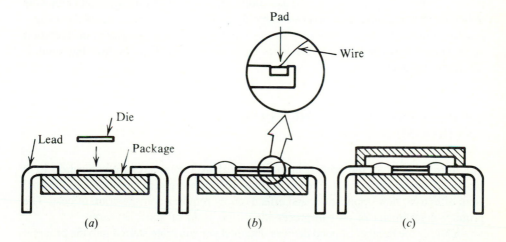

(a) (b) (c)

Figure 1-6 A cross-section of a generalized package, showing operations required: (*a*) bonding of the die to the package, (*b*) bonding of wires to connect pads on the die to leads in the package, and (*c*) encapsulation and sealing of the package.

solder bond to a metallized area of the package, or by using polymer adhesives. It is often necessary to have a conductive path to the body of the die, so that any polymer used must be made conductive. A second operation is *wire bonding* by running very small wires between each pad and the associated package *lead*. The bond between the wires and the package leads can be thermocompressive or ultrasonic. A variation is *tape automated bonding,* in which the leads, along with small copper fingers that can contact the pads, are formed into a continuous metallic tape. The tape is fed into an automated bonder, which bonds the fingers to the pads, and is later cut up to form the leads for individual packages. Tape automated bonding requires fabrication of specialized "bumps" on each pad.

After die and wire bonding, the package is closed and sealed to exclude moisture, dust, and contaminants. In plastic packaging, the package may be molded around the die and leads; ceramic and refractory packaging require sealing together of a top and bottom part. The completed package must meet several criteria. Environmental moisture can react with environmental chlorine, or with phosphorus present in CVD layers in the chip, to corrode chip metallization. Either moisture must be excluded, or the chip materials designed to prevent corrosion. Packaging materials, especially ceramics, can generate a low level of α particles. Alpha radiation causes "soft errors" in microelectronic devices, so it must be excluded or designed around. The electrical properties of the leads must also be compatible with device operation.

The packaging methods we have discussed are peculiar to the silicon integrated circuit. However, because they are all small and fragile, almost all microelectronic devices undergo some type of packaging or assembly after fabrication. Some examples are discussed further in Chapter 12. Additional information on packaging can be found in Refs. 6 and 7.

After assembly, devices are subjected to final testing. This testing serves to identify devices damaged during the assembly operation. In addition, certain elements of device performance, such as speed, can only be measured in the completed package. For some devices, final testing may be preceded by a *burn-in*—a period of operation under stress—designed to cause failure of weak devices and thus improve the reliability of the product.

1-5 PROCESSING AND THE ECONOMIC ENVIRONMENT: YIELD

The ultimate goal of the processing professional in a commercial organization is to produce a product at a profit. Even in the most research-oriented environment, the goal is an effective, new application, and effectiveness requires consideration of cost. The primary determinant of cost in processing is the *yield* of good devices.

Yield is the number of good devices obtained per substrate started into the process. It is the product of three factors, as follows:

$$Y = (LY)\,(DY)\,(FY) \tag{1-7}$$

Here LY is the *line yield,* which is the percentage of substrates starting the process that survive until the end. Line-yield losses may result from breakage, errors, or failures in equipment or process reliability. The overall line yield may be expressed in terms of the yield of each particular process step, $(LY)_i$, as

$$LY = \prod_i (LY)_i \qquad (1\text{-}8)$$

If a process has, say, 50 steps, the yield of each step must be 99.42% to obtain an overall line yield of just 75%.

The term FY in Eq. (1-7) represents the *final test yield,* the percentage of assembled die passing final test. If the prepackaging test program is sound and the packaging process reliable, this yield should be high. In our discussion here, we will not explore it further.

The term DY is the *die yield:* the number of dice that pass the wafer level electrical testing after processing is completed. This number may also be called *electrical test* or *E-test yield, sort yield,* etc. It is the primary measure of whether the processing engineer has successfully translated the device design into physical reality.

Defective dice and thus lowered die yield can be caused either by arealike or pointlike mechanisms. An area of the substrate may fail to yield due to areas of contamination or damage, or due to process control failures. An area of damage, for example, may exist around the periphery of the substrate, where it is repeatedly contacted during transport and handling. Process control failures may lead to areas that are overdoped or underdoped, have incorrect pattern dimensions or layer thicknesses, etc. These effects are similar to the process factors involved in line yield, so that tighter control improves both types of yield.

Pointlike defects result chiefly from particulate contamination. A single particle of size comparable to the feature dimensions of the design can disrupt the design and cause die failure. The percentage of failures due to point defects is dependent on die size, because one defect, however small, totally incapacitates one die, however large. Figure 1-7 illustrates this concept, by showing the effect of 30 identically distributed point defects on two substrates containing devices of different size. In Fig. 1-7a, there are 34 possible good dice. The 30 defects destroy 22 of them, or 65%. (Two dice contain two defects each, and six defects lie in partial dice at the edge, which are incomplete and thus nonfunctional.) In Fig. 1-7b, the die area is reduced by a factor of 4, and only 28 of 144 dice (19%) are defective. Reducing the die size has dramatically improved the yield.

Die yield can be predicted in terms of the density of defects by an expression such as

$$DY = (DY)_o a \exp(-AD) \qquad (1\text{-}9)$$

where $(DY)_o$ is the number of possible good dice on the wafer, a is the percentage of area free from area-type defects, A is the area occupied by a die, and D is the point defect density per unit area. If we ignore the complications of partial dice at the edge, the number of good dice is given by the wafer area divided by the die area, and

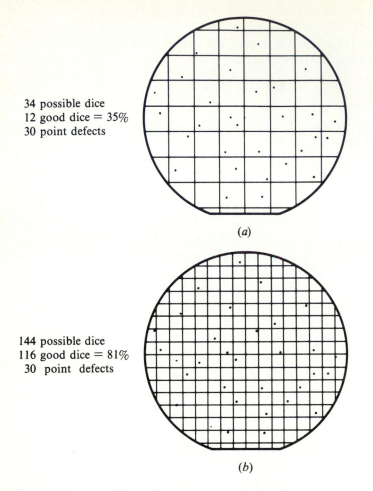

34 possible dice
12 good dice = 35%
30 point defects

(a)

144 possible dice
116 good dice = 81%
30 point defects

(b)

Figure 1-7 The effect of die size on yield loss due to pointlike defects. The distribution of defects, shown by black dots, is identical in cases (a) and (b), but the die size in case (b) is 25% of that in case (a). The percentage yield is improved by a factor of 2.3.

$$DY \cong \frac{\pi d^2}{4A} a \exp{(-AD)} \tag{1-10}$$

The exponential relationship between defects and yield is precisely true for a Poisson distribution of defects, and approximately true in many other cases.[8]

In applying such expressions, we should keep in mind that D and a depend on the typical dimensions in the design and are thus independent of A only for a given set of design rules. For a given typical device dimension x, particles below some limit (perhaps $0.1–0.25x$) can be considered insignificant. As x is shrunk to produce smaller devices, small particles become significant, and the effective defect density rises even though process conditions remain constant. Similarly, tighter tolerances require better process control to ensure that all parameters are maintained over the entire wafer area.

Thus shrinking dimensions may not cause yields to rise as expected unless there are simultaneous improvements in process quality. Nevertheless, the benefit of small dimensions is obvious.

Equation (1-10) explains the economics behind the principal desirables of microelectronics: smaller dimensions, larger substrates, and cleaner processing. Cost of production is roughly constant, being a fairly weak function of wafer size. Revenue depends on number of dice sold and is thus proportional to yield. Reducing die size, by making features smaller and tolerances tighter, means an exponential increase in profit. Similarly, increasing d, the wafer diameter, increases available dice by a factor of d^2 while only moderately increasing cost. The economic leverage of improved yield also provides for a large staff of process professionals, who can push a towards unity, and D towards zero, by improving process control and cleanliness.

It follows that yield enhancement deserves attention throughout this book. High yield requires tight process control and a low-particle environment. Process control is considered worthy of a separate section at the end of almost every chapter. Environmental cleanliness depends chiefly on the physical environment of the processing facility, which we consider next.

1-6 PROCESSING AND THE PHYSICAL ENVIRONMENT: CLEAN ROOMS

Microelectronic processes can be carried on effectively only in a tightly controlled environment from which various types of contamination are rigorously excluded. Such an environment is called a *clean room*. In the clean room, great care is taken to suppress particles in the air, in process gases and liquids, and from people in the environment.

The first line of defense against particles is the air-conditioning system of the room. All air entering the room is temperature- and humidity-controlled and is filtered using *high efficiency particulate air (HEPA)* filters. To be so designated, a filter must be 99.97% effective in removing particles 0.3 μm in size. In addition, the room is pressurized relative to its surroundings by at least 0.10 in of water (about 0.52 psi or 24.9 Pa). This is to ensure that all leakage is in the *out* direction and that dust does not enter through cracks, ceiling tiles, etc. Attention is given to the location of air supplies and returns, to minimize turbulence and eddies. Air-conditioning may also include charcoal filters, designed to remove airborne oxidant chemicals (i.e., smog), which can interfere with process chemistry.

These precautions can reduce air particle counts from the natural level (well over 100,000) to below 1000. Particle counts are given in units of particles of 0.5-μm size or larger, per cubic foot. A room that can consistently attain a count below 1000 ft^{-3} is called a *Class 1000* room, etc. Various federal standards have been established regulating particle counting and the classification of clean rooms. Class 100 or Class 1000 performance may be satisfactory for the overall atmosphere of a clean room, but is not adequate for the protection of devices under manufacture.

A second level of particle removal is required to achieve satisfactory counts at the working surface on which devices are processed. This involves covering all work spaces with *laminar flow hoods (LFH)*. A laminar flow hood, which may be of the vertical (VLF) or horizontal (HLF) type, is designed to produce a laminar (i.e., nonturbulent) flow of very clean air over the work area. Figure 1-8 shows a VLF hood positioned over a worktable. Air is forced through a HEPA filter and then through a matrix of baffles that induce laminar flow. Since turbulence can be induced by any physical barriers to the flow (shelves, fixtures, people), care must be given to the "furniture" in the work stations as well as to the return path for the air flow. A well-designed laminar flow work station should achieve particle counts below 10 ft^{-3} at the work surface.

Not only the air, but other materials that contact microelectronic devices must be particle-free. After air, water is the most common material to touch substrates during processing: copious amounts are used for rinsing and cleaning. Water used in microelectronics is *deionized* to remove all dissolved ions, producing *deionized ("D.I.")* water. Ion removal raises the resistivity of water, allowing water purity to be given in terms of megaohms of resistance. Pure water has a resistivity of 18 MΩ and good process water will be maintained at least at 12—often higher. Water is also filtered and is passed under ultraviolet lights to kill bacteria. Particle counts and bacterial counts are monitored frequently.

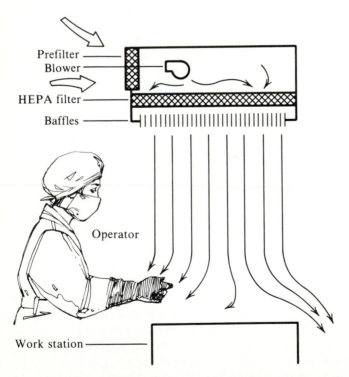

Prefilter
Blower
HEPA filter
Baffles
Operator
Work station

Figure 1-8 Use of a VLF hood to prevent particles from reaching the work surface in a microelectronic clean room. Air is highly filtered and passed through a baffle to create smooth, nonturbulent flow, isolating the work surface from dirtier room air.

Purity is also required in the many gases and chemicals used in processing. Process gases are usually filtered near the point of use. Processing chemicals such as acids and photoresists are increasingly being filtered, and vendors are now actively marketing low-particulate chemicals. Equipment that handles substrates and equipment for handling process chemicals are also capable of adding particles, an area that is receiving increasing attention from machinery suppliers.

Without a doubt, the biggest source of particles affecting the clean room is *people*. People, being organic rather than mechanical systems, produce a constant barrage of particles from skin, clothing, and breathing. People also use things that are copious sources of particles behaviors, such as paper, makeup, and cigarettes. Of course, smoking (along with eating and drinking) is strictly forbidden in clean room, but smokers exhale particles for up to one hour after their last cigarette, spreading their coffee-break and lunch-hour smoke throughout the day.[9]

People-generated particles are controlled by proper garb for clean-room workers, and by tightly controlling behavior in the clean room. Workers are required to dress in lint-free garments that cover most of their body. Depending on the requirement of the technology, such garb can range from smocks and gloves that expose most of the face, to "bunny suits" with goggles and surgical masks, completely covering the body. To keep the protective clothing from becoming dirty itself, it is stored under clean-room conditions in a separate changing room, which serves as a vestibule to the clean room. Cleanliness can be further improved by requiring personnel to pass through air showers (blasts of air that remove particles) and over adhesive mats (flypaperlike mats on the floor which clean the shoe soles) on their way into the clean room.

Particle-generating materials, such as wood, paper, and pencils are strictly banned from clean rooms. Where writing is required, special "clean-room paper" materials are used that do not generate lint. Makeup use is also controlled. Where the face and head are not totally covered, particular care must be given to control of hair. Since the conflict between processing quality and personal appearance can become a sensitive one, each facility will have its own rules covering these matters.

The generation of particles by people will be a problem as long as personnel work within the clean envelope containing processed substrates. An obvious solution would be to create a smaller clean envelope, containing only the wafers and the working parts of the machinery, with the operators working by remote control, as is done in nuclear technology. Surprisingly little progress has been made in this area, but as devices become more demanding, it seems inevitable that this method will be adopted. In the meantime, the processing professional will remain concerned with the control of the clean-room environment. More information on particulate control is given in Refs. 9–14.

PROBLEMS

1-1 A single-crystal silicon boule is grown from a melt containing 1 ppm oxygen and 0.1 ppm boron. Growth of the crystal consumes 90% of the melt. Find the oxygen and boron concentrations at the seed end and at the tang or butt end of the boule.

1-2 Find the maximum pull rate for a silicon crystal under the following conditions:

$$L = 264 \text{ cal g}^{-1}$$

$$\kappa = 0.05 \text{ cal sec}^{-1} \text{ cm}^{-1} \text{ °C}^{-1}$$

$$\frac{dT}{dx} = 100\text{°C cm}^{-1} \text{ (experimentally determined)}$$

$$d = 2.3 \text{ g cm}^{-3}$$

At this rate, how long will it take to pull a 2-m-long crystal?

1-3 A newly released manufacturing process for integrated circuits has a step line yield of 99% at each of 50 steps and a defect density of 1.0 defects per square centimeter. The die area is 1 cm², and there are 115 possible dice per substrate. (*a*) What is the line yield? (*b*) What is the die yield, assuming $a = 100\%$? (*c*) What is the overall yield, assuming final yield = 90%? (*d*) As engineering manager, you feel you have the resources either to (1) reduce step line yield losses 20% at each step, that is, improve step line yield to 99.2%, or (2) reduce defect density by 20%. Which alternative do you elect?

1-4 Repeat Problem 1-3 for a more mature process with a step line yield at each step of 99.7% and a defect density of 0.3 defects per square centimeter.

1-5 Through aggressive design and the improvement of processing methods, the die size of the device in Problem 1-3 is reduced to 0.8 cm². What is the expected change in yield?

REFERENCES

1. Leon D. Crossman and John A. Baker, "Polysilicon Technology," in *Semiconductor Silicon 1977,* Howard R. Huff and Erhard Sirtl (eds.), The Electrochemical Society, Princeton, NJ, 1977.
2. W. R. Runyan, *Silicon Semiconductor Technology,* McGraw-Hill, New York, 1965.
3. C. W. Pearce, *Crystal Growth and Wafer Preparation,* in *VLSI Technology,* S. M. Sze (ed.), McGraw-Hill, New York, 1984.
4. B. R. Pamplion, *Crystal Growth,* Pergamon, New York, 1983.
5. Sorab K. Ghandhi, *VLSI Fabrication Principles,* Wiley, New York, 1983.
6. Ernel R. Winkler, *Solid State Technol.* **25**:94 (June 1982).
7. C. A. Steidel, "Assembly Techniques and Packaging," in *VLSI Technology,* S. M. Sze (ed.), McGraw-Hill, New York, 1984.
8. W. J. Bertram, "Yield and Reliability," in *VLSI Technology,* S. M. Sze (ed.), McGraw-Hill, New York, 1984.
9. Mary L. Long, *Solid State Technol.* **27**:159 (March 1984).
10. Michael K. Kilpatrick, *Solid State Technol.* **27**:151 (March 1984).
11. J. M. Duffalo and J. R. Monkowsi, *Solid State Technol.* **27**:109 (March 1984).
12. Stuart A. Hoenig and Steven Daniel, *Solid State Technol.* **27**:119 (March 1984).
13. Don L. Tolliver, *Solid State Technol.* **27**:129 (March 1984).
14. Mauro A. Accomazzo, Kenneth L. Rubow, and Y. H. Benjamin, *Solid State Technol.* **27**:141 (March 1984).

THE PHYSICS AND CHEMISTRY OF SEMICONDUCTORS

The vast majority of microelectronic devices involve semiconductors. Therefore, we begin our study of microelectronic processing with a two-chapter section devoted to semiconductor properties. In this chapter, we will review the electronic and physical characteristics of semiconductors. In Chapter 3, we will examine techniques for modifying semiconductor properties through dopant diffusion and oxidation.

This chapter gives a brief overview of various properties of semiconducting materials. The intent is not to give a detailed development of the subject, but to introduce the basic concepts that provide the motivation for microelectronic processing methods. Students who have previously studied semiconductor properties will want to treat this chapter as a review. On the other hand, those new to the subject should concentrate on understanding general principles for later application. The reader interested in a detailed understanding of device physics should consult one of the texts mentioned in the references.

The first four sections deal with electronic properties. In Section 2-1, we use the band theory of conduction to explain conductivity in semiconductors. Semiconductor conductivity can be manipulated either by chemical means or by the application of fields. In Section 2-2, we see how conductivity is affected chemically, by the addition of trace impurities, while in Section 2-3 we deal with field effects. In Section 2-4, we summarize how these properties are exploited to make useful electronic devices.

The rest of the chapter deals with the physical properties of real semiconducting materials. Since most useful semiconductors are crystalline, in Section 2-5 we examine properties common to crystals. In Section 2-6, we look more closely at the most common elemental semiconductor, silicon. In Section 2-7, we examine an increasingly common compound semiconductor, gallium arsenide.

2-1 THE BAND THEORY OF CONDUCTION

What is a *semiconductor?* Originally, the term implied simply a material whose conductivity lay between the highly conductive metals and the insulators. We now know that conduction in semiconductors involves a different mechanism than in metals, with important technological implications. To understand why, we must begin with the modern understanding of electrical conduction in solids, summarized in the band theory. This theory explains the resistivity of solids in terms of the quantum-mechanical behavior of electrons as they move through a regular array of atoms.

In the quantum-mechanical view, elementary particles may exist only in discrete energy states. These states are assigned *quantum numbers,* which designate the states according to potential energy, spin, etc. In the absence of other factors, particles occupy the lowest available energy states. However, for electrons, the Pauli exclusion principle requires that no two electrons may occupy the same state. This implies that, given a number of available states and a number of electrons, the electrons will "fill up" the available states in order of ascending energy until all electrons have been placed. Since electrical conduction is chiefly the result of electron movement, conductivity becomes a matter of the number and energy distribution of available electron states.

In only a few cases can electron states be calculated completely. One of these is the hydrogen atom, with a single electron orbiting a positive nucleus. In this case, there is a series of distinct states or *orbitals,* separated by wide bands of energy in which no states are found. This type of state distribution is typical of isolated atoms. With only limited approximations, we can also calculate the result when two hydrogen atoms are brought together to form a hydrogen molecule, H_2. The result is shown in Fig. 2-1. Constructive and destructive interference occurs between orbitals. As the atoms approach, each pair of atomic orbitals evolves into two molecular orbitals, one higher and one lower in energy than the originals.

Mathematically, the molecular orbitals are linear combinations of the atomic orbital *wave functions,* $\psi_{a,b}$, one of form $\psi_a + \psi_b$ and one of form $\psi_a - \psi_b$. The average potential energy of the pair of molecular orbitals, taken together, is the same as the energy of the two atomic orbitals, but the lower-energy orbital of the pair is preferentially occupied by electrons. This lowers the potential energy of the molecule with respect to the atom, stabilizing and thus bonding it.

In more complicated molecules, a similar situation occurs. Atomic orbitals again combine into molecular orbitals, some of higher and others of lower energy. With increased numbers of atoms, interference also occurs between molecular orbitals, spreading them out into groups of orbitals with distinct, but similar energies.

A solid crystal can be considered as a huge molecule and the effects on orbital energy are qualitatively similar to smaller molecules. From the multitude of atomic orbitals, broad bands of states are formed. One band is lower in energy than the separated atoms, and one is higher. The various states within each band are so many and so close in energy that they are described, not by a separate count, but by a *density of states function,* $N(E)$, such that $N(E)\,dE$ is the number of states in an energy increment dE. Figure 2-2 illustrates these concepts. It should be noted that, while we

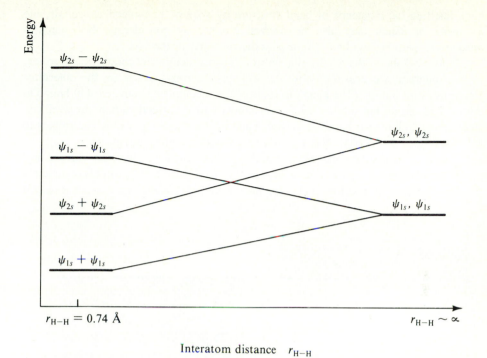

Figure 2-1 Evolution of hydrogen atom atomic orbitals into orbitals of the H_2 molecule, as the distance between atoms is reduced. (The diagram is schematic and is not to scale.)

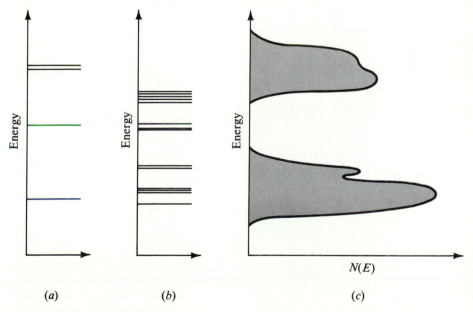

Figure 2-2 Typical state distributions for: (a) an atom, (b) a moderately complex molecule, and (c) a crystal.

have justified the existence of band structure by analogy to molecular orbitals, the existence of bands may also be derived directly by considering the quantum-mechanical particle-in-a-box with a periodic potential in the box.[1]

Figure 2-3 shows the relationship between band structure and conduction in solids. In this diagram, we deal only with the *valence* electrons, which are the outermost electrons of the atoms. (The inner electrons occur in complete shells, closely bound to the nuclei, and do not interact with other atoms.) As described earlier, the available states separate into two bands, a *valence* band of lower energy, which contributes to the bonding of the crystal, and a *conduction* band of higher energy.

Figure 2-3a displays the band structure of an insulator. Here the valence and conduction bands are separated by a large energy gap. The valence band is completely filled with electrons, and the conduction band is empty. Conduction can occur only if

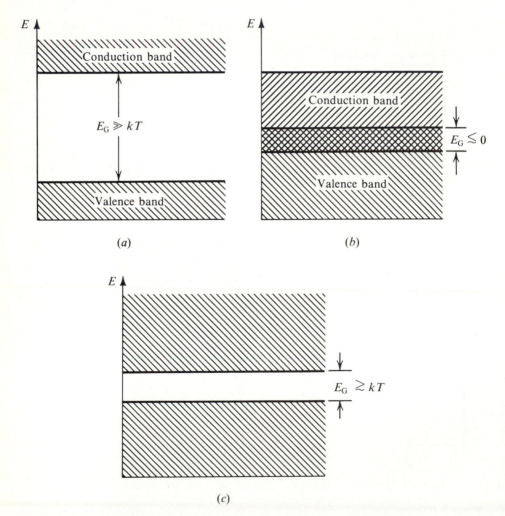

Figure 2-3 The band structure of (*a*) an insulator, (*b*) a conductor, and (*c*) a semiconductor.

electrons are free to move. Movement is possible only if there are unoccupied states to move into, and electrons to move into them. In the valence band, conduction can occur only if some states are made empty to allow movement. In the conduction band, conduction requires that electrons be supplied to move among the many empty states. If the valence band is filled and the conduction band empty, conduction can occur only if an electron is promoted to the conduction band, creating an empty state in the valance band and a free electron in the conduction band. This requires energy, either from heat or absorption of a photon. The large band gap of an insulator makes promotion unlikely, and thus insulators do not conduct electricity.

In Fig. 2-3*b*, we see the band structure typical of metallic conductors. Here the valence and conduction bands overlap, to form one continuous range of allowable electron states. Conduction can thus take place by the absorption of an infinitesimal amount of energy and is virtually unlimited. Indeed, metals have been described as an array of nuclei immersed in an "electron sea." In metallic conduction, the only resistance to conduction is the scattering of electrons from crystal defects, nuclei, etc. This scattering increases with increasing temperature, and thus the resistivity of metallic conductors increases with temperature.

In a semiconductor, the band structure of Fig. 2-3*c* is found. Here there is a definite separation between bands, but the separation is fairly small. (Here "large" and "small" are defined relative to the thermal energy of approximately kT, where T is the temperature and k is Boltzmann's constant of 8.62×10^{-5} eV $°K^{-1}$.) There is thus a reasonable probability, at room temperature, that a valence band electron will absorb enough thermal energy to jump to the conduction band, thus becoming a free charge carrier. This event, in turn, leaves behind a vacancy in the valence band, into which another electron can move. By the successive movement of electrons, charge can also be transported in the valence band, making use of the vacancy left by the promoted electron. Thus promotion of a single electron creates mobile charge in both bands.

Conduction in the valence band involves a cascade of electron movements that we prefer not to keep track of. Therefore, we speak of valence band conduction in terms of current-carrying "*holes*." By this abstraction, the vacant electron site becomes a positively charged "particle" which serves as a charge carrier. This imaginary particle has the charge of an electron and can be assigned an effective mass and other particle properties. The potential energy of a hole increases as it moves downward in the band diagram, away from the valence band edge. This visualization is precisely analogous to focusing our attention on the motion of a bubble in a fluid while being aware that the liquid in fact moves around the bubble.

We may say that in a semiconductor, an *electron-hole pair* is created by the absorption of energy, allowing the movement of charge in both valence and conduction bands. As a result, the conductivity of a semiconductor, unlike a metal, increases with temperature.

Further insight into semiconductivity results if we quantify the energy relationships we have been discussing. We begin with the Fermi-Dirac distribution function, which deals with the distribution of electrons among orbitals or states of varying energies. The probability of any state being occupied by an electron is

$$f(E) = (1 + e^{(E - E_F)/kT})^{-1} \tag{2-1}$$

where E is the energy of the state, T the temperature in degrees Kelvin, k is Boltzmann's constant, and E_F is a parameter known as the *Fermi level*.

The Fermi level is an important and useful parameter of semiconductor behavior. Thermodynamically, it is defined as the chemical potential of an electron in the semiconductor. In statistical-mechanical terms, it is the energy at which the probability of a state being occupied by an electron is precisely one-half. These two definitions are technical enough that they sometimes obscure the basic simplicity of the concept.

A useful analogy can be drawn between the Fermi level and another well-known benchmark, sea level. In thermodynamic (hydrodynamic?) terms, sea level is a measure of the gravitational potential, or simply the height, of the ocean. Any real ocean has waves, however, and sea level is therefore a mathematical construct to which no real ocean conforms. At any given time, there is a probability that a particular element of space above sea level is in fact occupied by water, and a similar probabiltiy that a volume below sea level is unoccupied by water. This probability will vary with the size of the waves and decreases sharply a few meters from sea level. It is thus also valid to define sea level as that plane at which the probability of a space being occupied by water is one-half. In a similar sense, the Fermi level marks the "ideal top of the "electron sea" in a solid crystal.

This analogy can be carried further. At equilibrium, the surface of the sea is everywhere at the same level, and there is no net flow of water. Sea level represents the zero point for the gravitational potential of water. If water is to flow, the water level must be no longer flat, and it is thus a necessary and sufficient condition for equilibrium that the sea level be the same everywhere. (For instance, the level of the sea falls by perhaps 3 ft as the Gulf Stream, an impressive example of flowing water, rounds the tip of Florida.) Similarly, for a semiconductor at equilibrium, the Fermi level is constant everywhere, and the flow of current both requires and implies that the Fermi level be no longer flat.

With a better understanding of the parameter E_F, we will now proceed to manipulate Eq. (2-1). First we develop two useful approximations. Usually, as we shall see, the edges of the conduction and valence bands are several kT from the Fermi level. Under these conditions, we can simplify Eq. (2-1) to give

$$f(E) \doteq e^{-(E-E_F)/kT} \qquad\qquad E \gg E_F \qquad\qquad (2\text{-}2a)$$

$$f(E) \doteq 1 - e^{-(E_F-E)/kT} \qquad\qquad E \ll E_F \qquad\qquad (2\text{-}2b)$$

Usually, we are more interested in the number of charge carriers than in the occupation of states. If we know the density of states $N_c(E)$ in the conduction band and the density of states $N_v(E)$ in the valence band, then the number of carriers will be given by the state density times the probability of occupation.

Such calculations have been performed, and the results are usually summarized as

$$n = N_c e^{-(E_c-E_F)/kT} \qquad\qquad (2\text{-}3a)$$

$$p = N_v e^{-(E_F-E_v)/kT} \qquad\qquad (2\text{-}3b)$$

where n and p represent the number of holes and electrons per unit volume. N_c and N_v are the *effective* densities of states, and E_c and E_v are the energies of the upper edge of the valence band and the lower edge of the conduction band, respectively. [Although Eq. (2-3) appears to follow from Eq. (2-2), it is worth remembering that there are some assumptions hidden in the term "effective energy of states."]

By multiplying Eqs. (2-3a) and (2-3b), we obtain

$$pn = N_c N_v \exp \frac{-E_G}{kT} \tag{2-4}$$

which gives the product of the carrier concentrations in terms of densities of states, temperature, and the energy gap E_G. The energy gap is the energy separating E_c and E_v. The pn product in turn defines an *intrinsic carrier density*, n_i, such that

$$n_i^2 = pn \qquad \text{at equilibrium} \tag{2-5}$$

or

$$n_i^2 = N_c N_v \exp \frac{-E_G}{kT} \tag{2-6}$$

The only quantities that determine n_i are the band gap, the density of states, and the temperature. At a given temperature, band gap and state density are functions of the material only. (We should note that state density does depend on temperature and is proportional to $T^{3/2}$.) Thus the carrier concentration in a given material at equilibrium is a function of temperature only, and Eq. (2-6) is a powerful tool in calculating carrier concentration.

It is possible, in nonequilibrium situations, to have carrier concentrations in excess of n_i. Excess carriers may be *generated* by illuminating the material with photons of an energy greater than the band gap. The result is the formation of electron-hole pairs, due to promotion of conduction-band electrons into the valence band. Alternatively, excess electrons or holes may be *injected* into the semiconductor by making it part of an electrical circuit. In either case, left to themselves, the excess carriers will recombine with carriers of the opposite sign until equilibrium is restored. Alternatively, carrier concentration may be reduced below equilibrium, for example, by sweeping carriers away with an electrical field. In this case, spontaneous generation of electron-hole pairs will tend to restore equilibrium.

Although we have discussed the Fermi level, we have not fixed its position. Intuitively, we might expect it to lie in the center of the band gap. Since the band gap E_G is much greater than kT, the exponential form of Eq. (2-3) suggests that very few valence states will be vacant and very few conduction sites filled. Also, holes are formed only by the promotion of valence band electrons, which ensures that $n = p$. To confirm this notion, we can set $n = p$ and eliminate both quantities from Eq. (2-3) to obtain

$$E_i = \frac{1}{2}(E_c + E_v) + \frac{1}{2} kT \ln \frac{N_v}{N_c} \tag{2-7}$$

Here the notation E_i is used to distinguish the *intrinsic* Fermi level, (i.e., the value of E_F in the pure equilibrium semiconductor), because Eq. (2-7) is not true under nonequilibrium conditions. Intrinsic Fermi level varies from the band gap center by a very small factor, which reflects any difference in the effective densities of conduction and valence band states.

With these concepts, the resistivity of an intrinsic semiconductor can be derived. The flow of current is given by the product of the number of charge carriers, their charge, and their velocity. Equation (2-3) gives us the number of charge carriers available; their charge is the electron charge q. Charge is measured in units of coulombs, abbreviated C. In a solid, the velocity of a moving charge carrier is related to the applied field $\mathscr{E}$ by

$$v = \mu\mathscr{E} \tag{2-8}$$

where μ is the mobility of the carrier in the solid. (This equation is justified more fully in Chapter 3, where we investigate the analogous equations for ions diffusing in an electric field.) Since the mobilities of holes and electrons differ, we must obtain the total current density $\mathscr{J}$ by summing hole and electron components:

$$\mathscr{J} = qn\mu_n\mathscr{E} + qp\mu_p\mathscr{E} \tag{2-9}$$

where μ_n and μ_p are the electron and hole mobilities. The resistivity p is given by $\mathscr{E}/\mathscr{J}$, so

$$\rho = \frac{1}{q(n\mu_n + p\mu_p)} \tag{2-10}$$

Table 2-1 gives the properties of common semiconductors.[2]

Table 2-1 Important properties of common semiconductors

	Ge	Si	GaAs
Atomic or molecular weight (amu)	72.6	28.09	144.63
Atoms or molecules (cm^{-3} × 10^{22})	4.4	5.00	2.21
Lattice constant (nm)	0.566	0.543	0.565
Density (g cm^{-3})	5.32	2.33	5.32
N_c (× 10^{17} cm^{-3})	104	28	4.7
N_v (× 10^{17} cm^{-3})	60	104	7.0
E_G (eV)	0.67	1.11	1.40
n_i (cm^{-3})	2.4 × 10^{13}	1.45 × 10^{10}	9 × 10^6
Intrinsic lattice mobilities (cm^2 V^{-1} sec^{-1})			
Electrons	3900	1350	8600
Holes	1900	480	250
Dielectric constant	16.3	11.7	12
Melting point (°C)	937	1415	1238
Vapor pressure (torr)	10^{-7} (@ 880°C)	10^{-7} (@ 1050°C)	1 (@ 1050°C)
Specific heat (J g^{-1} °C^{-1})	0.31	0.7	0.35
Thermal conductivity (W cm^{-1} °C^{-1})	0.6	1.5	0.81
Linear coefficient of thermal expansion (ppm)	5.8	2.5	5.9

Source: Reference 2, used by permission. Copyright 1967, John Wiley & Sons.

Example Find the resistivity of room temperature, undoped silicon. In this case, n and p are each equal to the intrinsic carrier density n_i, which, from Table 2-1, is 1.45×10^{10} cm^{-3}. The electron and hole mobilities are 1350 and 480 cm^2 V^{-1} sec^{-1}, respectively, and so

$$\rho = [(1.60 \times 10^{-19} \text{ C})(1.45 \times 10^{10} \text{ cm}^{-3})$$

$$\times (1350 \text{ cm}^2 \text{ V}^{-1} \text{ sec}^{-1} + 480 \text{ cm}^2 \text{ V}^{-1} \text{ sec}^{-1}]^{-1}$$

$$= 235,000 \text{ } \Omega \text{ cm}$$

Thus the resistivity of intrinsic semiconductors is fairly high.

2-2 CHEMICAL EFFECTS ON CARRIER DENSITY: DOPING AND JUNCTIONS

The conduction mechanism of semiconductors leads to interesting results when the density of charge carriers is varied. One means of manipulated charge carrier density is by the addition of small amounts of impurities, a process known as *doping*.

Consider, for example, a crystal of silicon. Each silicon atom is bound to four nearest neighbors and shares a pair of electrons with each. In this way, the four valence electrons of each silicon atom are employed in bonding. This is a stable situation that accounts for the rarity of conduction-band electrons in intrinsic silicon. Figure 2-4a shows a two-dimensional schematic of the silicon crystal.

Now consider the replacement of a single silicon atom with an atom of arsenic, as shown in Fig. 2-4b. Arsenic lies in the same row of the periodic table as silicon, in column VB. It is thus about the same size, but has one additional electron and an additional unit of nuclear charge. The arsenic atom should thus fit neatly into the silicon lattice, using four of its five valence electrons to form bonds with the silicon. However, the fifth electron has no place in the bonding scheme of the silicon crystal and should be easily separated. In fact, the *ionization energy*, or energy for separation, of the fifth electron turns out to be 0.049 eV, or about $2kT$ at room temperature. Thus arsenic in silicon becomes almost completely ionized. The extra electron enters the conduction band and leaves behind a positively charged arsenic ion bound into the silicon lattice. Addition of arsenic atoms to silicon results in a one-to-one increase in the number of conduction-band electrons.

Addition of an atom of a Group III element, such as boron, has similar results. Boron replaces silicon in the lattice, but can contribute only three bonding electrons. The result is a vacancy in the valence band, which allows mobility to adjoining electrons. This is a hole; thus addition of boron produces holes in the conduction band and produces negatively charged bound ions in the lattice. Figure 2-4c shows the effect of added boron.

Impurities that have this type of effect on the carrier distribution in a semiconductor are known as *dopants*. Elements such as arsenic that donate electrons are known as *donors*, or *n-type* (for negative) *dopants*. A hole-generating dopant like boron is an

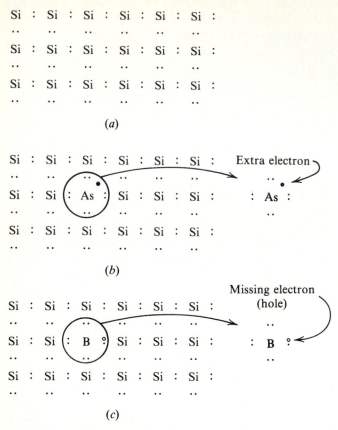

Figure 2-4 The effect of introducing dopants into a semiconductor lattice. (*a*) The undoped silicon lattice (conceptual representation). (*b*) Effect of substituting an arsenic atom into the lattice. (*c*) Effect of substituting a boron atom into the lattice.

acceptor, or *p-type dopant.* In an energy-band diagram such as Fig. 2-5, these atoms contribute donor and acceptor levels very close to the band edges. Table 2-2 gives the energies of these levels for common dopant-semiconductor combinations.[3]

The addition of small amounts of dopant have dramatic effects on the semiconductor's resistivity. For dopants such as arsenic and boron in silicon, ionization is almost complete. This allows us to write

$$n \sim N_\mathrm{D} \tag{2-11a}$$

for an *n*-doped semiconductor, and

$$p \sim N_\mathrm{A} \tag{2-11b}$$

for a *p*-doped semiconductor, where N_A and N_D are the acceptor and donor atom concentrations. Dopant addition not only adds *majority* charge carriers (e.g., electrons in *n*-doping), but also affects the population of *minority* charge carriers (e.g., holes in *n*-doping). This is because Eq. (2-5) continues to hold at equilibrium and thus

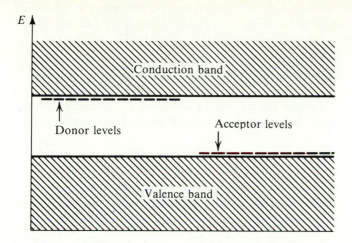

Figure 2-5 An energy-band diagram for a doped semiconductor, showing position of the donor and acceptor levels very near the conduction- and valence-band edges.

Table 2-2 Ionization energies for dopants in silicon

Dopant	p- or n-type	Ionization energy (eV)
P	n	0.044
As	n	0.049
Sb	n	0.039
B	p	0.045
Al	p	0.057
Ga	p	0.065
In	p	0.16

Source: Reference 3, used by permission. Copyright 1983, John Wiley & Sons.

$$p = \frac{n_i^2}{N_A} \qquad n\text{-doped} \qquad (2\text{-}12a)$$

$$n = \frac{n_i^2}{N_D} \qquad p\text{-doped} \qquad (2\text{-}12b)$$

Figure 2-6 relates doping to resistivity for n- and p-type silicon.[4]

Example Arsenic is added to silicon at a concentration of 10^{16} atoms per cubic centimeter. Find (*a*) the arsenic concentration in parts per million (ppm), (*b*) the resistivity, (*c*) the minority carrier concentration.

From Table 2-1, there are 5×10^{22} atoms per cubic centimeter in silicon and thus the added arsenic concentration is 2.0 ppm. From Fig. 2-6, the silicon resistivity is 1 Ω cm. The hole concentration is given by Eq. (2-12a), as

$$p = \frac{(1.45 \times 10^{10})^2}{10^{16}} = 2.1 \times 10^4$$

Thus an addition of 2 ppm of arsenic, barely detectable by chemical analysis, reduced silicon resistivity by a factor of 10^6 and reduces hole concentration drastically.

The Fermi level is a measure of the occupation of states in a semiconductor, and it follows that doping, which changes the ratios of charge carriers, must move the Fermi level away from the center of the band gap. From Eq. (2-3), we can write for an n-type semiconductor

$$N_D \doteq N_c e^{-(E_c - E_F)/kT} \tag{2-13}$$

For the undoped case, $n = n_i$ and $E_F = E_i$, so Eq. (2-3) would give

$$n_i = N_c e^{-(E_c - E_i)/kT} \tag{2-14}$$

Combining these expressions gives a value for the difference between the Fermi level of a doped and undoped semiconductor:

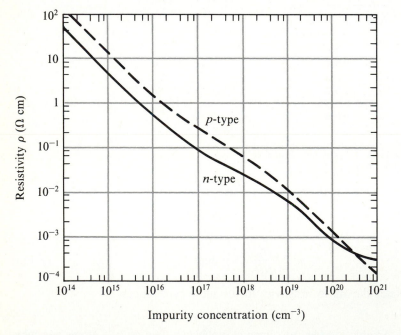

Figure 2-6 Resistivity versus dopant concentration for silicon. (*After Ref. 4. Reprinted with permission from the* Bell System Technical Journal. *Copyright 1962, AT&T.*)

$$N_D \doteq n_i e^{(E_F - E_i)/kT} \tag{2-15}$$

We can define the *Fermi potential* ϕ_F as the displacement of the Fermi level from the intrinsic Fermi level, so that

$$\phi_F = E_F - E_i = kT \ln \frac{N_D}{n_i} \qquad \text{\textit{n}-doped} \tag{2-16a}$$

For the p-doped case, by a similar logic,

$$\phi_F = E_i - E_F = kT \ln \frac{N_A}{n_i} \qquad \text{\textit{p}-doped} \tag{2-16b}$$

The addition of dopants shifts the Fermi level closer to the band containing the majority carriers. If the doping is particularly strong, the Fermi level can approach within a few kT of the band edge, and the approximation used to derive Eq. (2-3) breaks down. Such a semiconductor is called *degenerate*.

Deep-Lying Impurities

Impurities in semiconductors may serve other roles besides donors or acceptors. Some additives give rise to levels near the middle of the band gap, as shown in Fig. 2-7. These levels are called *deep-lying* levels to distinguish them from the shallow impurity levels of dopants. Such levels can accept an electron from the conduction band, later transferring it to the valence band. (This process may also be perceived as the transmission of a hole to the conduction band, followed by its replacement from the valence band.) Since the probability of electron transition between states depends on

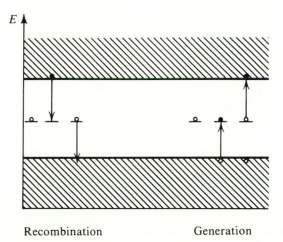

Recombination   Generation

Figure 2-7 The effect of deep-lying dopant states in facilitating generation and recombination of electron-hole pairs.

the energy ΔE between them as $\exp(-\Delta E/kT)$, halving the energy difference can dramatically increase the transition rate. In bipolar devices, in which excess charge carriers must be neutralized in order to switch a circuit, the rate of recombination is a major determinant of circuit speed. Therefore, gold, a source of deep-lying states, is sometimes used to dope bipolar devices. Table 2-3 presents the energy levels of some deep-lying dopants.[5]

The *pn* Junction

The effect of dopants upon bulk resistivity is interesting, but the real potential of semiconductors is seen only when *p*- and *n*-doping are combined within a single crystal. Figure 2-8*a* shows an idealized situation, in which a zone of *n*-doped silicon, of constant dopant concentration, adjoins a similarly homogeneous zone of *p*-doped silicon. The doping densities, in atoms per cubic centimeter, are N_A and N_D, respectively. In isolation, the excess majority carriers in each region would equal the doping, according to Eq. (2-11).

However, with the two zones in contact, this situation changes. On one side of the boundary is a large excess of electrons; on the other side, almost no electrons. The result is an immediate diffusion of electrons across the boundary, from the electron-rich *n*-doped region to the electron-poor *p*-doped region. Simultaneously, holes diffuse in the opposite direction. As the carriers migrate, they leave behind an increasing number of uncompensated, fixed charges on the dopant ions bound into the crystal lattice. The result is an electric field in the semiconductor, which opposes the further diffusion of charge carriers. Eventually, a new equilibrium is attained, at which the density gradient of charge carriers, which encourages migration, is exactly balanced by the electric field, which retards migration. As shown in Fig. 2-8*b*, there is now a volume of material at the junction between the two regions that is depleted in charge

Table 2-3 Ionization energies of deep-lying impurities in silicon

Impurity	Acceptor level, distance from valence band (eV)	Donor level, distance from conduction band (eV)
Ag	0.89	0.79
Au	0.57	0.76
Cu	0.24, 0.37, 0.52	—
Mo	0.3	—
Ni	0.21, 0.76	—
O	—	0.16
Pt	0.42, 0.92	0.85
Tl	0.26	—
Zn	0.31, 0.56	—

Source: Reference 5, used by permission. Copyright 1983, John Wiley & Sons.

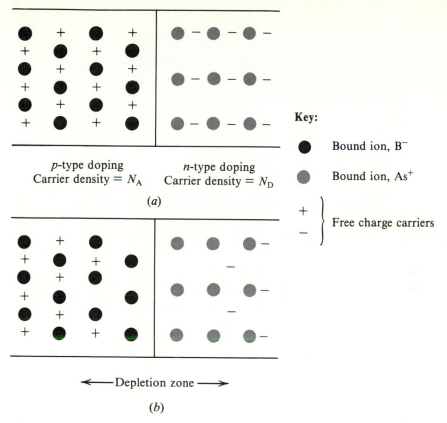

Key:

● Bound ion, B⁻

● Bound ion, As⁺

+ } Free charge carriers

(a)

(b)

Figure 2-8 An idealized *pn* junction: (*a*) Initial condition, prior to diffusion of charge carriers. (*b*) After diffusion of charge carriers to form a depletion zone.

carriers and contains an electric field. The junction of the zones is a *pn junction,* and the zone surrounding it is a *depletion zone.*

We know that at equilibrium the Fermi level is everywhere constant. We also know that in an *n*-doped semiconductor, the Fermi level approaches the conduction band; the reverse is true in *p*-doped material. Combining these facts allows us to sketch the band structure near a *pn* junction as in Fig. 2-9a. We can be more quantitative with the aid of some basic electrical relationships. One is Poisson's equation, which connects the electric potential ϕ to the charge density function $p(x, y, z)$. In one dimension, this equation is

$$\frac{d^2\phi}{dx^2} = -\frac{\rho}{\epsilon}$$

(2-17)

where ϵ is the permittivity of the material. The electric potential is the potential energy per unit charge of a charge carrier in the semiconductor, relative to some arbitrary zero. It is convenient to use the intrinsic Fermi level E_i as this zero, obtaining the relationship

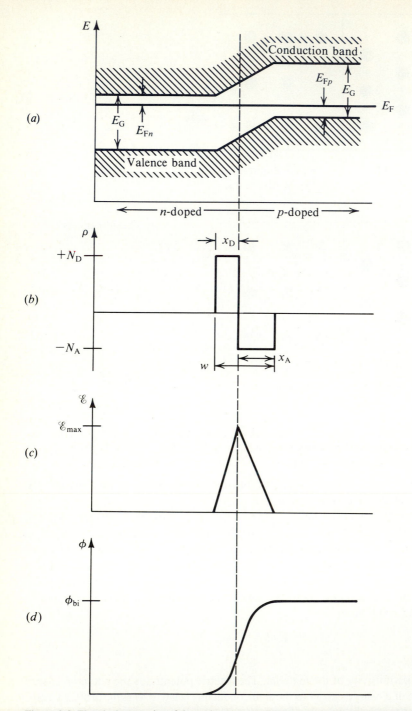

Figure 2-9 Electrical properties of the *pn* junction. (*a*) Band structure. (*b*) Charge distribution. (*c*) Electric field. (*d*) Electric potential.

$$\phi = -\frac{E_i}{q} \qquad (2\text{-}18)$$

The electric field is in turn the derivative of the potential with respect to distance:

$$\mathcal{E} = -\frac{d\phi}{dx} \qquad (2\text{-}19)$$

It follows that

$$\frac{1}{q}\frac{d^2 E_i}{dx^2} = \frac{d\mathcal{E}}{dx} = \frac{\rho}{\epsilon} \qquad (2\text{-}20)$$

These relations allow us to determine the width, maximum field, and potential drop across the depletion zone.

We begin with an assumption about the charge distribution in the depletion zone. The simplest assumption to make is the one illustrated in Fig. 2-9b. Here we assume that the depletion zone consists of a totally depleted area of width x_D on the n-doped side of the junction and a similar area of width x_A on the p-doped side. The electric field $\mathcal{E}$ in the depletion zone can be obtained by integrating the charge distribution

$$\mathcal{E} = \frac{1}{\epsilon}\int_{-x_D}^{x_A} \rho(x)\, dx \qquad (2\text{-}21)$$

Since in this case the charge distribution is constant within each region, the integral consists of two line segments of slopes qN_D and $-qN_A$, as shown in Fig. 2-9c. The maximum value of the field is at the apex of this triangle; it can be obtained by working in either direction to give

$$\mathcal{E}_{max} = \frac{q}{\epsilon}N_D x_D = \frac{q}{\epsilon}N_A x_A \qquad (2\text{-}22)$$

By integrating the field, we can obtain the potential difference, ϕ_{bi}, across the zone. In this case, integration may be done by inspection. The area of the triangle in Fig. 2-9c is

$$\phi_{bi} = \tfrac{1}{2}\,\mathcal{E}_{max}\,w \qquad (2\text{-}23)$$

where $w = x_A + x_D$ is the depletion zone width.

The value of the depletion zone width may be found using the fact that the total charge over the entire depletion zone must be zero. As a result $x_A N_A = x_D N_D$ and so

$$w = x_A + x_D = x_D\left(1 + \frac{N_D}{N_A}\right) \qquad (2\text{-}24)$$

Substituting Eqs. (2-24) and (2-22) into Eq. (2-23) gives

$$\phi_{bi} = \frac{q}{2\epsilon}N_D x_D^2\left(1 + \frac{N_D}{N_A}\right) \qquad (2\text{-}25)$$

Solving for x_D gives

$$x_D^2 = \frac{2\epsilon}{q}\phi_{bi}\frac{1}{N_D(1 + N_D/N_A)} \qquad (2\text{-}26)$$

and, from Eq. (2-25),

$$x_{\mathrm{D}}^2 = \frac{w^2}{(1 + N_{\mathrm{D}}/N_{\mathrm{A}})} \tag{2-27}$$

so that

$$w = \left[\frac{2\epsilon}{q} \phi_{\mathrm{bi}} \frac{N_{\mathrm{A}} + N_{\mathrm{D}}}{N_{\mathrm{A}} N_{\mathrm{D}}} \right]^{1/2} \tag{2-28}$$

The depletion zone width w is now known in terms of the potential drop ϕ_{bi}. Figure 2-9a shows that the relationship between potential drop and Fermi level displacement is

$$\phi_{\mathrm{bi}} = \frac{1}{q} E_{\mathrm{F}n} - \frac{1}{q} E_{\mathrm{F}p} \tag{2-29}$$

where $E_{\mathrm{F}n}$ and $E_{\mathrm{F}p}$ are given by Eq. (2-16). The potential drop is thus a function of the doping levels only (for a given material). It is called the *built-in potential,* because this potential appears automatically as a result of forming the junction. The potential increases logarithmically with increasing doping. However, the $(N_{\mathrm{A}} + N_{\mathrm{D}})/(N_{\mathrm{A}} N_{\mathrm{D}})$ term of Eq. (2-28) increases linearly with doping, and thus the depletion zone becomes narrower as the doping is increased.

> **Example** Find the built-in potential of a junction formed by doping both the n- and p-sides to a concentration of 10^{18} cm^{-3}.
>
> From combining Eqs. (2-29) and (2-16), we have
>
> $$\phi_{\mathrm{bi}} = \frac{kT}{q} \ln \frac{N_{\mathrm{A}} N_{\mathrm{D}}}{n_i^2}$$
>
> $$= \frac{(0.0259 \text{ eV})(1.6 \times 10^{-12} \text{ erg eV}^{-1})(10^{-7} \text{ J erg}^{-1})}{1.6 \times 10^{-19} \text{ C}}$$
>
> $$\times \ln \frac{(10^{18} \text{ cm}^{-3})^2}{(1.45 \times 10^{10} \text{ cm}^{-3})^2}$$
>
> $$= 0.935 \text{ V}$$

To see the usefulness of a pn junction, we must put it into an electrical circuit and apply voltages to it. Now the junction is not in equilibrium, and the Fermi level will no longer be constant. Instead, the Fermi level will vary to match the applied potentials at either end of the semiconductor. Figure 2-10b shows what happens if a negative bias $-V_{\mathrm{b}}$ is applied to the p-doped end of the semiconductor. This is called *reverse bias.* The applied voltage attracts holes away from the junction on the p-doped side; electrons are drawn away from the junction on the n-doped side. The depletion zone increases in width. Equation (2-28) remains valid, except that the potential across the junction is now

$$\phi = \phi_{\mathrm{bi}} + V_{\mathrm{b}} \tag{2-30}$$

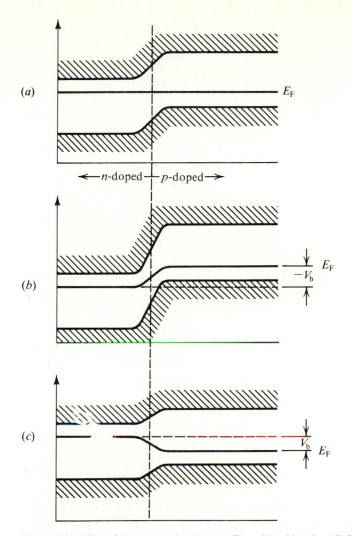

(a)

E_F

←—n-doped—┼—p-doped—→

(b)

$-V_b$ E_F

(c)

V_b E_F

Figure 2-10 Effect of bias on a pn junction. (a) The unbiased junction. (b) Reverse bias. (c) Forward bias.

Under these conditions, there is no flow of charge carriers across the junction and thus no current flow. A reverse-biased pn junction conducts no current and acts like an open circuit.

Now we apply *forward bias* to the junction by biasing the p-doped end positively with respect to the n-doped end. The potential drop decreases, and the depletion zone narrows. As the applied bias approaches the built-in bias, the depletion zone disappears, and large quantities of charge carriers flow across the junction. Any further increases in voltage will lead to massive increases in current, so that the junction exhibits very low impedance. A forward-biased pn junction conducts current freely.

We should note that the one-way-flow properties of the pn junction are not unlimited. Under moderate reverse bias, there is some current flow (usually nanoam-

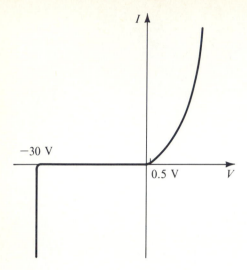

Figure 2-11 Current-voltage characteristics of a *pn* junction, showing some typical voltage values. (Note that the voltage scale is not linear.)

peres). This current results from the generation of electron-hole pairs in the depletion region. At sufficiently large reverse biases, there is a catastrophic increase in the current, known as *junction breakdown. Avalanche breakdown* results from acceleration of charge carriers above the ionization energy of the semiconductor by the field in the depletion zone. *Zener breakdown* results from the ionization of atoms near the junction by tunneling rather than impact. In both cases, breakdown occurs at a well-defined value of the electric field within the depletion zone and is avoided by keeping the reverse bias at reasonable levels. The voltage at which breakdown voltage occurs (denoted V_{BV} or BV) can be used as a check on the properties of junctions.

Figure 2-11 depicts the current-voltage characteristics of a *pn* junction. Current can flow through a junction in only one direction: the forward biased one. The *pn* junction is thus a *rectifier,* in itself a useful electronic device. The junction has two other important properties. Forward bias leads to the *injection* of charge carriers across the junction, where they become *excess minority carriers*. Reverse bias creates a depletion zone of variable width in which there are few carriers. When combined, these properties result in more versatile devices.

2-3 FIELD EFFECTS ON CARRIER DENSITY: METAL-OXIDE-SEMICONDUCTOR STRUCTURES

Doping a semiconductor is one way to change the concentration of charge carriers. Another is by applying an electric field. This is done by using the semiconductor as one electrode of a capacitor.

Figure 2-12a shows a conventional capacitor, constructed of two metallic conductors separated by an insulating layer of thickness *s*. We recall from elementary

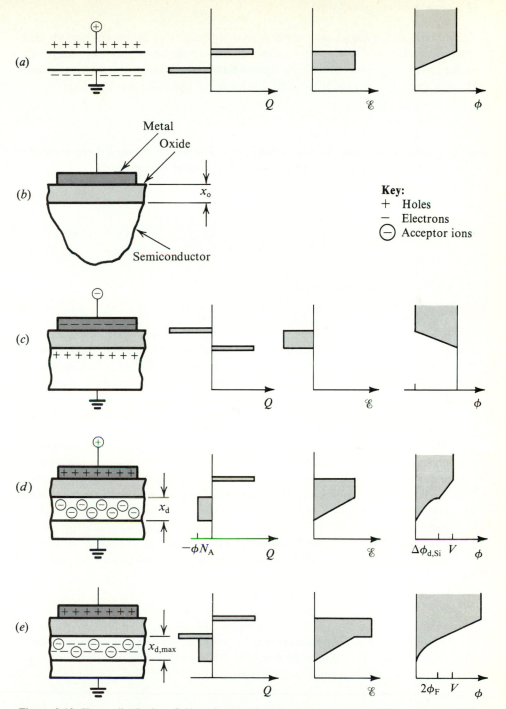

Figure 2-12 Charge distribution, field, and potential in classical and p-type MOS capacitors. (a) The classical capacitor. (b) Structure of an MOS capacitor. (c) Accumulation: charge stored as accumulated majority carriers. (d) Depletion: charge stored as a depletion zone. (e) Inversion: charge stored both in depletion zone and in minority carriers.

electrostatics that in such a case, application of a voltage V to the upper electrode leads to an accumulation of charge $+QA$ on the upper electrode and a matching charge $-QA$ on the lower electrode. Here, as throughout this book, Q represents a charge density per unit area; A is the electrode area. Between the electrodes exists a field:

$$\mathscr{E} = \frac{Q}{\epsilon} \tag{2-31}$$

where ϵ is the permittivity of the insulating layer. The potential drop $\Delta\phi$ experienced in crossing the field region is given by integrating $\mathscr{E}$ over s to give

$$\Delta\phi = \frac{Qs}{\epsilon} \tag{2-32}$$

or

$$\Delta\phi = \frac{Q}{C} \tag{2-33}$$

where C is the *capacitance per unit area* given by

$$C = \frac{\epsilon}{s} \tag{2-34}$$

The potential drop $\Delta\phi$ equals the applied voltage V, so that the second electrode remains at ground potential. In the classic capacitor, the metallic conductors are abundant sources of charge carriers, and the charge appears in a very thin sheet at the conductor surface. The figure shows the resulting charge distribution and electric potential.

Figure 2-12*b* shows a capacitor in which one electrode is a semiconductor. Such devices consist of a semiconductor, an insulating layer usually made of silicon oxide, and a metallic electrode. Therefore, they are called *MOS* (metal-oxide-semiconductor) or *MIS* (metal-insulator-semiconductor) devices. In the MOS capacitor, the supply of charge carriers is limited, and the compensating sheet of charge may appear at various depths within the semiconductor, depending on the applied voltage. Therefore, before studying them, we need to expand Eq. (2-33) to allow for more complex charge distributions.

If a sheet of charge is induced in the MOS device at a distance x from the metal-oxide interface, Eq. (2-31) still gives the field, and integration for $\Delta\phi$ gives

$$\Delta\phi = \int_0^x \frac{Q}{\epsilon(x')}\, dx' \tag{2-35}$$

The charge may also be spread into a more complex distribution, which we can describe by a charge density function $\rho(x)$. In this case, we consider the distribution as composed of sheets of charge of areal density $dQ = \rho(x)/dx$, and sum their effect. The field induced by the sheet of charge at x is

$$d\mathscr{E} = \frac{\rho(x)\, dx}{\epsilon} \tag{2-36}$$

and thus the potential drop experienced due to the sheet is

$$d\phi(x) = \int_0^x \frac{\rho(x)\ dx}{\epsilon(x')}\ dx' \tag{2-37}$$

Note that this integration is over x' and integrates the field between zero and x caused by the sheet of charge around x and no other charge. The overall potential drop is then found by integrating over x to sum up the charge distribution

$$\Delta\phi = \int_{\text{Distribution}} d\phi \tag{2-38}$$

We can now apply these formulas to obtain the charge distribution, capacitance, and potential behavior in an MOS device in which an effective voltage V' is applied to the metal electrode. Figure 2-12c shows the effect when V' is negative, and the semiconductor is p-type. (The n-type case is identical except that the sign of the voltage is reversed. The implications of the "effective" applied voltage will be developed in Chapter 3.) Note that this "forward biases" the device: the p-type material is more positive than the "no-type" metal electrode. The majority carriers, in this case holes, are attracted by the applied voltage, resulting in an *accumulation* of majority carriers at the semiconductor-insulator interface. In this *accumulation* regime, the MOS capacitor acts just like a conventional capacitor, with a layer of charge at the interface and a capacitance C_o given by

$$C_o = \frac{\epsilon_{ox}}{x_o} \tag{2-39}$$

where x_o is the thickness of the oxide layer.

Example Find the capacitance per unit area of an oxide layer 0.1 μm thick. The permittivity of silicon dioxide is found by multiplying its dielectric constant (found in Table 3-3) by the permittivity of free space to obtain

$$\epsilon_{ox} = 3.46 \times 10^{-13}\ \text{F cm}^{-1}$$

The capacitance is then

$$C_o = \frac{3.46 \times 10^{-13}\ \text{F cm}^{-1}}{10^{-5}\ \text{cm}}$$

$$= 3.46 \times 10^{-8}\ \text{F cm}^{-2}$$

The charge per unit area accumulated is given by

$$Q_{\text{acc}} = \frac{V'}{C_o} \tag{2-40}$$

When the MOS capacitor is "reverse-biased" by making V' positive, accumulation gives way to *depletion*, as shown in Fig. 2-12d. The field repels majority carriers and creates a *depletion region*, similar to the one formed in a *pn* junction, in which no mobile charge carriers are found. The charge density in the depletion region is given by

$$\rho = -qN_A \tag{2-41}$$

We can find the potential drop through the oxide and semiconductor by applying Eq. (2-37) to obtain

$$d\phi = \int_0^{x_o} \frac{-qN_A}{\epsilon_{ox}} \, dx' + \int_{x_o}^{x_o+x_d} \frac{-qN_A}{\epsilon_S} \, dx' \tag{2-42}$$

where x_d is the thickness of the depletion layer and ϵ_S is the permittivity of the semiconductor. Carrying out the integration results in

$$d\phi = \frac{-qN_A}{C_o} \, dx + \frac{-qN_A}{2\epsilon_S} x_d \, dx \tag{2-43}$$

Integration over all the charge in the depletion layer gives

$$\Delta\phi = \frac{-qN_A}{C_o} x_d + \frac{-qN_A}{2\epsilon_S} x_d^2 \tag{2-44}$$

where the first term reflects the potential drop through the oxide, and the second, the drop within the depletion layer itself. The figure illustrates the resulting charge distribution and potential variation for the MOS capacitor in *depletion*.

When the semiconductor is depleted, any additional charge on the metal electrode results in a widening of the depletion region, so that compensating charge appears at depth x_d in the silicon. As a result, the capacitance is the series capacitance of the oxide and the depletion layer:

$$\frac{1}{C} = \frac{1}{C_o} + \frac{1}{\epsilon_S/x_d} \tag{2-45}$$

Thus, at zero voltage, the capacitance merges smoothly with the value for accumulation; as the applied voltage becomes more negative, the depletion layer widens and the capacitance falls. The width of the depletion layer can be related to the applied voltage by solving Eq. (2-44), realizing that $\Delta\phi$ must equal V'.

Although the applied field repels majority carriers, it also attracts minority carriers. At sufficiently large negative fields, this effect predominates and the semiconductor enters *inversion,* shown in Fig. 2-12e. Inversion begins when there are as many electrons at the semiconductor surface as there were originally holes. At this point, the surface has become effectively *n*-type. This happens when the Fermi level at the surface is as close to the valence band as the bulk Fermi level is to the conduction band. We can express this condition in terms of the Fermi potential ϕ_F as

$$\Delta\phi_{d,Si} = 2\phi_F \tag{2-46}$$

Here $\Delta\phi_{d,Si}$ is the amount of band-bending, which is identical to the potential drop across the depletion layer, given by the second term on the right-hand side of Eq. (2-44). It follows that the maximum width of the depletion layer, before inversion occurs, is given by

$$x_{d,max} = \sqrt{4\epsilon_S\phi_F/qN_A} \qquad (2\text{-}47)$$

Since $\Delta\phi$ equals V', Eq. (2-46) defines the applied voltage at which inversion begins in terms of ϕ_F, which is determined by the doping density.

Once inversion begins, additional charge on the metal electrode is balanced by electron charge in the semiconductor. If capacitance is tested using a low-frequency signal, the electrons will have time to migrate to the silicon surface, and the effective insulator thickness will again be x_o. In this case, the capacitance C_{inv} in inversion will be C_o. At higher frequencies, electron charge appears at the edge of the depletion layer, and the capacitance is given by Eq. (2-45) with $x_d = x_{d,max}$. The electron charge Q_{inv} contributes a potential drop given by

$$\Delta\phi = \frac{Q_{inv}}{C_{inv}} \qquad (2\text{-}48)$$

Adding this potential drop to the depletion-layer contribution given by Eq. (2-43) and using Eq. (2-46) for the depletion layer width gives an expression for Q_{inv}:

$$Q_{inv} = -\frac{1}{C_o}[V' - 2\phi_F] - qN_Ax_{d,max} \qquad (2\text{-}49)$$

(where we have used the low-frequency value for C_{inv}).

To summarize, we have seen that the capacitance of an MOS capacitor varies with applied voltage because there are three separate modes of charge storage in the semiconductor. Accumulation occurs for negative voltages, depletion for moderate positive voltages, and inversion predominates at higher positive voltages. Figure 2-13 shows a *capacitance-voltage (C-V)* curve for a *p*-type MOS device, along with the voltage conditions, capacitance, charge distribution, and potential drop for each mode. As we shall see in Chapter 3, the characteristic shape of the *C-V* curve turns out to be very useful in monitoring the properties of oxides. It is also the basis for the MOS transistor, which we shall see in the next section.

2-4 APPLICATIONS OF SEMICONDUCTOR PROPERTIES

Three sections into this chapter, we have already seen all the properties that make semiconductors useful. Since this book deals with processing rather than device physics, we will not pursue this subject much further. But before leaving it, it seems worthwhile to give a brief survey of how semiconductor properties are applied that may serve as an introduction to device concepts for the student with no prior exposure to the subject.

What makes semiconductors unique and uniquely useful is the mechanism by which they conduct. Metals possess an abundance of charge carriers, and the study of metallic conductivity is the study of carrier *movement*. Semiconductors are characterized by a scarcity of charge carriers, and semiconductivity is dominated by factors

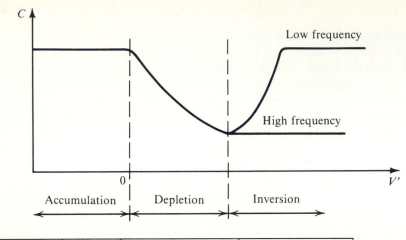

Figure 2-13 Capacitance-voltage characteristic of an MOS device with important properties of the accumulation, depletion, and inversion regimes.

Voltage limits	$V' < 0$	$V' > 0$	$\phi_{d,Si} \geq 2\phi_F$
Charge storage	Majority carriers	Dopant ions in depletion layer	Minority carriers
Capacitance	$C_o = \dfrac{\epsilon_{ox}}{x_o}$	$C = \left(\dfrac{1}{C_o} + \dfrac{1}{\epsilon_S/x_d}\right)^{-1}$	Low frequency: C_o High frequency: $\left(\dfrac{1}{C_o} + \dfrac{1}{\epsilon_S/x_{d,max}}\right)^{-1}$

determining carrier *supply*. By focusing on these factors, we have derived a number of semiconductor properties.

These properties include:

A conductivity mechanism dominated by electron transition between two bands, separated in energy

A resistivity that is adjustable by the addition of trace amounts of impurities

The formation of a rectifying junction between differently doped regions

The formation of a depletion zone at a unbiased or reverse-biased junction, or in a "reverse-biased" MOS device

Injection of charge across a forward-biased *pn* junction

Inversion of the surface under sufficient "forward" bias in an MOS device

This list of properties allows the construction of a number of electrical and partly electrical devices.

Semiconductors are used most widely in purely electrical devices, which manipulate only electrical energy. Traditionally, electrical circuits were constructed by combining resistors, inductors, capacitors, and two types of vacuum tube: the diode or rectifier and the triode. With the exception of the inductor, each of these elements can be duplicated by microscopic bits of silicon, which are then connected to make circuits of great complexity.

Resistors are easily formed in a volume of silicon, by doping the region to the appropriate resistivity. Either *n*- or *p*-type dopants will serve, and the required dopant level may be read off a graph like that of Fig. 2-6 or calculated from Eq. (2-10).

The MOS device can serve as a capacitor. A reverse-biased *pn* junction also has capacitance, because it consists of two conductive, carrier-rich regions, separated by a depletion zone that is devoid of free charge carriers and thus nonconducting. The capacitance of a reverse-biased *pn* junction can be used intentionally, or it may appear as an unwanted design constraint in a semiconductor circuit.

As for the diode or rectifier, the *pn* junction has exactly this property. The triode remains: its semiconducting incarnation is the transistor. Transistors, like triodes, are devices connecting two circuits, an input and an output. This is done using three terminals, one of which is common to the two circuits, as shown in Fig. 2-14. The essential property of the transistor is that a small signal applied to the input drives a large signal at the output. In an analog mode, this property is used for *amplification:* an input of arbitrary shape is reproduced, at higher amplitude, at the output. In digital electronics, a transistor is used as a *switch:* a low-power on-off signal at the input results in a on-off output of sufficient additional power to drive several additional switches.

Both *pn* junctions and MOS capacitors can be used to form transistors, resulting in two rival semiconductor technologies: the *bipolar* method and the *MOS* method. The bipolar transistor relies on the injection of carriers across a forward-biased junction. It contains two *pn* junctions back to back, as in Fig. 2-15. The middle region, common to both junctions, is the *base*. One junction, involving the base and *emitter*, is forward biased by the input. The emitter injects a large number of carriers into the base, with a magnitude strongly dependent on the emitter-base bias. (The base also injects into the emitter, but this current is not useful: it is minimized by a proper choice of dopant concentrations.) The other junction, consisting of the base and the *collector*, is reverse-biased and thus is swept clear of majority carriers. However, if the base is narrow enough, most of the injected *minority* carriers from the emitter will reach the

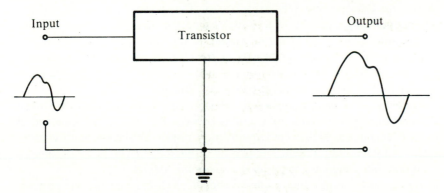

Figure 2-14 The concept of the transistor.

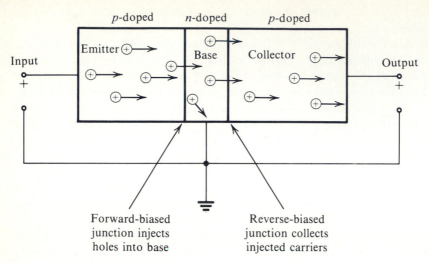

Figure 2-15 A bipolar transistor. A *pnp* transistor is shown: the operation would be the same if all biases and dopant polarities were reversed.

base-collector depletion zone, where the resulting field sweeps them *into* the collector to create an output current. The result is a large output current variation, mediated by a small voltage modulation at the input. The term bipolar refers to the use of both *n*- and *p*-type doping in this transistor. Chapter 10 is given over to bipolar fabrication methods.

The MOS transistor is based on the change in conductivity caused by the manipulation of carrier concentrations by the applied voltage. In the MOS transistor, an MOS capacitor is formed at the surface of the semiconductor, as shown in Fig. 2-16. If the device is driven into inversion, a "channel" rich in minority carriers appears at the surface, so that it is effectively *n*-type. An input signal applied to the metal electrode, or *gate,* of the MOS transistor, will modulate the carrier concentration and thus the conductivity of the channel. Power consumption in the input circuit will be negligible, because no current can flow across the oxide barrier. The output circuit is connected to two *n*-type regions on either side of the gate, called the *source* and *drain.* Output current now flows in the semiconductor, parallel to the gate. The resistance to the current is determined by the carrier density in the channel. A large output current variation is thus modulated by the gate voltage with consumption of very little input current. Chapter 11 is devoted to MOS transistor fabrication.

Semiconductors can also be applied to devices not purely electrical. We have repeatedly mentioned the transition of electrons between the valence and conduction bands of a semiconductor. This event requires the emission or absorption of energy: either thermal or light energy. If light energy is involved, the semiconductor becomes a tool for coupling light and electricity: an *optoelectronic* device.

A device designed for maximum recombination of electron-hole pairs will produce light from electrical energy. Examples are the *light-emitting diode* or *LED* and the

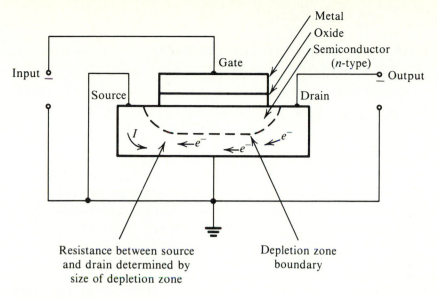

Figure 2-16 An MOS transistor.

semiconductor laser. Conversely, by deliberately using light energy for the generation of electron-hole pairs, we can produce electricity from light. Examples of this application are the space-age solar cell and the old-fashioned photocell light sensor. Chapter 12 covers optoelectronic applications in more depth.

A third semiconductor application is of growing importance: the semiconductor serves simply as a substrate, whose electronic properties are not used at all. In this way, existing microfabrication knowledge is applied to create devices that are small but not necessarily electronic. Devices as disparate as ink-jet print nozzles and Josephson junctions have been made using silicon because the material and the fabrication techniques were readily available.[6,7]

2-5 CRYSTALLINE SEMICONDUCTORS

Most semiconductor applications require the regular and precise structure found in a single crystal. A semiconductor substrate is commonly a slice, cut from a single large crystal, and the properties of the crystal significantly affect the behavior of the fabricated device.

All crystals are based on the repetition of a characteristic unit cell. Semiconductor crystals are members of the cubic crystal family and have a unit cell based on the cube. For such crystals there is a useful system of notation based on numbers known as *Miller indices*. Consider an arbitrary plane in three-dimensional space, with the equation:

$$\frac{x}{\alpha} + \frac{y}{\beta} + \frac{z}{\gamma} = 1 \tag{2-50}$$

Here x, y, and z are any points in the plane, and α, β, and γ are the intercepts formed by the plane with x, y, and z axes. This equation can be rewritten as

$$hx + ky + lz = 1 \tag{2-51}$$

where h, k, and l are the reciprocals of the intercepts: that is,

$$h = \frac{1}{\alpha} \tag{2-52}$$

etc. If the plane we are discussing is a plane in a cubic crystal, the numbers h, k, and l identify the plane uniquely, so the plane can be identified by the shorthand notation (hkl). Normal to the plane is a direction, whose projections onto the axes are h, k, and l. This direction is written $[hkl]$. Because a crystal is a repeating array, there will in fact be many planes parallel to our chosen one, with the same normal direction. In an ideal crystal, these planes are physically indistinguishable, and we can chose to represent all of them by reference to one plane whose Miller indices are a convenient combination of integers.

Figure 2-17 gives some examples of the application of this system. For a cube centered at the origin, the cube walls correspond to planes with intercepts such as $(1, \infty, \infty)$ or $(\infty, 1, \infty)$, resulting in indices such as (100) or (010). Negative directions are shown as overscores, so the cube wall in the negative x direction has the indices $(\bar{1}11)$. Minor diagonals will have indices such as (110), and the major diagonals have indices such as (111). If we consider planes involving multiple cells, we may obtain sets of intercepts such as $(2, 1, 0)$, implying indices such as $(\frac{1}{2}10)$: in such cases we conventionally use instead the indices (120). Frequently, we desire to speak of a family of planes of identical symmetry, such as the minor diagonals of the cube, without regard to the distinction between positive and negative directions, or x, y, and z axes. This set of planes is denoted using braces: $\{110\}$ includes the planes (110), $(\bar{1}10)$, (101), etc. The equivalent set of directions is identified with angle brackets: $\langle 110 \rangle$.

A few other properties of this system are worth noting. The angle between two directions $[h_1 k_1 l_1]$ and $[h_2 k_2 l_2]$ is given by

$$\cos \theta = \frac{h_1 h_2 + k_1 k_2 + l_1 l_2}{\sqrt{(h_1^2 + k_1^2 + l_1^2)(h_2^2 + k_2^2 + l_2^2)}} \tag{2-53}$$

The distance d between adjacent planes along the direction $[hkl]$ is given by

$$d = \frac{a}{\sqrt{h^2 + k^2 + l^2}} \tag{2-54}$$

where a is the length of a side of the unit cube. Two planes $(h_1 k_1 l_1)$ and $(h_2 k_2 l_2)$ intersect along a line of direction $[hkl]$, where

$$h = k_1 l_2 - k_2 l_1 \tag{2-55a}$$
$$k = h_1 l_2 - h_2 l_1 \tag{2-55b}$$
$$l = h_1 k_2 - h_2 k_1 \tag{2-55c}$$

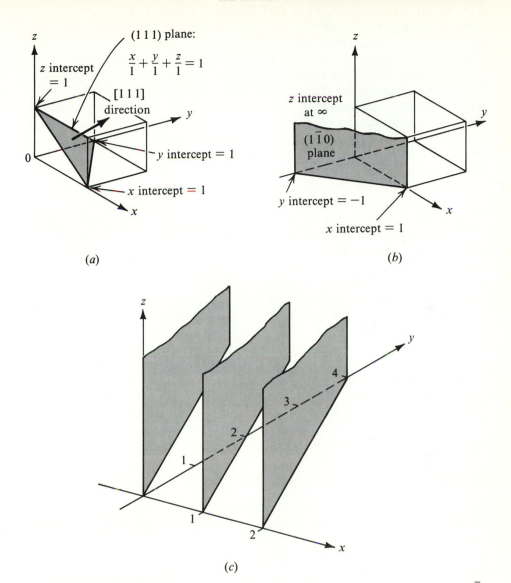

Figure 2-17 Some examples of Miller indices. (a) The (111) plane and the [111] direction. (b) The (1$\bar{1}$0) plane. (c) A set of (120) planes.

Example Find the distance separating atomic planes in the $\langle 111 \rangle$ direction is a silicon crystal ($a = 0.543$ nm), and the line of direction along which the $\langle 111 \rangle$ and $\langle 11\bar{1} \rangle$ planes meet.

From Eq. (2-54), $d = 0.543$ nm/$\sqrt{3} = 0.313$ nm. From Eq. (2-55), the line of direction is [$\bar{2}\bar{2}0$], which is equivalent of [$\bar{1}\bar{1}0$].

The simplest type of cubic lattic is, appropriately, the simple cubic lattice, shown in Fig. 2-18a. This lattice is rarely found in nature. More common are the *body-centered cubic (bcc)* and *face-centered cubic (fcc)* lattices, shown in Figs. 2-18*b* and *c*. In the body-centered lattice, the eight corner atoms are joined by one at the center of the unit cube; in the face-centered lattice, there are eight corner and six face atoms.

Common semiconductor crystals are based on an elaboration of the fcc lattice. In this structure, a second fcc lattice is built in the interstices of the first one, with an origin located at (¼ *a*, 1/4*a*, ¼ *a*). Figure 2-18*d* shows this lattice, with the "old" fcc lattice made of dark spheres and the "new" one of white. In Fig. 2-18*e*, the same lattice is shown demonstrating the bonding of each sphere with four nearest neighbors. When this lattice is built of alternating atoms of different types, as in Fig. 2-18*d*, it is called the *zincblende* lattice. Gallium arsenide has this crystal structure. If all atoms in the lattice are identical, the lattice is a *diamond* lattice. This is the crystal lattice of the Group IVB elements carbon (in its diamond form), silicon, and germanium. The diamond lattice can be broken up into subcells, one of which is shown in Fig. 2-18*f*.

The diamond lattice is fairly loosely packed. If each atom is considered a sphere, touching its nearest neighbors, the proportion of the unit cell occupied by atoms is around 34%, versus 74% for a bcc lattice. The coordination number, or number of nearest neighbors, is four. The distance between nearest neighbors is $(\sqrt{3}/4)\,a$, where *a* is the unit cube edge length. Along the $\langle 100 \rangle$ directions, crystal planes are separated by length *a*; along the $\langle 110 \rangle$ directions by $0.707a$, and in the $\langle 111 \rangle$ directions by $0.597a$. Thus the $\langle 111 \rangle$ direction is the most compact. Located within the diamond lattice are a number of cavities or interstitial voids. If we pack the lattice with hard spheres, each of these voids is sufficient to contain an additional atom, although there is not room for an atom to pass from one void to the next. One of these voids is at the center of the unit cell; four others are midway between the center and the cell corners. Additional voids are found at the center of each cube edge: there are twelve edges, but each void is shared by four unit cells, so the average number of voids per cell is three. This makes a total of eight voids per cell.

Crystal Defects

Real crystals never achieve complete regularity. Instead, various types of *defects* occur, partly in response to thermal energy and partly due to external stress. Defects can be divided into point and line defects.

Point defects involve only a single lattice location. There are three logical possibilities. A lattice site, supposed to be occupied, may be empty: this is a *vacancy* defect. Or a nonlattice location, supposed to be empty, may contain an atom: this is an *interstitial* defect. Finally, a lattice site may be occupied by an impurity atom: this type of disorder is a substitutional "defect." Figure 2-19 illustrates these defects. Some number of point defects is inevitable, due to the thermal energy in the crystal. Each defect has an energy of formation, and we can expect there to be an equation for the equilibrium density of such defects, containing a Boltzmann factor. Indeed, the equilibrium density of vacancy, or Schottky, defects is given by

$$n_S = N \exp \frac{-E_S}{kT} \tag{2-56}$$

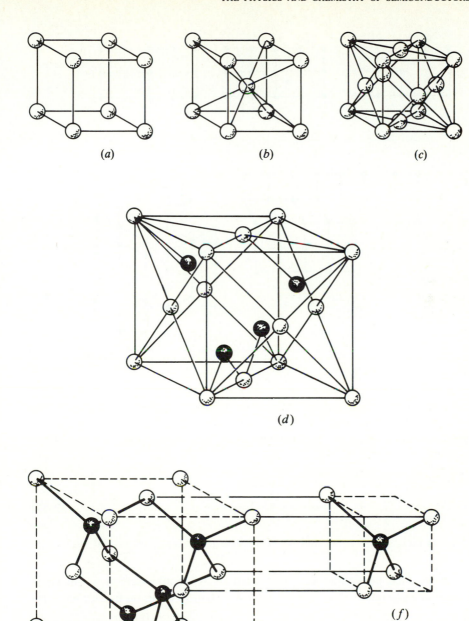

Figure 2-18 Cubic lattices. (*a*) The simple cubic lattice. (*b*) Body-centered cubic lattice. (*c*) Face-centered cubic lattice. (*d*) The zincblende lattice. (*e*) Bonding in the zincblende lattice. (*f*) A subcell of the diamond lattice (degenerate zincblende lattice).

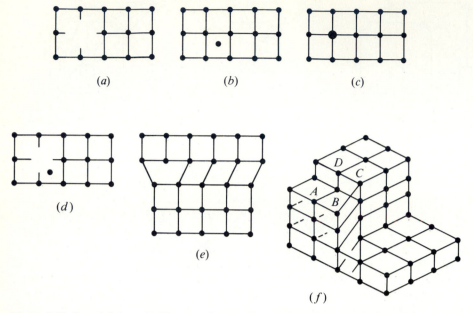

Figure 2-19 Crystal defects. (*a*) Vacancy. (*b*) Interstitial. (*c*) Substitutional. (*d*) Frenkel defect. (*e*) Line dislocation. (*f*) Screw dislocation.

where E_S, the energy of formation for Schottky defects, is 2.3 eV for silicon. Here N is the number density of crystal atoms.[8] Another type of defect with a name is the Frenkel defect, consisting of a vacancy and an adjacent interstitial. This pair of defects arises from pushing a lattice atom out of its place and has an equilibrium density

$$n_F = \sqrt{NN'} \, \exp \frac{-E_F}{2kT} \tag{2-57}$$

where E_F is around 0.5–1 eV, and N' is the concentration of interstitial sites.[8]

Example Find the equilibrium concentration of Schottky defects in silicon at (*a*) room temperature of 300°K; (*b*) 1200°K, a typical diffusion furnace temperature.

The value of kT at room temperature is 0.0259 eV, so at 300°K

$$N_S = (5.0 \times 10^{22} \text{ cm}^{-3}) \exp \frac{-2.3 \text{ eV}}{0.0259 \text{ eV}}$$

$$= 1.36 \times 10^{-16} \text{ cm}^{-3}$$

which is essentially zero. However, at 1200°K

$$N_S = (5.0 \times 10^{22} \text{ cm}^{-3}) \exp \frac{-2.3 \text{ eV}}{0.104 \text{ eV}}$$

$$= 1.14 \times 10^{13} \text{ cm}^{-3}$$

which is within a few orders of magnitude of common dopant concentrations.

Line defects involve stacking failures of the lattice as a whole, leading to lines along which lattice bonding is incomplete. Such defects, termed *dislocations,* include screw and edge dislocations. Figure 2-19*e* shows an edge dislocation, which occurs when there is an additional plane of atoms in the crystal structure. At the boundary of this plane there is a line where bonding is disturbed: this is the dislocation. Figure 2-19*f* illustrates a screw defect. Normally, following a path such as *ABCD* along the bonds in a crystal would return us to our starting point, but in a screw defect, the bonding is distorted so that the path *ABCD* ends up on a different plane. Following this path repeatedly would "advance" us through the crystal, just as a screw thread advances with rotation. The line defect lies along the axis of the "screw." We have illustrated defects in a conceptually simple lattice; dislocations in the diamond lattice are more complex and may involve combinations of edge and screw dislocations.

Defects have both electronic and chemical effects on the semiconductor crystal. Point defects involve either extra bonding electrons or unsatisfied bonds and thus provide electronic states not present in the ideal crystal. In addition, they add strain to the material. Point defects may interact so as to increase their stability. For example, two independent vacancies produce eight dangling bonds; two adjacent vacancies require only six. Thus defect clusters may be more stable and thus more common than individual defects.

Line defects possess all the characteristics of an array of point defects. In addition, they are sites of considerable strain and thus distort the energy-band structure of the crystal. The diffusion of impurities is enhanced near line defects, so that "pipes" of dopant may accumulate along them during dopant diffusion. They also tend to collect metallic impurities, which can cause circuit failure. All of these deleterious effects are amplified by the ability of line defects to move within the crystal. Figure 2-20, for example, illustrates the movement of an edge defect. By the successive making and

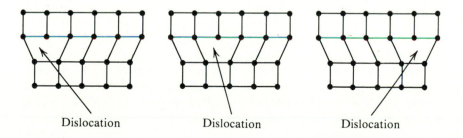

Dislocation   Dislocation   Dislocation

Figure 2-20 Movement of a line dislocation by slip. Note that only the dislocation moves: there is no motion of atoms.

breaking of bonds, the dislocation line travels through the crystal. Note that no atoms move during this process. This process is called *slip*. Another mechanism, called *climb*, allows dislocations to move in the perpendicular direction.

With all of these detrimental effects, it would seem that defects are always to be avoided. Indeed, defect prevention is an important consideration in processing. Defects are reduced by minimizing stress, whether from mechanical, thermal, or chemical sources. Mechanical stress results from such things as scribing the wafer surface, for example, to imprint an identification number. Thermal stress results from the heating and cooling of substrates. While some heating is unavoidable in processing, lower temperatures, slower heating and cooling rates, and more uniform application of heat all help prevent defects. Thermal stress is aggravated by larger and thicker wafers, so that each increase in wafer size requires more care in thermal processing. Impurities, either dopants or contaminants, also induce stress. Stress due to doping is a function of the misfit factor in size between substrate and dopant atoms, and is minimized by a careful choice of dopant. Stress-producing impurity precipitates can also form either during crystal growth or subsequent processing.

However, absolute prevention of defects is unattainable, if only because there is an equilibrium density for each type of defect. For this reason, defects are sometimes intentionally introduced into one part of a crystal in order to reduce their density elsewhere. In a process called *gettering,* strain is induced in an electrically inactive part of a substrate, in order to attract defects away from the electrically active region. One method of gettering is to diffuse a high concentration of phosphorus into the back of a substrate, creating defects and stress.[9] Another method uses the controlled growth of oxygen impurity precipitates, unavoidably present in silicon, to produce a *denuded zone* free of defects at the substrate surface.[10,11] Such techniques, while widely used, are not yet understood in detail.

To control defects, they must first be measured. For common use, the principle means of defect detection is chemical etching. Since a defect area is under higher stress, it will etch more quickly in a selective etchant than will the bulk semiconductor. The result will be a pit in the semiconductor surface, at the location of the defect. One common etchant is Sirtl etch, announced by E. Sirtl and A. Adler in 1961.[12] Other formulations include Seeco and Wright etch.[13,14] Table 2-4 gives the compositions of these etchants. They usually include chromium oxide, hydrofluoric acid, and nitric acid.

The etching of defects produces pits or mounds of distinctive shapes, depending on the etchant, the nature of the defect, and the orientation of the crystal. With silicon, for example, the {111} planes are the densest and thus the slowest etching. If the substrate surface is a {111} plane, etching of the surface will be slow and uniform. However, once a defect is encountered, faster etching takes place around the defect until the {111} surfaces within the crystal are exposed. These planes then etch uniformly, producing a pit whose well-defined shape reflects the way {111} planes meet the (111) surface. In this case the pit will be triangular, as shown in Fig. 2-21.

Preferential etching can be used for reasons other than defect detection. If etch rate depends on crystal orientation, preferential etching can create trenches or grooves on the surface of a substrate. These grooves will have well-defined orientations and

Table 2-4 Some common selective etchants for defect decoration in silicon

Sirtl etch:	50% by volume HF
	50% by volume 5M CrO_3 in H_2O
	Typical etch time: 5 min
Secco etch:	67% by volume HF
	33% by volume 0.15M $K_2Cr_2O_7$ in H_2O
	Typical etch time: 5 min
Wright etch:	60 ml HF
	30 ml HNO_3
	60 ml CH_3COOH (glacial)
	60 ml H_2O
	30 ml of a solution of 1 g CrO_3 per 2 ml H_2O
	2 g $(CuNO_3)_2 \cdot 3\, H_2O$

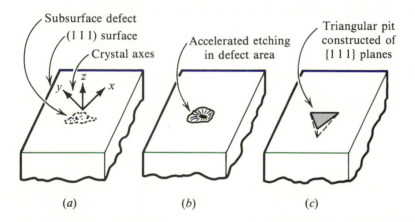

Figure 2-21 Decoration of a defect on $\langle 111 \rangle$ silicon by selective etching. (*a*) Prior to etching, showing the crystal axes and the buried defect. (*b*) Accelerated etching around the defect when it is exposed to the etchant. (*c*) Formation of a triangular etch pit.

shapes, lying along principal crystal planes. Several fabrication technologies have been devised that make use of this method.

2-6 SILICON: AN ELEMENTAL SEMICONDUCTOR

Silicon is the most widely used semiconductor today. Silicon is an element and is fairly inert at room temperature. It resists acids under normal conditions, but does react with dilute alkalis, halogens (chlorine, fluorine), and certain combinations of acids with strong oxidants. In the form of a single crystal, it has a grayish silvery color and polishes to a high metallic luster. Silicon is brittle in its crystalline form, and silicon wafers break easily into knifelike shards.

Silicon has a diamond lattice with a cube edge dimension a of 0.543 nm. The spacing between adjacent atoms is therefore 0.235 nm. In this lattice, the spacing is closest between {111} planes, so that the ⟨111⟩ direction shows the fastest growth during crystal formation. For the same reason, these planes are the slowest etching and the slowest to form alloys when heated in contact with metals. Thus silicon crystals oriented in the ⟨111⟩ direction are easily grown, polish and etch smoothly, and form smooth alloy junctions. These properties make *⟨111⟩ silicon* (that is, slices cut from a crystal oriented along a ⟨111⟩ direction) the material of choice for fabrication in the absence of special considerations. Most bipolar circuits are built on ⟨111⟩ silicon.

Silicon is strongest in the ⟨111⟩ direction and thus cleaves most easily perpendicular to that direction, along {111} planes. These planes intersect a (111) surface so as to form an equilateral triangle. This is an inconvenience in the manufacture of devices built on rectangular chips, since the chips must be cleaved from the wafer without breaking. In practice, the chip is oriented so that one pair of edges lie along a ⟨111⟩ intersection. Thus the wafer is easily cleaved into strips of chips. Scribes are then made between chips and the strips forced to break: the resulting edges are jagged composites of {111} planes.

If silicon is grown in the ⟨100⟩ orientation, some of the {111} planes meet the (100) surface at right angles to each other. It is thus straightforward to scribe and cleave rectangular chips. The (100) surface also has fewer dangling bonds projecting from the surface atoms, which leads to a lower level for some types of electrical noise.[15] ⟨100⟩ silicon is usually chosen for MOS circuits as well as noise-sensitive bipolar ones.

Silicon crystals cannot be grown in the ⟨110⟩ direction, but ⟨110⟩ slices can be produced for specialized applications by taking an oblique slice out of a ⟨111⟩ crystal.

To exploit orientation effects in cleaning or etching, it is necessary to know the orientation of the crystal slice being processed. This would be difficult with a slice cut from a perfectly cylindrical crystal, since the slice would be circular without distinguishing marks. Therefore, as we saw in Chapter 1, wafer manufacture includes a step in which a flat surface is ground into one side of the cylinder. Now the slices cut from it have a flattened side, or *flat,* as shown in Fig. 2-22. The flat shows the crystal orientation, and the grid of chips to be fabricated is oriented relative to the flat. In

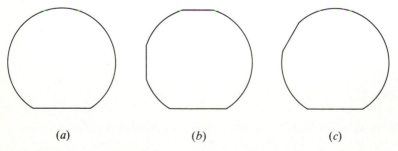

(*a*)   (*b*)   (*c*)

Figure 2-22 Wafers with flats. (*a*) Primary flat only. (*b*) Primary flat and two secondary flats, ⟨100⟩ silicon. (*c*) Primary flat and secondary flat, ⟨111⟩ silicon.

addition to this *major flat,* other, smaller flats are frequently added to identify wafers with differing resistivities or orientations. By this method, material type can be easily determined during fabrication.

We have already discussed doping, using silicon as an example. We saw that the common *n*-type dopants of silicon are taken from Group VB of the Periodic Table, and that the *p*-type dopants come from Group IIIB. Boron is the most common *p*-type dopant for silicon; phosphorus, arsenic, and antimony are all used as *n*-type dopants. We have also mentioned the use of gold as a deep-lying impurity, to increase the speed of devices.

Two other common impurities deserve mention: oxygen and carbon. Oxygen is inevitably present in silicon, due to the use of SiO_2 (*silica* or *fused quartz*) for crucibles for crystal formation. Carbon is a close chemical relative of silicon, and cannot be totally removed from it. Both these impurities, which are often present in concentrations of 10^{16} cm^{-3}, are electrically inert. However, they form microprecipitates in the silicon crystal, and these insulating precipitates cause curvature of field lines in depletion regions. As a result, local field gradients become higher than expected, and junction breakdown occurs at unexpectedly low applied voltages. Oxygen precipitates may also collect defects and impurities. There is increasing interest in manipulating the oxygen content of silicon crystals for the purpose of gettering.[10]

Silicon has become the premier industrial semiconductor due to its ease of fabrication. As we shall see, it conveniently forms a masking layer upon oxidation, facilitating selective doping. It is an element and hence not subject to decomposition. It has a low vapor pressure, a high melting point, and is chemically unreactive, so that no special precautions are required to process it. Silicon adheres well to a number of metals and insulating films, and alloys with aluminum, thus allowing electrical connection between circuit elements. For ease of fabrication, no material is likely to surpass silicon. Only where there are inherent physical limitations on the use of silicon are there incentives for the use of other semiconductors.

2-7 GALLIUM ARSENIDE: A COMPOUND SEMICONDUCTOR

Gallium arsenide is more difficult to process than silicon, but it can be used for purposes for which silicon is inappropriate. To understand why, we must delve deeper into band theory, and explore the notions of direct and indirect semiconductors. We will then look at how the crystal structure, chemistry, and doping behavior of gallium arsenide are affected by the fact that it is a compound rather than an element.

Direct and Indirect Band Gaps

In Fig. 2-3, we presented a simplified picture of the band structure of a semiconductor. The *y* axis represented energy, and the shading represented a crude density of states, but the *x* axis really represented nothing but some generalized distance. We thus implied that the state density of a semiconductor was a function of energy only. In fact, in a regular structure like a crystal, the direction of electron momentum is an important factor, and the state density is in reality a function of both energy and the momentum vector **k**. A detailed calculation produces a band diagram like Fig. 2-23.[16] Here the y axis still represents energy, but the x axis shows the direction of the momentum vector with respect to principal crystal axes.

What is most significant about Fig. 2-23 is that the minimum in the conduction-band energy and the maximum in the valence-band may occur at different values of momentum. Because the transition probability drops off exponentially with energy difference, electron transitions virtually always occur between the valence-band maximum and the conduction-band minimum. When these extrema coincide in momentum, the semiconductor is a *direct band gap semiconductor*. In gallium arsenide, the valence-band maximum and conduction-band minimum both occur at zero momentum, and the semiconductor is direct. Transitions between bands thus require no change in momentum. Irradiation with light of a proper wavelength will create electron-hole pairs without impediment, and excess charge carriers can easily recombine to produce light photons. Gallium arsenide is therefore an effective converter between light and electrical energy, and good material for the fabrication of optoelectronic devices.

In germanium and silicon, the band energy extrema occur at difference values of momentum. These elements are thus *indirect band gap semiconductors*. Creation or annihilation of electron-hole pairs requires an exchange of both energy and momentum. Light quanta carry no momentum, and thus an electron transition cannot involve a photon alone. An interaction with a lattice vibration (quantized as a phonon) or some other momentum source is also required. Such three-body interactions are inherently unlikely; thus neither silicon nor germanium is suitable for optoelectronic applications.

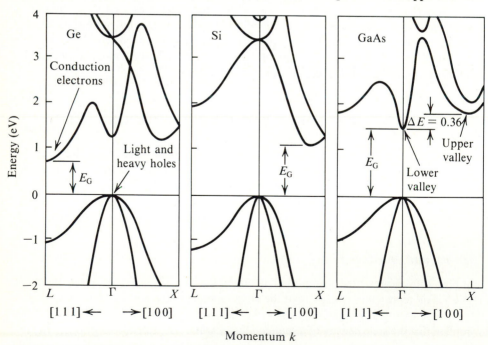

Figure 2-23 Band structures for common semiconductors. (*Reference 16, used by permission.*)

Another advantage of gallium arsenide can also be derived from Fig. 2-23. The slope of the band edge diagram is related to the effective charge carrier mass, m^*. In fact,

$$m^* = \frac{\hbar^2}{\partial^2 E / \partial k^2} \tag{2-58}$$

Here k is the magnitude of $\mathbf{k}$. Note that double lines are shown in some of the band diagrams. In this case, there are "hard" and "easy" directions for electron movement, and thus two effective masses are derived from the slopes. The general shapes and slopes of the valence bands are similar in all of the diagrams, and the effective masses for holes in both silicon and gallium arsenide are in the range $0.15–0.5m_0$ (where m_0 is the electron rest mass).

However, in the conduction band, the situation is different. The minimum for gallium arsenide is much more steeply sloped than that for silicon, and the effective mass for conduction band electrons is $0.068m_0$ in gallium arsenide versus $0.19–0.97m_0$ in silicon. As a result, the electron mobility in GaAs exceeds that in silicon by a factor of almost 6. In MOS devices, speed is determined primarily by the mobility of the majority carrier, and gallium arsenide devices are potentially much faster than silicon devices. As silicon devices approach their performance limits, interest in gallium arsenide increases.

Orientation Effects

Gallium arsenide is a compound, made of equal numbers of gallium and arsenic atoms, with the formula GaAs. Because it is a compound, its crystal is more complicated than that of silicon. In Fig. 2-24a, the GaAs unit cell is shown. White spheres represent gallium atoms, and black represents arsenic. Bonds between atoms are shown with solid lines while dashed lines indicated the outlines of the unit cube. The corners of the cube are labeled to facilitate reference. The GaAs lattice is a *zincblende* lattice: it is identical to the diamond lattice except that it involves two types of atoms.

Note the existence of a W-shaped chain of five atoms diagonally across the top of the cube, from corner *a* to *c*, and the similar "W" from *h* to *f*. On closer examination, other "W"s are found along diagonals of the other faces of the cube. Each atom in the lattice is bonded to four others in a tetrahedral arrangement, and forms part of four W-shaped chains. In Fig. 2-24a, only bonds lying within the cube are shown, and thus only the four gallium atoms are depicted with their full complement of bonds.

In Fig. 2-24b, we have abstracted a (100) face of the cube, selecting the one with corners *cdgh*. On this face, we find five arsenic atoms: four at the cube corners and one in the center. Also shown are the bonds made by these atoms to others within the unit cell. Dotted lines symbolize bonds receding behind the plane of the paper, and wedge shapes illustrate bonds projecting out of the paper. Since all of the bonds that are within the cube must be behind the face we selected, all bonds in Fig. 2-24b are receding.

In Fig. 2-24c, we have added all of the bonds to the atoms on the (100) face, so that each atom has all four of its bonds. We now see that a (100) plane of GaAs contains atoms all of one species (here As), with two bonds per atom extending down

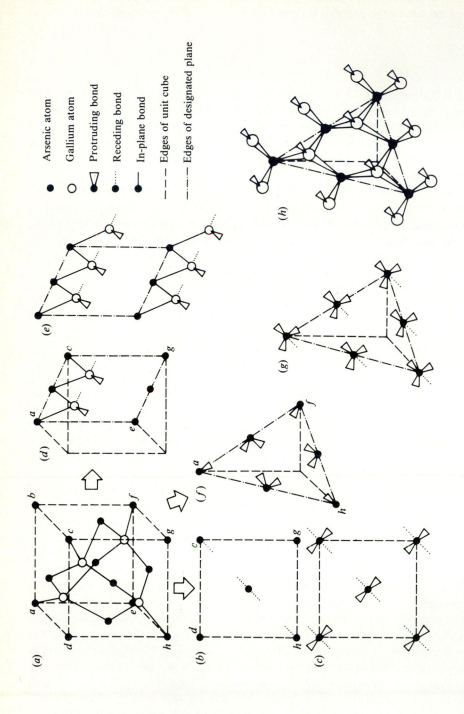

Figure 2-24 Crystal planes of gallium arsenide. (*a*) The GaAs unit cell. (*b*) The (100) plane, abstracted from the unit cell. (*c*) The (100) plane, showing bonds to adjacent atoms outside the unit cell. (*d*) The (100) plane. (*e*) The (100) plane, showing bonds to adjacent atoms. (*f*) The (111) plane. (*g*) The (111) plane, showing bonds to adjacent atoms. (*h*) A more realistic rendering of the (111) plane, showing bonding to the underlying As layer of Fig. 2-24*g*.

into the solid, and two *dangling bonds* reaching up from the surface. There is an average of two atoms per unit cell (one at the center, and four quarter-atoms at the corners). If we moved deeper into the cell to the location of the gallium atoms, we would again find two atoms per cell (both intact this time) and two dangling bonds per atom. (100) faces of GaAs are thus alternatively of all gallium, or all arsenic atoms, with two unsatisfied bonds per atom.

In Fig. 2-24*d*, the (110) plane *aceg* is abstracted. This plane lies along the "W"s in one direction, and thus two bonds per atom lie *in* the (110) plane. In this plane there are both arsenic and gallium atoms. In Fig. 2-24*e*, atoms and bonds outside the unit cell are shown: we now see more clearly that the plane contains equal numbers of gallium and arsenic atoms, with two bonds per atom in the plane and one behind it. There are four atoms per unit cell (two whole gallium atoms, four quarters plus two halves of arsenic atoms). Because of the bonding within the plane and the small number of dangling bonds, the {110} plane is the preferred cleavage plane for GaAs. A (110) plane meets other {110} surfaces at right angles and thus cleavage along {110} planes allows fabrication of precisely rectangular GaAs chips. This is vital for laser diode applications.

Some fascinating things happen in ⟨111⟩ gallium arsenide. In Fig. 2-24*f*, we look at the (111) face *afh*. This plane contains only arsenic atoms. In Fig. 2-24*g*, which displays bonds outside of the unit cube, we see that each atom has one bond extending directly into the paper and three dangling bonds obliquely out of the paper. The receding bonds connect to a (111) plane of gallium atoms located directly behind the plane of the paper and one full bond distance away. The protruding, obliquely angled bonds would connect to another gallium layer only a short distance in front of the paper. Therefore, in moving through GaAs along the ⟨111⟩ direction, we will pass pairs of planes close together, separated from adjacent pairs of planes by a full bond distance.

The (111) face we have drawn has three projecting dangling bonds, and only one satisfied bond, per atom. This situation is highly unfavorable energetically and, in fact, never happens. The face we have drawn will not remain the exposed face of ⟨111⟩ GaAs: it is unstable. Instead, the exposed face will be a gallium face, such as the one directly in front of the paper. Figure 2-24*h* shows this plane and the bonds it makes to the original arsenic plane. The gallium face has only one dangling bond per atom and is much more stable. As a result, whether in crystal growth, cleaving, or etching, atoms always leave a {111} face in pairs so as to leave a gallium plane at the surface.

Now take another look at Fig. 2-24*g*, but let the bulk of the crystal be above the paper and the exposed face toward the back. In this case, the arsenic plane is stable, with only one dangling bond for every three satisfied ones. At the "back" of the crystal, then, the arsenic (111) face is energetically more stable and the gallium face is never observed. The ⟨111⟩ direction has turned out to be polar direction for GaAs: the (111) face is different from the ($\bar{1}\bar{1}\bar{1}$) face. The (111) face is, naturally enough, called the *gallium face,* the ($\bar{1}\bar{1}\bar{1}$) face is the *arsenic face*. Growth and dissolution at these faces is always by matching pairs of atoms.

Although symmetrical from a geometric point of view, the gallium and arsenic faces do not behave identically because they are chemically different. Gallium, as a

Group IIIB element, has three valence electrons to employ in the three bonds it makes behind the surface. At the opposite face, the Group VB element arsenic has two leftover valence electrons after making its three bonds to the bulk. The gallium face is thus more chemically stable. It etches slower, mechanically laps slower, and takes a smoother polish. It also oxidizes and evaporates more slowly. Neither of the two faces is a good cleavage plane for GaAs: the electrostatic bonding between the adjacent planes inhibits smooth cleavage.

Compositional Effects

Being a compound, gallium arsenide can be decomposed into its constituent elements. This fact has implications for gallium arsenide processing. The vapor pressures of gallium and of various forms of arsenic vapor (As, As_2, As_4) over gallium arsenide as a function of temperature have been determined.[17] Even at 600°C, these pressures are appreciable. The vapor pressure curves are double-valued: the vapor pressure can have two values, depending on whether the GaAs is gallium-rich or arsenic-rich. Since at most temperatures the arsenic vapor pressure is higher, the GaAs will tend to become gallium-rich as decomposition proceeds. As a result, if GaAs is processed in a closed tube, arsenic will preferentially evaporate until the equilibrium vapor pressure is reached, resulting in gallium-rich GaAs. If this is not acceptable, arsenic vapor must be supplied by the evaporation of a charge of arsenic to stabilize the semiconductor. In any case, the evaporation of GaAs must be controlled, which rules out the simple open-tube processing methods common with silicon. Like most constraints, this one can also be turned to advantage: brief heating to around 600°C, where the gallium and arsenic pressures are nearly equal, is an effective method of cleaning GaAs.

Because GaAs is relatively volatile, vapor pressure effects dominate the defect distribution within the crystal. Of course, the nature of defects inside GaAs is inherently more complicated than with silicon. Vacancies may be either gallium or arsenic vacancies, and interstitials come in four kinds (gallium or arsenic interstitials, surrounded by either gallium or arsenic). Arsenic vacancies occur most readily at low arsenic vapor pressures and are suppressed by increasing vapor pressure; gallium vacancies show the opposite behavior. Thus processing history has an important effect on the nature and number of crystal vacancies.

Because gallium arsenide is a compound, it dopes differently from an elemental semiconductor. Impurities can substitute either into the gallium or arsenic lattices. It is the Group II impurities, with two valence electrons, which are the common *p*-type dopants of GaAs. They substitute into the gallium lattice and contribute holes. Group VI elements substitute into the arsenic lattice and act as *n*-type dopants. Because of the high electron mobility of GaAs, *n*-type dopant atoms have exceptionally low ionization energies (around 0.005 eV) and *n*-type doping is almost always degenerate.

Group IV elements can substitute into either the gallium lattice, where they act as *n*-dopants, or into the arsenic lattice, as *p*-dopants. Usually, the distribute themselves between the two lattices in a proportion governed by the vacancy distribution, and thus are partially self-compensating. Which type of doping predominates depends on the vacancy distribution, which in turn depends on processing history.

Zinc is the most common p-type dopant for GaAs. Beryllium has also been studied as a possible ion for implantation. Sulfur is a frequent n-type dopant; selenium and tellurium are also used. Of the Group IV elements, carbon and silicon are always present as contaminants and can act either as n- or p-type dopants. Tin, if present, almost always contributes as an n-dopant.

Three other contaminants deserve mention. Copper is a frequent contaminant that can be used to reduce the n-type diffusion length in GaAs. Chromium and oxygen are dangerous contaminants, because they both can make GaAs semi-insulating. Oxygen in particular is highly soluble in GaAs and can raise resistivity to around 10^8 Ω cm. To avoid the risk, oxygen needs to be rigorously excluded from GaAs processing environments.

With this description of gallium arsenide, we end our general exploration of semiconductors. Beginning with the band theory of conduction, we have looked at the nature of semiconductivity, the reasons for its sensitivity to impurities, and its potential for exploitation. Then we looked at the semiconductor crystal and two specific semiconducting materials. We now turn to ways of fabricating these raw materials into devices of practical usefulness.

PROBLEMS

2-1 The Fermi level has been defined as the energy at which the probability of occupation of a state is precisely one-half. Show that this definition is implicit in Eq. (2-1).

2-2 (a) Derive Eq. (2-2) from Eq. (2-1), under the approximation that $E - E_F \gg kT$.

(b) For silicon at 300°C, use both Eqs. (2-2) and (2-3) to calculate $f(E)$ at the two band edges, and comment on the validity of the approximation.

2-3 Find the location of the Fermi level, relative to the top of the valence band, for

(a) Intrinsic silicon at 300°K

(b) Intrinsic silicon at 500°K

(c) p-doped silicon, $N_A = 10^{16}$ cm^{-3}, 300°K

(d) n-doped silicon, $N_D = 10^{16}$ cm^{-3}, 300°K

(e) n-doped silicon, $N_D = 10^{23}$ cm^{-3}, 300°K

(f) Are any of these cases degenerate?

2-4 A certain integrated circuit is expected to function over a range from −40° to +100°C. Calculate the percentage change in resistivity of the intrinsic semiconductor, for (a) Si; (b) Ge; (c) GaAs; from the bottom to the top of this temperature range. Assume that N_C, N_V are proportional to $T^{3/2}$, and that μ_n, μ_p are proportional to $T^{-3/2}$.

2-5 Compute the number density of holes and electrons in silicon that is doped as follows: (a) intrinsic; (b) arsenic-doped with 10^{16} atoms per cubic centimeter; (c) boron-doped, 10^{16} atoms per cubic centimeter; (d) phosphorus-doped, 10^{23} atoms per cubic centimeter. Assume complete ionization of dopants.

2-6 A pn junction is formed between p-doped silicon, $N_A = 10^{18}$ atoms per cubic centimeter and n-doped silicon, $N_D = 10^{16}$ atoms per cubic centimeter. (a) Find the built-in voltage ϕ_{bi}. (b) Find $\mathscr{E}_{max}$. (c) Under reverse bias, avalanche breakdown will occur when the electric field exceeds some value $\mathscr{E}_{crit}$. Express $\mathscr{E}_{max}$ for reverse bias as a function of the bias V_b and the doping densities, and write an equation for the reverse bias when breakdown occurs, V_{BV}. (d) How will V_{BV} be affected by doubling N_A? (e) By doubling N_D?

2-7 An MOS capacitor is constructed using p-type silicon with $N_A = 10^{16}$ and $x_o = 0.1$ μm. Find: (a) V' at which accumulation gives way to depletion, (b) V' for the onset of inversion, (c) $x_{d,max}$. (d) C_o. $\epsilon_{ox} = 3.9\epsilon_o$. (e) The capacitance at the onset of inversion. (f) Sketch the capacitance-voltage curve, assuming that the capacitance in inversion has the high-frequency characteristic.

2-8 Use Eqs. (2.53)–(2.55) to confirm the following statements in the text about silicon. (a) The spacing is closest between {111} planes. (b) {111} planes intersect a (111) surface so as to form an equilateral triangle. (c) {111} planes meet the (100) surface at right angles to each other.

2-9 When chips are cleaved from a silicon crystal, orientation affects not only the angle between cleavage lines on the surface, but also the angle of the chip edge to the surface. Therefore, the "sides" of rectangular chips may not be vertical if the chip is laid on a flat horizontal surface, but may be beveled at some angle. Calculate the angle between the chip surface and the chip edge for cleavage along {111} planes in (a) ⟨111⟩ silicon; (b) ⟨100⟩ silicon.

2-10 Oxygen is usually present in silicon at a concentration around 10^{16} atoms per cubic centimeter. It has been reported that it forms SiO_2 precipitates of roughly 1 μm in diameter (Ref. 3, p. 25). Assume that the density of these precipitates is the bulk SiO_2 density of 2.27 g cm^{-3}, and estimate the mean spacing of the precipitates. The atomic weight of Si is 28 g $mole^{-1}$, that of O is 16 g $mole^{-1}$, and one mole is 6.03×10^{23} atoms.

2-11 How many As atoms per unit cell are present on the (111) face of the zincblende lattice in Fig. 2-24g? How many Ga atoms per cell are present in the plane directly above it, shown in Fig. 2-24h? These numbers are unequal: how can this be?

REFERENCES

1. Bodie E. Douglas and Darl H. McDaniel, *Concepts and Models in Inorganic Chemistry*, Blaisdell, Waltham, MS, 1965, p. 291.
2. A. S. Grove, *Physics and Technology of Semiconductor Devices*, Wiley, New York, 1967, p. 102.
3. Sorab K. Ghandi, *VLSI Fabrication Principles: Silicon and Gallium Arsenide*, Wiley, New York, 1983, p. 24.
4. J. C. Irvin, *Bell System Tech. J.* **41**:387 (1962).
5. Sorab K. Ghandi, *op cit.,* p. 25.
6. Juri Matisoo, "The Superconducting Computer," *Scientific American* **242**:50 (May 1977).
7. James B. Angell, Stephen C. Terry, and Phillip W. Barth, "Silicon Mechanical Devices," *Scientific American* **248**:44 (April 1983).
8. Sorab K. Ghandi, *op. cit.,* pp. 18–19.
9. O. Pax, E. Hearn, and E. Fayo, *J. Electrochem. Soc.* **126**:1754 (1979).
10. H. R. Huff, H. F. Schaake, J. T. Robinson, S. C. Baber, and D. Wong, *J. Electrochem. Soc.* **130**:1551 (1983).
11. L. Jastrzebski, R. Soydan, B. Goldsmith, and J. T. McGinn, *J. Electrochem. Soc.* **131**:2944 (1984).
12. E. Sirtl and A. Adler, *Z. Metalk,* **52**:529 (1961).
13. F. Secco d'Aragona, *J. Electrochem. Soc.* **119**:948 (1972).
14. M. Wright Jenkins, *Electrochemical Society, Spring Meeting 1976,* Abstract 118 (Electrochemical Society, Princeton, N.J., 1976).
15. Sorab K. Ghandi, *op cit.,* p. 12.
16. M. L. Cohen and T. K. Bergstresser, *Phys. Rev.* **141**:789 (1966).
17. J. R. Arthur, *J. Phys. Chem. Solids* **28**:2257 (1967).

THREE

DOPANT DIFFUSION AND
RELATED OPERATIONS

In Chapter 2, we discussed the electrical behavior of semiconductors and the effects of small amounts of impurities. We saw that, by the addition of small amounts of impurities to selected parts of a semiconducting substrate, electrical circuits of great complexity can be fashioned. This chapter deals with methods of selectively incorporating controlled amounts of impurities into a semiconductor.

Elements intentionally added to a semiconductor to modify its electrical behavior are termed *dopants*. The most common method of adding dopants is by *diffusion*. In this method, the surface of the semiconductor is heated together with a compound containing the desired dopant. At sufficiently high temperature, the dopant will diffuse into the semiconductor. Dopant diffusion thus requires a *source* of the doping element (along with a technique for transporting it to the substrate surface) and a *furnace* to apply the required heat. In addition, if the doping is to be selective, affecting only portions of the substrate, the remainder of the surface must be *masked* to exclude the dopant. In Section 3-1 we look at furnaces and related equipment. In Section 3-2 we develop the general laws affecting diffusion, while in Section 3-3 we discuss dopant sources. In Section 3-4 we explore the *oxidation* of silicon, which is the commonest way to mask the surface. In Section 3-5 we summarize principles and tools for the control of diffusion processes.

3-1 EQUIPMENT FOR DIFFUSION AND RELATED OPERATIONS

Figure 3-1 depicts a typical furnace used for diffusion operations. It consists of a quartz tube surrounded by electrical heating elements. At the temperatures used for diffusion, metals diffuse rapidly into silicon, usually with catastrophic effects on circuit perfor-

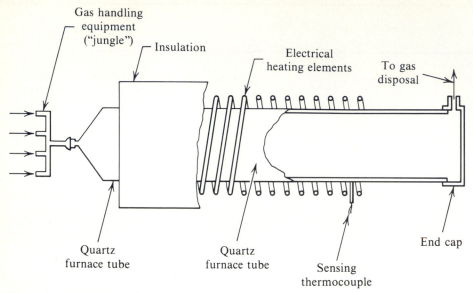

Figure 3-1 Furnace for use in dopant diffusion.

mance, so metals are inappropriate for the construction of furnace tubes. Also detrimental to the device are the alkalis (sodium, potassium, and other Group IA elements), which are found in most glasses. This leaves quartz (fused silicon dioxide, or *silica*) as the construction material of choice. The long, cylindrical shape of the tube is convenient for the loading of numerous circular wafers and allows a smooth flow of gases to envelop the substrates.

Substrates are held vertically in the tube by quartz holders termed *boats, sleds,* or *cassettes*. The substrates slide into slots in the rods composing the boat: Fig. 3-2 shows a typical arrangement. A single long boat may hold some 200 wafers in a length of about 2 ft. Another method uses standardized cassettes about 6 in long, which hold 50 wafers and are strung together into a 200–250 wafer load.

Electrical heating elements surround the tube and heat it to the desired temperature. Usually at least three distinct elements are used: one in the center to heat the reaction zone plus one on each end to buffer the center zone against large temperature fluctuations. This arrangement allows control to ±1°C, over a range from 800 to 1200°C, throughout the center 2–3 ft of the tube. This area, where the wafers are positioned and the reaction occurs, is called the *flat zone*.

At the inlet or *source* end, the tube narrows to allow connection to the *jungle*, the array of tubing and controllers that meters the incoming gas flow. At the outlet or *handle* end, the tube is closed with a removable quartz plug called an *end cap*, which channels gases into disposal ducting. Since many diffusion sources are highly toxic and flammable, complete gas removal is given careful attention. Traditionally, boats or sleds of substrates are slid into the furnace from the handle end by removing the end cap and pushing the sled down the tube with a quartz *push-pull* rod. Figure 3-3 shows

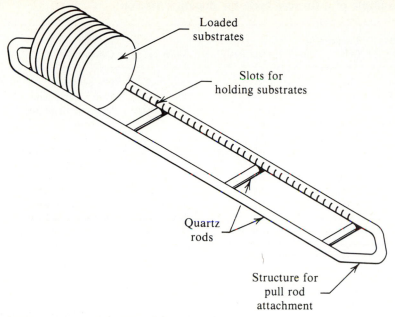

Figure 3-2 Quartz boat for holding substrates.

the insertion of a loaded diffusion boat into a furnace tube, using a push-pull rod. The boat has been loaded into a quartz tube extension (often called a *white elephant* or simply *elephant*) that is fitted onto the tube end during loading to prevent contamination. After loading, the elephant will be replaced by the end cap. Because the push-pull method of loading involves abrasion during sliding, which creates contaminating

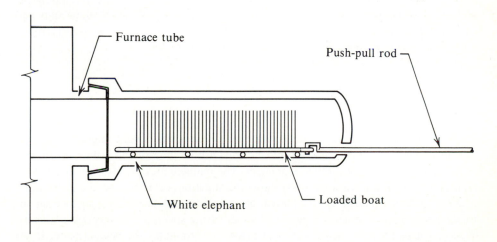

Figure 3-3 Insertion of a loaded diffusion boat into a furnace. Note the use of the push-pull rod and the removable extension called a *white elephant*.

Table 3-1 Example of a furnace cycle for dopant diffusion

Start (min)	Duration (min)	Temperature (°C)	Gas flow	Other	Comments
0	0	800	N_2	—	Idle condition
0	5	800	N_2	Pusher on	Push run into tube
5	†	800	N_2	Sound alarm	Disconnect push rod
5	60	Ramp 4°C min^{-1}	N_2	—	Ramp up
65	5	1040	N_2	—	Source
70	60	1040	N_2, O_2, source		Dopant predeposition
130	30	1040	N_2, O_2	—	Dry oxidation
160	120	Ramp −2°C min^{-1}	N_2	—	Ramp down
280	†	800	N_2	Sound alarm	Connect push rod
280	5	800	N_2	Puller on	Pull run from tube
285	†	800	N_2	Sound alarm	Disconnect pull rod
285	—	800	N_2		Idle

† Program stops and waits for operator intervention.

particles, there is increasing interest in alternative loading methods, such as the *cantilever method,* in which a cantilever beam suspends the boat clear of the tube during loading and unloading.

Semiconductor substrates that experience large, abrupt temperature changes will warp or develop crystal defects due to thermal strain. Therefore, substrates are usually not pushed or pulled with the furnace operating at full temperature. Instead the wafers are loaded at an *idle temperature* (800°C is a popular choice) and the furnace then *ramps up* to the desired temperature. After diffusion, the furnace is *ramped down* to the idle temperature prior to substrate removal. Modern furnaces are highly automated, and allow software control over the entire reaction cycle, including ramp rates and temperature calibration. Table 3-1 illustrates a typical sequence of events in an automated furnace cycle.

3-2 LAWS OF DIFFUSION

Dopants diffuse into semiconductor substrates because the dopant migrates from the dopant-rich surface of the substrate into the dopant-poor interior and is thus an example of diffusion due to a concentration gradient. The mathematics of diffusion under these conditions is well known. Figure 3-4 shows the situation, on a microscopic scale, when a concentration gradient exists. Shown are two adjacent thin slices of material, of thickness λ. Slice 1 contains n_1 atoms of the diffusing species, and slice 2 contains n_2. Plane P_1 separates the two slices, and plane P_2 terminates the second slice. If we consider a cross-sectional area A, we can write the densities of the impurities in each slice as

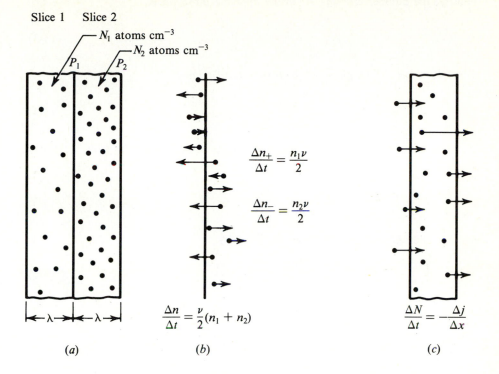

$$\frac{\Delta n_+}{\Delta t} = \frac{n_1 \nu}{2}$$

$$\frac{\Delta n_-}{\Delta t} = \frac{n_2 \nu}{2}$$

$$\frac{\Delta n}{\Delta t} = \frac{\nu}{2}(n_1 + n_2)$$

$$\frac{\Delta N}{\Delta t} = -\frac{\Delta j}{\Delta x}$$

(a) (b) (c)

Figure 3-4 Diffusion under a concentration gradient: (a) two slices with different concentrations. (b) Focus on plane P_1 showing change in concentration. (c) Focus on slice 2 showing flux.

$$N_1 = \frac{n_1}{A\lambda} \tag{3-1a}$$

$$N_2 = \frac{n_2}{A\lambda} \tag{3-1b}$$

Atoms throughout a solid are constantly vibrating, and at a given temperature, each atom in a slice has a probability of "breaking loose" and jumping to a site in the adjacent slice. Let the frequency of this occurrence be ν.

Now examine the flow of atoms across the plane P_1. In time Δt, the number of atoms leaving slice 1 is the product of the jumping frequency ν and the number of atoms n_1. Since half of these atoms jump in either direction, the number crossing P_1 in the positive direction is

$$\frac{\Delta n_+}{\Delta t} = \frac{n_1 \nu}{2} \tag{3-2}$$

Similarly, the number crossing P_1 in the opposite direction is

$$\frac{\Delta n_-}{\Delta t} = \frac{n_2 v}{2} \tag{3-3}$$

The net flow across the plane is given by the difference

$$\frac{\Delta n}{\Delta t} = \frac{v}{2} (n_1 - n_2) \tag{3-4}$$

as illustrated in Fig. 3-4b. Substituting the density terms N_1 and N_2 from Eq. (3-1) gives

$$\frac{\Delta N}{\Delta t} = \frac{(N_1 - N_2) A \lambda v}{2} \tag{3-5}$$

Since the concentration gradient can be given by

$$\frac{\Delta N}{\Delta x} = \frac{N_2 - N_1}{\lambda} \tag{3-6}$$

we can write

$$\frac{\Delta N}{\Delta t} = \frac{-A \lambda^2 v}{2} \frac{\Delta N}{\Delta x} \tag{3-7}$$

Flux density j is defined as the number of atoms crossing P_1 per unit time per unit area. Dividing by area gives

$$j = \frac{1}{A} \frac{\Delta N}{\Delta t} = -\frac{\lambda^2 v}{2} \frac{\Delta N}{\Delta x} \tag{3-8}$$

If we define a diffusion constant D such that

$$D \equiv \frac{\lambda^2 v}{2} \tag{3-9}$$

we obtain the expression

$$j = -D \frac{\Delta N}{\Delta x} \tag{3-10}$$

In partial differential terms, this is Fick's first law:

$$j = -D \frac{\partial N}{\partial x} \tag{3-11}$$

which relates flux of atoms across a boundary to the concentration gradient of those atoms.

Fick's Second Law

In dopant diffusion, we are interested in changing the impurity density. Thus we next examine how the impurity density function, $N(x)$, changes with time. Consider slice 2 in Fig. 3-4c. Since matter is conserved, the time change in concentration must be equal to the input across P_1 minus the outgo across P_2, or

$$\frac{\Delta N}{\Delta t} = j_1 - j_2 \tag{3-12}$$

Now

$$\frac{j_1 - j_2}{\lambda} = \frac{\Delta j}{\Delta x} \tag{3-13}$$

and therefore

$$\frac{\Delta N}{\Delta t} = -\frac{\Delta j}{\Delta x} \tag{3-14}$$

In differential terms, this is the *transport equation*

$$\frac{\partial N}{\partial t} = -\frac{\partial j}{\partial x} \tag{3-15}$$

Substituting Eq. (3-11) into (3-15) gives

$$\frac{\partial N}{\partial t} = D\frac{\partial^2 N}{\partial x^2} \tag{3-16}$$

This is Fick's second law, a differential equation for the time-dependent concentration function $N(x,t)$. Since it is identical in form with the equation for heat conduction, differing only in the constant D, we can apply the large existing body of work on heat flow to the solution of diffusion problems.

Diffusion in an Electric Field

Electrical fields as well as concentration gradients can drive solid state diffusion, and, although dopant diffusion is primarily driven by concentration, electrical fields within the semiconductor can also have an effect. A charged dopant atom, or any other charged particle in an electric field, experiences a force F_{acc}:

$$F_{acc} = Z\mathscr{E} \tag{3-17}$$

where Z is the charge on the particle and $\mathscr{E}$ the field. Thus a particle of mass m undergoes an acceleration

$$\frac{dv}{dt} = \frac{Z\mathscr{E}}{m} \tag{3-18}$$

In a solid, the accelerating atom quickly encounters another atom that scatters it elastically, randomizing its direction of motion. This scattering can be represented as a restraining force F_r proportional to the velocity[1]:

$$F_r = m\alpha v \tag{3-19}$$

where α is a proportionality factor with units of $\sec^{-1}$, related to the collision frequency. Thus

$$\frac{dv}{dt} = \frac{1}{m}(F_{acc} - F_r) = \frac{Z\mathscr{E}}{m} - \alpha v \tag{3-20}$$

Setting $dv/dt = 0$, we can obtain the steady state velocity, or *drift velocity*, v_d, as

$$v_d = \frac{Z\mathscr{E}}{m\alpha} \tag{3-21}$$

We can define a *mobility* μ by the equation

$$\mu = \frac{Z}{m\alpha} \tag{3-22}$$

and thus write

$$v_d = \mu\mathscr{E} \tag{3-23}$$

Now the flux j_d due to a uniform motion of impurity atoms is just the velocity times the number density N, so we can write

$$j_d = \mu N\mathscr{E} \tag{3-24}$$

This is the flux due to diffusion induced by the electric field. Equations (3-11) and (3-24) can be summed to give an overall flux for diffusion under both electrical and concentration gradients:

$$j_{total} = -D\frac{\partial N}{\partial x} + \mu N\mathscr{E} \tag{3-25}$$

Diffusion in a Semiconductor Crystal

To diffuse within a semiconductor crystal, impurities must make their way either among or around the atoms already occupying the crystal lattice. Thus diffusion can be either interstitial or substitutional. In interstitial diffusion, the diffusing species occupy interstitial sites in the lattice. The barrier to motion between such sites is fairly low, and the number of sites is high, resulting in quite rapid diffusion. Many metals, most notably gold, diffuse in silicon by this mechanism.

However, the important *p*- and *n*-dopants, such as phosphorus, boron, and arsenic, diffuse substitutionally, occupying a lattice site and moving from one lattice site to the next. In fact, we assumed in our discussion of the effects of doping that the impurities replaced and bonded like the substrate atoms. Substitutional diffusion is a slower process. For one thing, an impurity atom cannot move from one lattice site to another until a vacancy occurs at the new site. The energy for vacancy formation is several electron volts; hence they occur rarely. In addition, the substitutional impurity is bound into the lattice and must break those bonds to diffuse.

We can quantify these concepts by considering the concept of an activation energy E_A, which is the energy required to move from one site to an adjacent one. We can view diffusion as the process of penetrating an energy barrier, of height E_A, which separates the two sites. In such a case, atoms react with a frequency given by the product of two terms. One term is the frequency with which the atom collides with the barrier; the second is the probability of getting over the barrier on each collision. The atom vibrates within the crystal with a frequency ν_0, which in silicon is about 10^{13}–10^{14} sec^{-1}. At each vibration, it oscillates within the potential well and thus assaults the potential barrier: therefore, ν_0 is the number of "attempted escapes" per unit time. Due to the thermal energy in the crystal, there is at any time a proportion of atoms with sufficient thermal energy to surmount the barrier: This probability is given by the Boltzmann factor exp $(-E_A/kT)$, which is a function of the activation energy and the temperature. Thus the rate of successful barrier crossings is

$$\nu = \nu_0 e^{-E_A/kT} \tag{3-26}$$

Since the atom can move to any adjacent site, we should multiply by the number of adjacent sites, which is four for both substitutional and interstitial sites in the diamond lattice of crystalline silicon. Therefore, the frequency of atomic movement for interstitial sites can be written

$$\nu_{\text{int}} = 4\nu_0 e^{-E_A/kT} \tag{3-27}$$

For substitutional diffusion, we must add an additional factor for the probability of a vacancy in adjacent site. If E_v is the energy for vacancy formation, the resulting expression is

$$\nu_{\text{subst}} = 4\nu_0 e^{-(E_A + E_v)/kT} \tag{3-28}$$

Experimentally determined activation energies (including the vacancy formation component) are around 3 eV for substitutional diffusers and from 0.5 eV up for interstitial ones.[2]

It is now possible to express the diffusion constant in terms of the properties of the material. For ν in Eq. (3-9), we can substitute the value from Eq. (3-27) or (3-28). In the diamond lattice, the spacing between atoms λ is $d/\sqrt{3}$, where d is the lattice distance. Thus we obtain

$$D = \frac{4\nu_0 d^2}{6} e^{-E_A/kT} \tag{3-29}$$

for interstitial diffusion and

$$D = \frac{4\nu_0 d^2}{6} e^{-(E_A + E_v)/kT} \tag{3-30}$$

for substitutional diffusion.

Example Estimate the diffusion coefficient D for boron diffusing in silicon at 1000°C. Since boron is a dopant and a substitutional diffuser, we apply Eq. (3-30), approximating $(E_A + E_v)$ as 3.0 eV and ν_0 as 10^{14} sec^{-1}. The lattice distance a is found in Table 2-1 as 0.566 μm. The resulting value is

$$D = \frac{4}{6} (10^{14} \text{ sec}^{-1})(0.543 \times 10^{-7} \text{ cm})^2 \exp\frac{-3 \text{ eV}}{(1273°\text{K})(8.62 \times 10^{-5} \text{ eV } °\text{K}^{-1})}$$

$$= 2.63 \times 10^{-13} \text{ cm}^2 \text{ sec}^{-1}$$

$$= 0.0947 \ \mu\text{m}^2 \text{ h}^{-1}$$

We can compare the resulting value of $\sqrt{D} = 0.031 \ \mu$m h$^{-1/2}$ with the actual value, from Fig. 3-5, of 0.07, an agreement within a factor of 2, which is satisfactory given the approximations we made. Note that, without further refinement, the number we have obtained will be the same for any interstitially diffusing species.

The actual diffusion of dopants in silicon is complicated by the existence of multiple diffusion mechanisms, electrical field effects due to charges on dopants and vacancies, and the sequential nature of real diffusion. R. B. Fair has developed expressions for the diffusion of common semiconductor dopants that have been successful in accounting for many of the observed nonidealities in real diffusions.[3] Figure 3-5 shows the diffusion constants of a number of common dopants as a function of temperature.[4]

The well-established mathematics of diffusion has helped make computer modeling of diffusion operations highly accurate. Therefore, it is now becoming customary to model the effects of diffusion and related processes, not only during process design, but when considering possible process modifications. References 5 and 6 give further information on process modeling.

Application of the Equations

In executing a diffusion operation, the process engineer desires to control the total amount of impurity introduced and the depth to which the impurity penetrates into the substrate. A frequently used strategy to achieve this control is a two-step diffusion consisting of a *predeposition* and a *drive-in* step. The predeposition step, done in the presence of a source, introduces the correct amount of dopant into the substrate. The drive-in, done with no source present, diffuses the dopant introduced during predeposition to the desired depth.

In predeposition, a source of constant concentration is supplied to the substrate surface. To improve the reproducibility of this operation, advantage is taken of the concept of *solid solubility*. At a given temperature, there is a sharply defined maximum level of dopant that the silicon crystal will absorb. Therefore, as long as excess dopant is present at the surface, the silicon immediately below the surface will be doped to the limit of solid solubility, independent of small variations in the dopant supply. Figure 3-6 gives solubilities for common dopants.[7] Under these conditions, we can solve Eq. (3-16) using the boundary and initial conditions

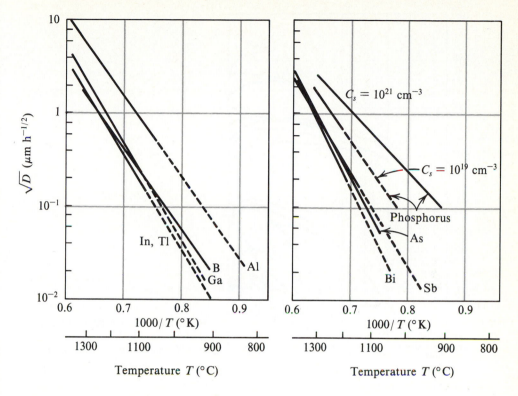

Figure 3-5 Diffusion coefficients of common dopants in silicon. (*Reference 4, used by permission. Copyright 1967, John Wiley & Sons.*)

$$N(0, t) = N_s \qquad (3\text{-}31a)$$

$$N(\infty, t) = 0 \qquad (3\text{-}31b)$$

$$N(x, 0) = 0 \qquad (3\text{-}31c)$$

Here N_s is the surface concentration, which should equal the solid solubility. The solution is the *complementary error function*[8]

$$N(x, t) = N_s \, \text{erfc}\left(\frac{x}{2\sqrt{Dt}}\right) \qquad (3\text{-}32)$$

The complementary error function is defined by

$$\text{erfc}(u) = 1 - \text{erf}(u) \qquad (3\text{-}33a)$$

$$= 1 - \frac{2}{\sqrt{\pi}} \int_0^u e^{-a^2} \, da \qquad (3\text{-}33b)$$

The purpose of a predeposition is to fix the total amount of dopant in the substrate. The total dopant per unit area of surface Q can be derived using the property that

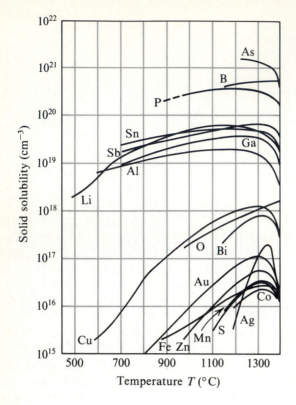

Figure 3-6 Solid solubilities of various impurities in silicon. (*Reference 7. Reprinted with permission from the* Bell System Technical Journal. *Copyright 1960, AT&T.*)

$$Q = \int_0^\infty \text{erfc}(u) \, du = \frac{1}{\sqrt{\pi}} \tag{3-34}$$

This implies

$$Q = \int_0^\infty N(x, t) \, dx = \frac{2N_s}{\sqrt{\pi}} \sqrt{Dt} \tag{3-35}$$

Thus the total doping per area is a straightforward function of the solid solubility, the diffusion constant, and the diffusion time. (Note that Q is used to denote both charge per unit area and doping per unit area. This notation, while unfortunate, is conventional: the distinction is usually quite clear from the context.)

It is convenient to have an approximation for the complementary error function. In most real conditions we can assume that

$$\frac{x}{\sqrt{Dt}} \gg 1 \tag{3-36}$$

This is particularly true of predepositions, which are designed to concentrate dopant near the surface. Under these conditions,

$$\text{erfc}(u) \sim \frac{1}{\sqrt{\pi}} \frac{e^{-u^2}}{u} \qquad u \gg 1 \tag{3-37}$$

so that

$$N(x, t) \sim \frac{2N_s\sqrt{Dt}}{\sqrt{\pi}} \frac{e^{-x^2/4Dt}}{x} \tag{3-38}$$

During drive-in, no source is applied to the substrate, and the total amount of dopant in the wafer remains unchanged. If the predeposition is kept short, the drive begins with most dopant quite near the substrate surface. These conditions approximate another known solution of Eq. (3-16). For a fixed amount of dopant Q, originally distributed as a delta function at $x = 0$, the distribution takes the form

$$N(x, t) = \frac{Q}{\sqrt{\pi Dt}} \exp \frac{-x^2}{4Dt} \tag{3-39}$$

This is the well-known Gaussian, or normal, distribution.[8] Note the occurrence of the term $\sqrt{Dt}$ in Eqs. (3-35), (3-38), and (3-39). This "root-D-t" term forms a convenient basis for comparing diffusions.

By proper selection of temperature and duration, the drive is designed to achieve the desired depth of doping. This depth is usually expressed as a *junction depth*, x_j. Since the Gaussian distribution only asymptotically approaches zero at large x, there is no clearly defined limit to the dopant depth in an originally undoped crystal. In actuality, however, there is always a background doping level in practical semiconductors, due either to crystal growth or a previous diffusion. Thus the junction depth is defined as the value of x at which $N(x)$ falls to the background level. If the background dopant and the added dopant differ in polarity, an electrical junction will occur at this point. Figure 3-7 shows the shape of the Gaussian and erfc curves, and illustrates the formation of a junction with the background doping.

Example A boron predeposition at 1100°C gives a dopant dose Q of 10^{16} cm^{-2}. It is followed by a drive-in diffusion of one hour at 1000°C. Find: (a) the appropriate predep time, (b) the surface concentration after drive-in $N(0)$, (c) the junction depth x_j if the background doping $N_B = 10^{15}$.
(a) From Eq. (3-35), we have

$$t = \frac{\pi Q^2}{4C_s^2 D}$$

with C_s from Fig. 3-6 as 4×10^{20} cm^{-3}, and D from Fig. 3-5 as 0.04 μm^2 h^{-1}, we obtain

$$t = 1.4 \text{ h}$$

(b) For the drive-in, D from Fig. 3-5 is 0.0049. From Eq. (3-39), with $x = 0$, we obtain

$$C_s = Q(\pi Dt)^{-1/2}$$

$$= 8.06 \times 10^{16} \text{ cm}^{-3}$$

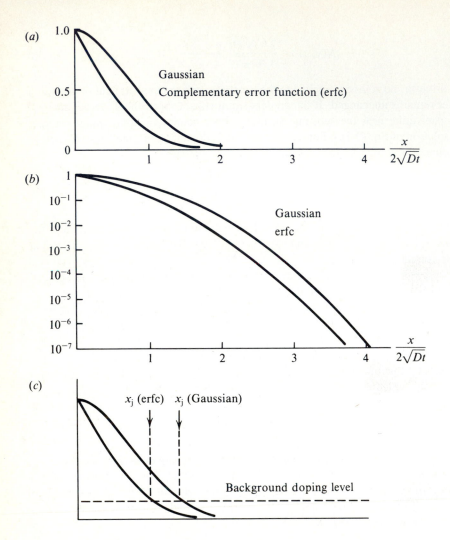

Figure 3-7 Gaussian and error functions, illustrating the formation of a junction with background substrate doping. (*a*) Linear scale. (*b*) Semilog scale. (*c*) Location of junction formed with constant background doping.

(c) The junction depth occurs at x such that $N(x) = N_B$. Solving Eq. (3-39) for these conditions gives

$$x_j^2 = -4Dt \ \ln\frac{N_B}{C_s}$$

$$x_j = 0.293 \ \mu m$$

Diffusion through a Mask

We have treated diffusion as a one-dimensional process. However, dopants are usually diffused through a window in a masking layer, and diffusion near the edges of the window is three-dimensional. Nevertheless, it is usually satisfactory to assume that diffusion proceeds equally in all directions from the edge of the window, resulting in a cylindrical or spherical distribution of dopant. Figure 3-8a illustrates the concept. Figure 3-8b shows the shape of some theoretically predicted diffusions near windows: note the essentially circular profile.[9]

3-3 DOPANTS AND DOPANT SOURCES

Dopant diffusion requires that a dopant *source* be placed in intimate contact with the semiconductor surface. Occasionally the source material is applied to the substrates before they enter the furnace, but more commonly a chemical reaction within the furnace is used to deposit a dopant-containing compound on the surface. In any case, the term *source* usually includes not only the chemical source, but the entire system involved in supplying dopant to the wafer. In this section, we discuss some common source systems for silicon doping.

Silicon dopants come from either Group IIIA or Group VA of the periodic table. As we saw in Chapter 2, the IIIA elements form *p*-type dopants, generating excess positive holes, while the VA elements are *n*-type, generating electrons. Boron (B) is usually chosen for *p*-doping of silicon although aluminum and gallium are sometimes used. Either phosphorus (P), arsenic (As), or antimony (Sb) may serve as *n*-type dopants, a choice usually determined by the diffusion coefficient and misfit factor. Arsenic and antimony are larger atoms than phosphorus and diffuse more slowly. When several diffusions are to be performed in sequence, the impact on earlier diffusions as a result of the later ones is minimized by using slow diffusers early in the process and faster ones later. The *misfit factor,* which is the percentage deviation of the dopant radius from that of the semiconductor atom, must be considered to avoid excessive crystal strain. The arsenic atom is exactly the same size as silicon, while antimony is 15.3% larger.[10] High concentrations of antimony can significantly strain the crystal, so arsenic is preferred in such applications. Table 3-2 gives misfit factors for common dopants and semiconductors.[10]

The source compound formed on the wafers' surface is usually the oxide of the doping element. The most direct method of application is to spray or spin-coat a slurry of the oxide onto the substrates. Alternatively, the solid powdered oxide can be allowed to evaporate in a closed or open tube, coating the wafers. A third method is to form the oxide by chemical reaction within the furnace, so that it accumulates on the substrates. In this case, the oxide is usually formed by oxidation of a halide, oxyhalide, or hydride. The hydrides are usually gases and can be directly admitted into the furnace tube. The halogen-containing compounds are usually liquid and are delivered to the furnace as vapors entrained in a carrier gas such as nitrogen, oxygen, or argon.

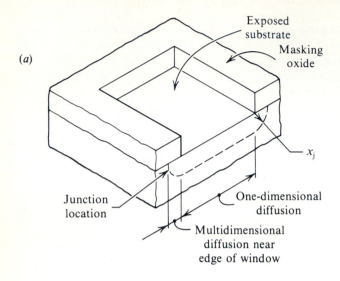

(a)

Exposed substrate

Masking oxide

x_j

Junction location

One-dimensional diffusion

Multidimensional diffusion near edge of window

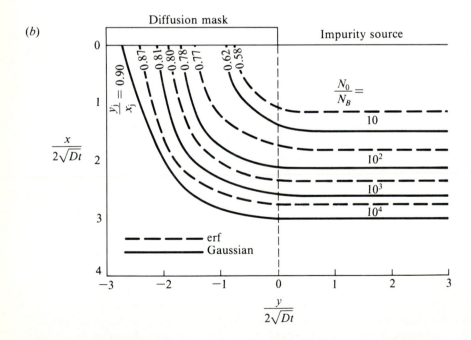

(b)

Figure 3-8 Pattern of diffusion near a masking window. (a) View of the window. showing one- and multidimensional diffusion zones. (b) Calculated two-dimensional diffusion profiles near a window. (*Figure 3-8b is from Ref. 9, copyright 1965 by International Business Machines Corporation; reprinted with permission.*)

Whatever method is chosen for dopant application, the number of dopant atoms per unit area of the surface must be controlled. If a predeposition is being run at the solid solubility limit, we must at least ensure that the dopant concentration is sufficient to saturate the silicon. For high resistivity depositions below the solubility limit, source concentration will directly determine doping level. If direct application of source is used, concentration control is determined by dopant concentration in the coating and the coating thickness. For evaporation from a solid source, the source vapor pressure determines concentration. Liquid sources vaporized into a carrier have a concentration dependent on vaporization temperature while gas source concentration can be directly metered with flowmeters.

Other considerations influence the choice of a dopant source. The properties of an ideal doping source have been summarized as including[11]

A concentration or vapor pressure invariant over the life of the source
Ability to mix with common carrier gases
Concentration or vapor pressure variable at will
Reaction products that do not attack the semiconductor or the masking layer
Minimum undesirable deposits on tube walls, boats, etc.
Chemical stability at storage and working temperatures
Availability at reasonable cost
Easily handled, nontoxic and nonexplosive

Unfortunately, this ideal source has yet to be found. As we examine some common sources, we shall find that they all fail to meet at least one of the preceding criteria.

Table 3-2 Tetrahedral radii and misfit factors for silicon and various dopants

Element	Tetrahedral radius	Misfit factor
Si	1.18 Å	0
P	1.10	0.068
As	1.18	0
Sb	1.36	0.153
B	0.88	0.254
Al	1.26	0.068
Ga	1.26	0.068
In	1.44	0.22
Au	1.5	0.272

Source: Reference 10, used by permission.
Copyright 1983, John Wiley & Sons.

Boron Sources

Boron is the most common p-type dopant for silicon. It is used over a wide range of concentrations from junction isolation at resistivities of 5 Ω per square to base diffusions at 250 Ω per square or more. Boron is usually deposited on the substrate as boric oxide (B_2O_3), which is liquid in the furnace but forms a glassy solid at room temperature. After predeposition, this glass must be removed, both to prevent further doping during drive-in and to allow photoresist adhesion at subsequent steps. A dilute solution of hydrofluoric acid (HF), which attacks the glass but does not etch silicon, can be used to remove this layer. Since HF also removes the masking oxide, the acid dip must be kept short.

When heated in contact with silicon, boron oxide forms a *boron-rich layer,* a few tens of nanometers thick, on the silicon surface. This layer is insoluble in hydrofluoric acid and is known to include such silicon-boron compounds as SiB_6. This layer can be converted to HF-soluble oxides by a *low-temperature oxidation* at about 600°C in either oxygen or steam. A complete boron diffusion process thus includes a predeposition, an HF acid wash to remove B_2O_3, a low-temperature oxidation (LTO) to oxidize the boron-rich layer, another HF wash, and finally a drive-in step.

Boron can be applied to the wafer directly by dissolving B_2O_3 in a suitable solvent such as ethylene glycol monomethyl ether (EGME). This method is difficult to control, being sensitive to mixing, storage, application and environmental conditions. A second doping method involves the oxidation of the gas *diborane:*

$$B_2H_6 + 3O_2 \rightarrow B_2O_3 + 3H_2O \qquad (3\text{-}40)$$

This reaction occurs in the source end of the furnace tube at 300–400°C (Refs. 12 and 13). Although gas sources are attractively easy to meter and control, diborane is a poisonous, explosive compound that is safely handled only when diluted below 0.1% in a carrier such as argon. For this reason, other boron sources have been sought.

Either boron trichloride or boron tribromide can be oxidized under furnace conditions to form boron oxide:

$$2BX_3 + \tfrac{3}{2}O_2 \rightarrow B_2O_3 + 3X_2 \qquad X = \text{Cl,Br} \qquad (3\text{-}41)$$

This reaction is fairly slow, and practical predepositions require temperatures above 950°C (Ref. 14). Chlorine etches silicon at these temperatures, so BBr_3 is the more common source. Boron tribromide is a liquid at room temperature, with a vapor pressure of 40 torr at 15°C (Ref. 15). (Boron trichloride boils at 13°C. It is usually supplied to the furnace in the form of vapor in a carrier gas.) Figure 3-9 shows a typical vaporization device, colloquially called a *bubbler.* The carrier gas (usually N_2 or O_2) is bubbled through a container of the liquid source, becoming saturated with the vapor. The saturated vapor is diluted further with additional carrier gas as desired: final concentrations of 0.5% or so are typical. Vapor concentration depends on the vapor pressure in the bubbler, so the source temperature is thermostatically controlled.

Several precautions must be observed. Boron trihalides react with water vapor in the air to form corrosive, toxic acids:

$$2BX_3 + 3H_2O \rightarrow B_2O_3 + 6HX \qquad X = \text{Cl,Br} \qquad (3\text{-}42)$$

Therefore the system must be leak-free, and the bubbler filled under protected conditions. (Chemical vendors will supply sealed, prefilled bubblers ready for connection and use.) The saturated vapor will condense if cooled, so the downstream portion of the gas system must be kept warmer than the bubbler.

The difficulty of handling BBr_3 has encouraged the development of boron nitride (BN) as a doping source. BN is an inert solid material which can be machined into source wafers of the same shape and diameter as the silicon substrates being doped. Semiconductor and source wafers are then loaded alternately in a quartz boat and placed in the furnace, as shown in Fig. 3-10. Under furnace conditions, the BN is oxidized to form a B_2O_3 glass on the source wafer; this layer then serves to transfer dopant to the silicon wafers next to it.

The BN source can be used in one of two ways. One is simply to oxidize the wafers, insert the loaded boat into the furnace, and allow boric oxide to diffuse under its own vapor pressure from source to substrate. This method is simple, but has two disadvantages. One is that, unlike a gas or liquid source, there is no way to "turn on" or "turn off" the BN wafers other than by the heat of the furnace. This means that the doping of substrates depends on their loading position: the first wafers in are the last out and are doped for a longer time. Second, this "dry oxidized" BN process proves extremely sensitive to ambient moisture. The reason for this sensitivity is the key to an alternative method of using BN.

In the presence of water vapor, boron oxide reacts to give boric acid:

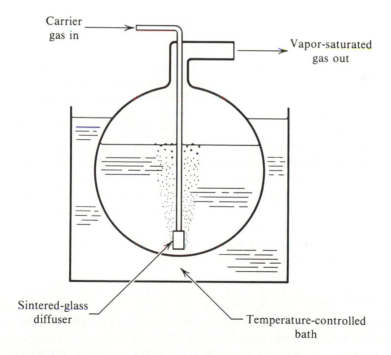

Figure 3-9 Vaporizer or "bubbler" for use with liquid sources.

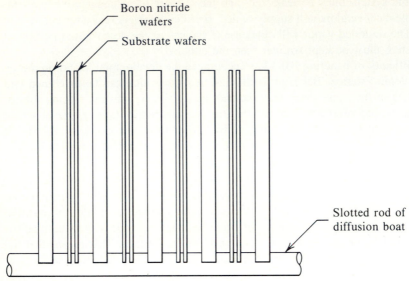

Figure 3-10 Method of use of solid-source wafers. Substrate wafers are placed parallel to solid-source wafers.

$$H_2O + B_2O_3 \rightleftharpoons 2HBO_2 \qquad (3\text{-}43)$$

Figure 3-11 shows the equilibrium constant of this reaction, and the resulting partial pressures of the two species. The vapor pressure of boric acid is roughly two orders of magnitude higher than that of the oxide.[16] As a result, a small amount of moisture in the furnace dramatically increases the vapor pressure of boron-containing gas in the tube. The reaction reverses at the silicon surface, depositing boron oxide at an accelerated rate. Thus even slight moisture adsorption by "dry oxidized" BN wafers leads to anomalously fast dopant transfer.

This effect is used intentionally in the *hydrogen-injected* BN process. After the loaded BN boat reaches furnace temperature, a short burst of hydrogen gas is admitted (e.g., 60 sec of 4% H_2). The hydrogen is oxidized to water and generates boric acid, resulting in rapid transfer of considerable boron to the silicon substrates. The hydrogen is then turned off, essentially ending dopant transfer, and the predeposition diffusion continues with the deposited oxide as the source. The dopant transfer during the hydrogen burst far exceeds the slow, continuing boric oxide evaporation during the remainder of the cycle. This process allows the BN to be "activated" on cue, like a gas or liquid source, and improves process controllability.[17]

Arsenic and Antimony Sources

Arsenic and antimony are slow-diffusing *n*-type dopants and are thus usually used for early diffusions such as bipolar buried collectors and CMOS tub diffusions. Common sources include the gaseous hydrides *arsine* (AsH_3) and *stybnine* (SbH_3), the solid oxides As_2O_3 and Sb_2O_3, and direct application using a solution of the oxides.

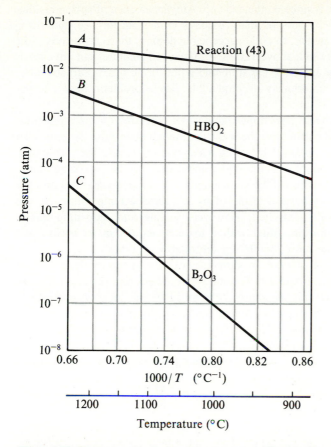

Figure 3-11 The boric oxide–boric acid–water system. Line A is the value of the equilibrium constant for the reaction of Eq. (3-43). Lines B and C are the partial pressures of HBO_2 and B_2O_3, respectively. (*Reference 16. Copyright 1974 by International Business Machines Corporation; reprinted with permission.*)

Gaseous sources are always attractive because of the ease of metering them. The hydrides of arsenic and antimony are gases and can react with oxygen to form the oxides:

$$2DH_3 + 3O_2 \rightarrow D_2O_3 + 3H_2O \qquad D = As,Sb \qquad (3\text{-}44)$$

However, these compounds are flammable and toxic: arsine, with an acceptable exposure level of 0.05 ppm, is among the most toxic gases known.[18] Therefore extreme care must be taken to avoid the escape of unreacted arsine or stybnine from the tube by ensuring an excess of oxygen. If insufficient oxygen is present, the hydride will decompose to give the metallic element:

$$2DH_3 \rightarrow 2D \text{ (solid)} + 3H_2 \qquad (3\text{-}45)$$

which will appear as a black residue on the end cap. Any appearance of such a residue or the detection of the garlicky odor of arsine requires immediate furnace shutdown. An additional disadvantage is that, to obtain acceptable doping, these diffusions must be performed at temperatures well above 1200°C.[13] These temperatures not only strain the crystal, but dangerously approach the softening point of the quartz furnace tube. Arsine depositions are also notoriously irreproducible. These problems have encouraged the use of alternative sources.

Arsenic and antimony oxides have appreciable vapor pressure at furnace temperatures and can be directly deposited onto the wafer. Although open-tube systems have been used, better control is obtained by using a *capsule* or *bomb*. The substrates, together with a measured amount of solid oxide, are sealed into an evacuated quartz capsule that is placed in the furnace. The known amount of oxide evaporates into the vacuum at a known vapor pressure determined by the furnace temperature, so the dopant transfer is quite well controlled. However, sealing and reopening quartz containers under vacuum that are filled with a toxic material can be tricky, and imploding capsules make the term "bomb" uncomfortably accurate.

A final method is *spin-on doping*. A slurry or solution of the oxide is applied directly to the substrate surface before diffusion. This method, less popular for other dopants, is appealing for arsenic or antimony because of the lack of satisfactory alternatives. To be successful, spin-on doping must result in a layer of uniform thickness and reproducible chemistry. The thickness requirement is met by borrowing the technique of *spin-coating* from photoresist technology. We shall study this method in Chapter 4. Several vendors offer proprietary spin-on solutions that, it is claimed, have the requisite chemical predictability.

Phosphorus Sources

Phosphorus is the fastest diffusing *n*-type dopant and is most commonly used for the later diffusions in a process. In bipolar technology, it is a common choice for emitter diffusion, at resistivities down to a few ohms per square. Three properties, found to some extent in most dopants, are aggravated in high-concentration phosphorus diffusions.

First is the existence of *electrically inactive* dopant. We have so far assumed that all of the dopant physically present in the crystal is available for the generation of electrically active charge carriers. However, in practice only a portion of the dopant may be electrically active. The remainder is either in nonlattice sites in the crystal or is accumulated in precipitates and inclusions. With phosphorus, a particularly high proportion of the dopant may be inactive. Figure 3-12 shows the inactive percentage of several dopants as a function of dopant concentration.[19]

Second, phosphorus significantly attacks masking oxide layers. All dopants react to some extent with the masking layer of SiO_2, slowly converting it to a doped glass. Thus the mask layer must be sufficiently thick that substantial unreacted oxide remains at the end of the diffusion. Phosphorus is a very demanding dopant in this regard, penetrating some 200 nm of oxide in 60 minutes at 1000°C. Figure 3-13 shows minimum masking thicknesses as a function of time and temperature of boron and

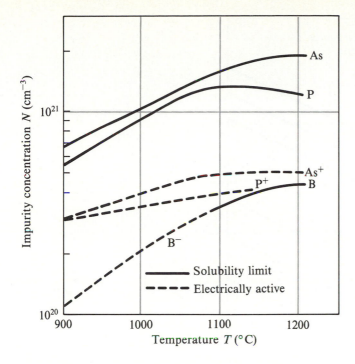

Figure 3-12 Amount of inactive dopant for common dopants. Solid lines are solid solubilities; dashed lines are limits of electrically active dopant concentration. (*Reference 19, used by permission. This figure was originally presented at the Spring 1977 Meeting of The Electrochemical Society, Inc. held in Philadelphia, Pennsylvania.*)

phosphorus.[20] Phosphorus also accelerates the oxidation of silicon and increases the etch rates of the resulting oxide. Therefore, standard oxidation and etch rate graphs may not be accurate when applied to phosphorus-rich layers.

Example A 0.2-μm layer of oxide covers a silicon substrate. Is this sufficient to mask against a 1-h diffusion at 1000°C of (a) phosphorus? (b) boron?

From Fig. 3-13, a boron diffusion under these conditions requires 0.02 μm of oxide, and thus the given layer is sufficient. A phosphorus diffusion requires about 0.25 and thus the masking oxide will not suffice. Using at least 0.3 μm would seem advisable to give some safety margin.

Third, the short duration and high doping level of many phosphorus diffusions magnifies the influence of furnace quartzware on doping levels. Both furnace tubes and boats used for source predeposition accumulate a coating of the dopant oxide glass. In some applications, these sources can contribute as much phosphorus to the wafer as does the intended source mechanism. As a result, depositions can become insensitive to changes in source parameters and can respond dramatically

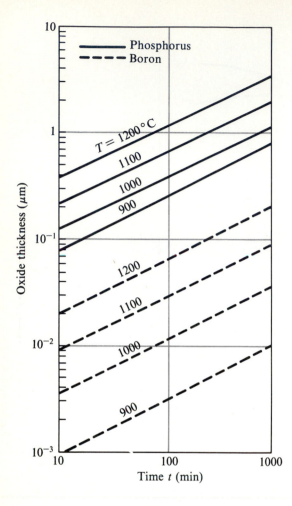

Figure 3-13 Minimum masking oxide thickness for common dopants. (*Reference 20, used by permission. Copyright 1983, John Wiley & Sons.*)

to cleaning or replacement of quartzware. Phosphorus tubes for heavy doping should either be cleaned often or be predoped to a steady-state dopant level before use.

Phosphorus sources fall into the categories discussed previously. *Phosphine* (PH_3) can be used as a gaseous source, while phosphorus oxychloride ($POCl_3$) is a possible liquid source. Wafers containing SiP_2O_7 can be used for a planar solid source similar to boron nitride.

Phosphine is a toxic and explosive gas, with a fishy odor. It is used in dilutions of 0.1–1% in a carrier, usually argon. Phosphine should be used with the same precautions mentioned for arsine: a dark red phosphorus deposit signals incomplete oxidation. Phosphine reacts with oxygen to form phosphorus oxide[21]:

$$2PH_3 + 4O_2 \rightarrow P_2O_5 + 3H_2O \qquad (3\text{-}46)$$

The uniformity of deposition from phosphine, and indeed of all gas or vapor sources, is affected by the patterns of gas flow surrounding the wafers. Flow is restricted between the closely spaced wafers in the furnace, and dopant concentration may significantly reduced near the wafer centers. Sometimes wafers are loaded parallel to the flow, which reduces but does not eliminate flow effects. Figure 3-14 shows contours of resistivity for phosphine depositions with parallel loading, showing the impact of flow on uniformity.[22]

Phosphorus oxychloride (POCl$_3$), colloquially called "pockle," is a popular liquid source for phosphorus. It is introduced into the furnace as a vapor in a carrier gas, using a bubbler like that in Fig. 3-9. The vapor reacts to give the oxide and chlorine:

$$2POCl_3 + 3O_2 \rightarrow P_2O_5 + 3Cl_2 \tag{3-47}$$

Although the conditions of use are similar to those for boron tribromide, POCl$_3$ reacts more rapidly and can be used at temperatures down to 850°C (Refs. 23 and 24). Solid source wafers for phosphorus deposition have been constructed using SiP$_2$O$_7$ in an SiO$_2$ matrix. The reaction involved is

$$SiP_2O_7 \rightarrow SiO_2 + P_2O_5 \tag{3-48}$$

This process is similar to the "dry oxidized" process using boron nitride.[25] There is no method for "turning on" solid phosphorus sources analogous to hydrogen injected boron nitride deposition. Spin-on applications for phosphorus have also been reported.[26]

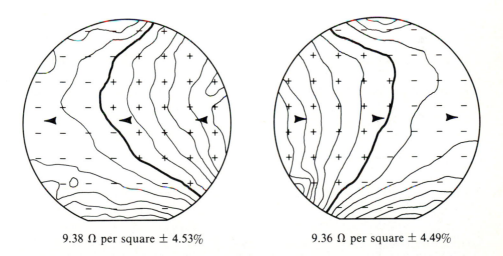

9.38 Ω per square ± 4.53%   9.36 Ω per square ± 4.49%

Figure 3-14 Doping patterns of phosphine under different flow conditions. Substrates were loaded parallel to the flow; arrows show flow direction. (*Reference 22. Reprinted with permission of* Solid State Technology, *published by Technical Publishing, a company of Dun & Bradstreet.*)

3-4 OXIDATION OF SILICON

Dopant diffusion is most useful when it is selective, so that various dopants can be introduced into different parts of the substrate. Selective diffusion requires a masking layer to block dopant from the substrate. A pattern of holes can then be etched into the mask layer to allow dopant to diffuse into selected areas. The masking layer must adhere firmly to the substrate without cracks or pores and must effectively block the diffusion of the dopant. To allow the patterning of the required holes, there must be a practical etchant for the mask material. Ideally, the substrate would be unetched by this etchant, so that overetching would not damage the substrate. Silicon oxide has all of these properties. This single fact does much to explain the preeminence of silicon in microelectronic fabrication. Table 3-3 gives some important properties of silicon dioxide.[27]

When heated in an oxidant such as oxygen or steam, silicon forms a uniform, nonporous, adherent oxide layer by the reactions

$$Si + O_2 \rightarrow SiO_2 \tag{3-49}$$

$$Si + 2H_2O \rightarrow SiO_2 + 2H_2 \tag{3-50}$$

It is therefore easy to create a masking layer for silicon simply by heating it with an oxidant.

Silicon oxidation is done in the same type of furnace used for diffusion. Oxidation in steam is much more rapid than in oxygen, so steam is usually used for thick oxide layers. Water vapor can be supplied using a bubbler (Fig. 3-9), filled with water near its boiling point. This simple method has some drawbacks. Either condensate formation or rapid boiling can carry liquid water into the tube, where it causes nonuniform oxidation. Continuous evaporation in the bubbler also concentrates any impurities present in the water.

These disadvantages are avoided by the use of *hydrogen injection*. Figure 3-15 illustrates the method: a hydrogen injector wand is placed at the entrance end of the tube, in a temperature zone hot enough to ignite the hydrogen smoothly, but suf-

Table 3-3 Important properties of silicon dioxide

Molecular weight (amu)	60.08
Molecules (cm^{-3} × 10^{22})	2.3
Density (g cm^{-3})	2.27
Resistivity	$\geq 10^{16}$ Ω cm @ 300°K
Dielectric constant (ϵ/ϵ_0)	3.9
Melting point (°C)	~1700
Specific heat (J g^{-1} °C^{-1})	1.0
Thermal conductivity (W cm^{-1} °C^{-1})	0.014
Linear coefficient of thermal expansion (ppm)	0.5

Source: Reference 27, used by permission. Copyright 1967, John Wiley & Sons.

ficiently cool to avoid erosion of the wand. The hydrogen burns to form water vapor, which serves as a pure oxidant. Care must be taken that explosive concentrations of hydrogen do not accumulate. Therefore, hydrogen is used in low concentration (4% is typical), and a thermocouple in the wand ensures that ignition temperature is maintained.

Kinetics of Oxidation

During oxidation, an oxide film covers the surface of the silicon. Continued growth requires either that the oxidizing species migrate through the oxide to react with silicon or that silicon atoms move through the oxide to the surface. In fact, the former mechanism occurs: silicon oxidation takes place at the silicon-oxide interface.

Figure 3-16 allows us to examine the growth process in detail.[28] Three events must occur during oxidation: Oxidant must enter the oxide surface from the gas ambient, with a flux j_1. The oxidant then diffuses across the oxide layer, which has a thickness denoted x_o; this flux is j_2. Finally, oxidant is consumed by reaction with the silicon, at a rate (in atoms cm^{-2} sec^{-1}) of j_3. For steady-state growth, these fluxes must be equal. Let the concentration of oxidant in the oxide be $N(x)$, where x is the distance from the silicon interface. The flux of oxidant into the substrate will depend on the difference in concentrations between the gas (N_G) and the surface of the oxide where $x = x_o$, so

$$j_1 = h[N_G - N(x_o)] \qquad (3\text{-}51)$$

where h is a proportionality constant. Within the oxide, Eq. (3-11) predicts a flux of

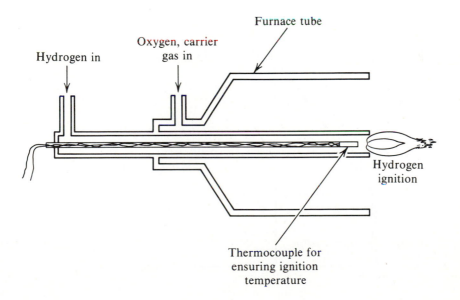

Figure 3-15 Hydrogen injection system for steam oxidation.

$$j_2 = D \frac{N(x_o) - N(0)}{x_o} \tag{3-52}$$

where D is the diffusion coefficient for the oxidant in the oxide. The rate of oxidant consumption by reaction is given in terms of the reaction rate constant k_s by

$$j_3 = k_s N(0) \tag{3-53}$$

By requiring $j_1 = j_2 = j_3$, we can solve for $N(0)$ and $N(x_o)$ to obtain

$$N(0) = \frac{N_G}{1 + k_s/h + k_s x_o/D} \tag{3-54}$$

$$N(x_o) = \frac{N_G(1 + k_s x_o/D)}{1 + k_s/h + k_s x_o/D} \tag{3-55}$$

Now the reaction rate is related to the increase in oxide thickness dx_o/dt by the expression

$$\frac{dx_o}{dt} = \frac{k_s N(0)}{\gamma} \tag{3-56}$$

where γ is the number of oxidant molecules per unit volume of oxide. (This number is 2.3×10^{22} for O_2 and twice that number for H_2O.[29]) This allows us to write a differential equation for $x_o(t)$:

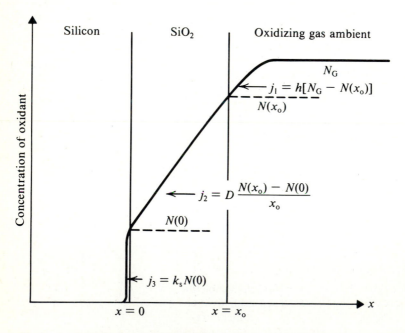

Figure 3-16 Oxidation of silicon, showing concentration of oxidant as a function of distance from the oxide-substrate interface. (*After Reference 28, used by permission. Copyright 1967, John Wiley & Sons.*)

$$\gamma \frac{dx_o}{dt} = \frac{k_s N_G}{1 + k_s/h + k_s x_o/D} \tag{3-57}$$

Deal and Grove[30] solved Eq. (3-57) under the initial condition that $x_o(0) = x_i$. This condition makes the solution applicable to both bare and previously oxidized surfaces. The resultant expression has been termed Grove's law[31]:

$$x_o^2 + A x_o = B(t + \tau) \tag{3-58a}$$

where

$$A = 2D \left(\frac{1}{k_s} + \frac{1}{h} \right) \tag{3-58b}$$

and

$$B = \frac{2DN_G}{\gamma} \tag{3-58c}$$

and

$$\tau = \frac{x_i^2 + A x_i^2}{B} \tag{3-58d}$$

Since τ is simply the time required to grow the initial oxide layer of thickness x_i, the term $(t + \tau)$ implies that for sequentially applied oxidation times t_1, t_2:

$$x_o = x_o(t_1 + t_2) \tag{3-59}$$

Note that oxidation thicknesses are *not* additive. Two limiting cases of Eq. (3-58) are important. At long times, with $t \gg A^2/4AB$, we have the *parabolic rate law:*

$$x_o^2 \simeq Bt \tag{3-60}$$

where B is the *parabolic rate constant.* For short times, with $(t + \tau) \ll A^2/4AB$, the *linear rate law* results:

$$x_o \simeq \frac{B}{A}(t + \tau) \tag{3-61}$$

with B/A termed the *linear rate constant.* Equation (3-58) proves highly reliable in predicting oxidation rates of silicon, deviating from observation only for very thin oxides.[32,33] Figure 3-17 shows observed oxide thicknesses as a function of oxidation time and temperature, for oxygen and steam ambients.[34]

Example A substrate is oxidized for 2 h at 1000°C in steam. Windows are then etched in the oxide, clearing parts of the surface back to the bare substrate. At the next step, the substrate is exposed to a steam ambient at 1100°C for 1 h. What is the oxide thickness (a) in the areas which were etched clear? (b) In the unetched areas?

(a) The etched areas have only the oxide grown due to 1 h of steam at 1100°C. From Fig. 3-17, the thickness is 0.65 μm.

(b) In the unetched areas, the second oxidation proceeded at a slower rate due to the presence of the first oxide. Since the times at a given temperature are additive, we can find the thickness by first finding, from Fig. 3-17, the thickness produced by the first oxidation. This is 0.6 μm. We next find the equivalent oxidation time at 1100°C to produce the same thickness; this is about 70 min. The total oxide in the unetched areas grown in the unetched areas corresponds, then, to that grown in 2 h and 10 min at 1100°C. This thickness is 1.0 μm. Note the way that subsequent oxidations tend to wash out differences in original thickness.

Implications of the Oxidation Mechanism

Several important effects follow from the fact that silicon oxidation occurs at the oxide-silicon interface. One is that oxidation effects the distribution of dopants already in the silicon. Dopants can either be accepted into the growing oxide layer as the interface moves or be pushed ahead into the silicon. Which event occurs depends on the relative stability of the dopant in the oxide and in silicon. In general, the *p*-type dopants tend to be taken up by the oxide while the *n*-type dopants are rejected. The resulting doping profile will also be affected by the diffusion rate of the dopant in oxide. Figure 3-18 illustrates the four possible results.[35]

The oxidation mechanism also implies that numerous oxygen atoms are available at the silicon surface. While most of these react, some continue into the silicon, adding

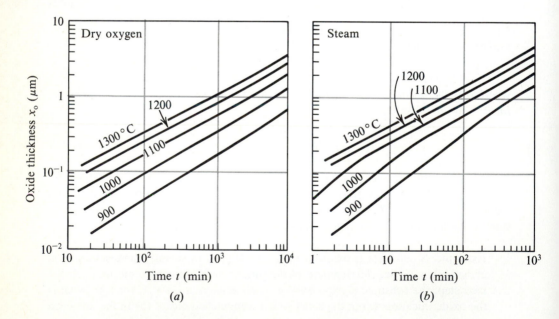

Figure 3-17 Oxide thickness as a function of time and temperature, in: (*a*) dry oxygen; (*b*) Steam. (*Reference 34, used by permission. Copyright 1967, Prentice-Hall.*)

Oxide *takes up* impurity $(m < 1)$

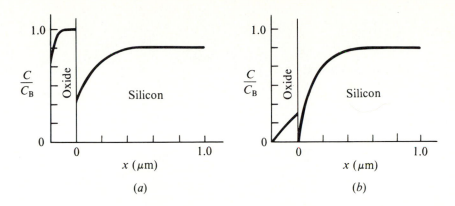

(a)   (b)

Oxide *rejects* impurity $(m > 1)$

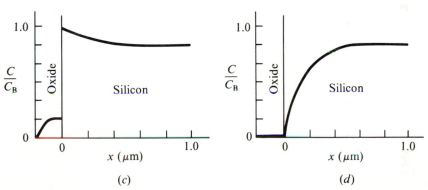

(c)   (d)

Figure 3-18 Dopant redistribution due to oxidation. (*a*) Diffusion in oxide slow (e.g., boron). (*b*) Diffusion in oxide fast. (*c*) Diffusion in oxide slow (e.g., phosphorus). (*d*) Diffusion in oxide fast (e.g., gallium). (*Figures are reprinted from Ref. 28, p. 70, by permission of the publisher. Copyright 1967, John Wiley & Sons. Original credit is given to Ref. 35. Reprinted by permission of the publisher, The Electrochemical Society, Inc.*)

to the amount of dissolved oxygen already present from crystal growth. It is possible to saturate the crystal with oxygen, leading to oxygen precipitation and the formations of crystal stacking faults. These *oxidation-induced stacking faults (OSFs)* are also encouraged by the mobile vacancies propagated into the crystal as silicon atoms rearrange at the oxide front. One way of avoiding OSFs is to conduct oxidation above 1230°C, so that the oxygen remains in solid solution.[36] Another is the addition of chlorine to the oxidation ambient, for reasons we shall see shortly.

At the silicon–silicon oxide interface, there will be silicon atoms partly bonded to the crystal lattice and partly to the oxide. Figure 3-19 shows this concept schematically. The result of partial bonding is the accumulation of a sheet of charge at the

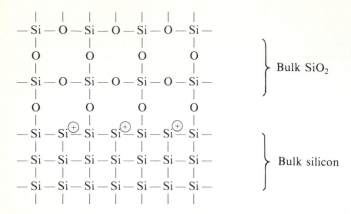

Figure 3-19 Partial bonding at the silicon–silicon oxide interface. (This figure is a conceptual presentation only and is not meant to portray the actual three-dimensional bonding.)

interface. The charge per unit area of this sheet is usually designated Q_{ss}. (Remember that it is a charge density in coulombs per square centimeter and not a charge in coulombs.) This charge can significantly affect the operation of some devices.

Both Q_{ss} and OSFs can be reduced by oxidation in the presence of chlorine. The chlorine may be introduced to the furnace as a gas (usually HCl) or as a chlorinated organic compound like trichloroethane. The chlorine diffuses through the oxide and forms a chlorine-rich layer at the silicon interface, where it reduces the number of partially bonded silicon atoms and limits the flow of oxygen and crystal vacancies into the crystal.[37,38] Chlorine also reduces the amount of ionic charge in the oxide layer. This charge results from alkali ions (e.g., Na^+) present as impurities: chlorine presumably reacts with the alkalis to form uncharged or immobile products. The use of chlorine in high-quality oxidations is standard in modern fabrication. It causes slight variation in the oxide growth rate, which may depend somewhat on the chlorine source.[39]

Since oxidation requires the flow of oxidant to the silicon, oxidation can be made *selective* if oxidant motion can be blocked. An effective oxidation masking material is *silicon nitride* (Si_3N_4), which can be deposited on substrates by methods explored in Chapter 7. By depositing and patterning a nitride layer, oxide can be grown only on selected areas of the substrate surface. Selective oxidation is one of the techniques that made possible metal-oxide-silicon technology, the subject of Chapter 11.

Selective oxidation is complicated by the *"bird's beak"* phenomenon, which is a consequence of the consumption of silicon during oxidation. The growth of oxide consumes silicon from the surface: for each micrometer of oxide grown, 0.48 μm of silicon is removed. The result is a movement of the silicon–silicon oxide interface inward from the original wafer surface. If oxidation were perfectly selective, the result would be an abrupt step in the silicon surface at the edge of the oxide. In fact, there is a transition area, where the silicon surface slopes downward. At the same location, the surface of the oxide rises, since the remaining 52% of the oxide thickness projects above the original silicon surface. Figure 3-20 illustrates the boundary of a selectively oxidized area. From the figure, the origin of the term *bird's beak* is clear. If the selective oxidation is thick, as in most MOS devices, control of this feature is important. If it extends too far into the masked area, the device may not function properly, and an abrupt step causes step coverage problems at later processing steps. For this reason, controlling the shape and reproducibility of the bird's beak was a significant obstacle in the initial implementation of MOS technology.

Implicit in the rate expressions for oxidation are alternative methods of controlling oxidation rate. Traditionally, oxidation rate has been controlled, once the oxidant was chosen, by varying temperature. However, to grow a thick oxide in a convenient time requires high temperatures, which can cause excessive thermal stress in large substrates. On the other hand, the increasingly thin gate oxides used in MOS applications require shorter and shorter oxidation times, which are difficult to control. As a result, it can be useful to vary oxidation rate independent of temperature. One way to do this is to perform the oxidation at other than atmospheric pressure, thereby changing the oxidant gas density, N_G, and thus the oxidation rate.

For example, a 0.5-μm layer of oxide can be grown in steam at 10 atm at 950°C in 30 min. At 1 atm, either a temperature of 1150°C or a time of over 2 h would be required. High-pressure oxidation thus allows a reduction in process temperature and/or a reduction in process time. Lower temperature leads to less thermal stress on the wafer and less diffusion of previously deposited dopant distributions. Beneficial effects on OSFs, dopant redistribution, and other oxide parameters have also been reported. Commercial equipment is currently available to perform oxidation at pressures of tens of atmospheres. Since VLSI devices increasingly require minimizing of

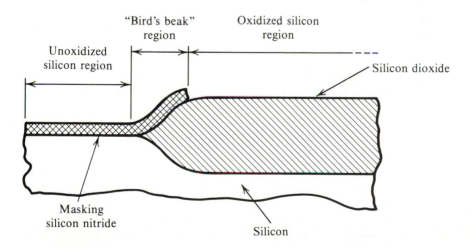

Figure 3-20 Effect of selective oxidation on wafer topography: formation of "bird's beak" in region at boundary of nitride mask.

process temperatures, oxidation at elevated pressures will become increasingly common in the future.[40]

Oxide Behavior in Metal-Oxide-Semiconductor Devices

In Chapter 2 we studied some of the properties of metal-oxide-semiconductor (MOS) devices. The most challenging application of silicon oxidation is in the fabrication of MOS insulator oxides, because the oxide layer serves not only as a diffusion mask but as a critical part of the device. Therefore, the electrical properties of the device depend on the properties of the oxide, which must be carefully controlled. Conversely, a sensitive way of measuring the properties of any oxide is to incorporate the oxide into a simple MOS device.

We saw in Chapter 2 that the capacitance of an MOS device varied with the effective applied field V' in a characteristic way. It was equal to the oxide capacitance C_o when V' was negative (for p-type silicon) and declined with increasing voltage as V' became positive. We will now see how oxide properties affect the relationship between *effective* applied voltage V' and the real voltage on the metal electrode V. First, however, we will see how MOS properties are used to make measurements on oxides.

Given an oxidized semiconductor substrate, an array of simple MOS capacitors can be made by depositing a number of metal dots on the surface. Each dot then becomes the metal electrode of an MOS capacitor. Contact is made to a dot with a probe, and a dc bias is applied. This bias determines the field penetrating the silicon. We then apply a small additional ac signal, dV, to the electrode and measure the resulting change in charge per unit area, dQ. The measured capacitance per unit area is then

$$C = \frac{dQ}{dV} \tag{3-62}$$

A graph of this capacitance versus the applied DC voltage is a *capacitance-voltage* or *C-V plot*. The measurement frequency is normally chosen so that a high-frequency response is observed in the inversion region. The result is a characteristic, sigmoid curve.

Figure 3-21 shows a typical *C-V* plot for a p-type MOS device. Figure 3-21a shows the ideal case we discussed in Chapter 2. The boundary between accumulation and depletion occurs at $V = V' = 0$. This voltage is called the *flat-band voltage*, V_{fb}, because with this applied voltage the valence and conduction bands are flat throughout the semiconductor. The flat-band voltage, V_{fb}, will be zero, and V will equal V', only if the oxide is a perfect, charge-free insulator. In practice, this is not the case, and therefore the value of V_{fb} is a sensitive probe of the properties of the oxide.

The flat-band voltage for real oxides can be obtained by summing up the effects of any oxide charge. We have already mentioned the sheet of surface charge, of density Q_{ss}, which occurs at the oxide-silicon surface due to dangling bonds. This charge creates an electric potential relative to the metal electrode, which we can evaluate using Eq. (2-35). The resulting potential contribution $\Delta\phi$ is

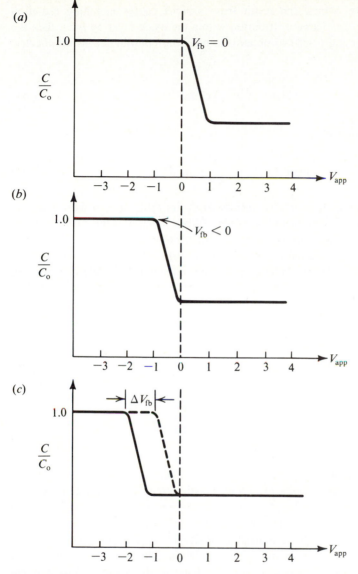

Figure 3-21 Capacitance-voltage characteristic of a p-type MOS capacitor. (a) Ideal case: $V_{fb} = 0$, depletion approximation. (b) Nonideal case. (c) V_{fb} shift after high-temperature bias.

$$\Delta\phi(Q_{ss}) = \frac{1}{C_o} Q_{ss} \qquad (3\text{-}63)$$

where C_o is the oxide capacitance. In addition, there is usually charge distributed within the oxide. One source of this charge is contamination by alkali metal ions (sodium, potassium), which are ubiquitous in nature. Ionic charge is moveable charge:

the ionic carriers can diffuse within the oxide at moderate temperatures. There can also be fixed oxide charge, which can result from various causes including radiation damage. If we denote the charge distribution within the oxide layer by $\rho(x)$, then we can apply Eqs. (2-37) and (2-38) to obtain the resulting contribution to the electric potential. We find that

$$d\phi(\rho) = \frac{x_o}{\epsilon_{ox}} \rho(x) \, dx \tag{3-64}$$

and, using the definition of C_o,

$$\Delta\phi(\rho) = \frac{1}{C_o} \int_0^{x_o} \frac{x}{x_o} \rho(x) \, dx \tag{3-65}$$

Another adjustment to applied voltage results from the difference in *work function* between the metal and semiconductor electrodes. Work function is the potential energy in a conductor with respect to the energy of an electron in free space. In general, this quantity differs from material to material and results in an electric potential between two dissimilar electrodes. We designate this potential Φ_{MS}. We can now relate the actual applied voltage on the metal electrode, V, to the effective applied voltage, V', by combining the effects of oxide charge and work function:

$$V' = V - \Phi_{MS} + \Delta\phi(Q_{ss}) + \Delta\phi(\rho) \tag{3-66a}$$

$$= V - \Phi_{MS} + \frac{1}{C_o}\left(Q_{ss} + \int_0^{x_o} \frac{x}{x_o} \rho(x) \, dx \right) \tag{3-66b}$$

The flat-band voltage is the applied voltage at which V' is zero and is thus

$$V_{fb} = \Phi_{MS} - \frac{1}{C_o}\left(Q_{ss} + \int_0^{x_o} \frac{x}{x_o} \rho(x) \, dx \right) \tag{3-67}$$

A change either in magnitude or position of the oxide charge will therfore change V_{fb}.

Figure 3-21*b* shows a *C-V* plot of a device with nonideal V_{fb}. Comparison with Figure 3-21*a* reveals that oxide charge shows up as an overall shift of the *C-V* curve to the left. The magnitude and reproducibility of this shift are of vital concern for the manufacture of MOS devices. MOS transistors utilize a voltage on the metal electrode to switch the underlying silicon in and out of depletion. Therefore, the switching or *threshold* voltage for MOS devices is closely related to the value of V_{fb}, and unpredicted changes in oxide charge can make a device inoperative.

The greatest danger to the device is posed by ionic charge, because it can move about under normal device temperatures and voltages. The resulting change in threshold voltage can make the device unreliable. Mobile oxide charge is monitored using the high-temperature bias technique of *C-V* measurement.[41]

In this method, an initial *C-V* plot is made using an oxide film, and the sample is then heated to about 300°C while a positive voltage (perhaps 10 V) is applied to the metal electrode. At this temperature, alkali ions are mobile, and they are driven to the silicon-oxide interface by the applied field. After perhaps 15 min the sample is cooled and the *C-V* characteristic redetermined. All ionic charge is now concentrated at $x = x_o$, and thus, according to Eq. (3-67), V_{fb} has become more negative. The result is a shift to the left in the (*p*-type) *C-V* curve. Figure 3-21*c* shows the application of this method. The difference in flat-band voltage, ΔV_{fb}, is a sensitive detector of mobile ion contamination and an early warning of potential device unreliability.

Example A 0.1-μm-thick oxide is *C-V* plotted and found to have $V_{fb} = -1.0$ V. After high-temperatute bias, $V_{fb} = -2.0$ V. Assume that, prior to bias, the ionic charge is uniformly spread throughout the oxide with a charge distribution ρ, and that after bias, the ionic charge is all localized at the oxide interface. Find ρ and the number of ions cm^{-3} in the oxide.

By integrating Eq. (3-65) with a constant $\rho(x)$, the ionic contribution to V_{fb} is

$$\frac{-\rho x_o}{2C_o}$$

With the ionic charge concentrated in a sheet of charge $Q = \rho x_o$ at the interface, the ionic contribution is

$$\frac{-\rho x_o}{C_o}$$

and so the ΔV_{fb} of -1 V corresponds to

$$\frac{1}{2}\frac{\rho x_o}{C_o}$$

Thus

$$\rho = \frac{2(1 \text{ V})(3.46 \times 10^{-8} \text{ F cm}^{-2})}{10^{-5} \text{ cm}}$$

$$= 6.92 \times 10^{-3} \text{ C cm}^{-3}$$

(The capacitance of a 0.1-μm oxide layer was calculated in an example in Chapter 2.)

The equivalent ionic density is found by dividing by the electronic charge q to give 4.32×10^{16} ions per cubic centimeter, about a 2-ppm concentration of ions in the oxide.

With further refinement, additional oxide and silicon parameters can be deduced from *C-V* behavior, making it a diagnostic tool of unusual sensitivity and power. Therefore, *C-V* plotting of oxide films is a routine part of oxide process control, whether or not the oxide is intended for use in a MOS device. Unexpected values of V_{fb} or ΔV_{fb}, or changes in the shape of the *C-V* curve, signal problems in the oxidation process that are frequently undetectable by other means.

3-5 DIFFUSION PROCESS CONTROL

The primary task of the microelectronic process engineer, either in development or manufacturing, is to design and maintain a controlled process. For this reason, in most chapters we will devote a section to tools and philosophies of process control. In general, control over a process is demonstrated by predictability of the relevant output variables for the process, e.g., oxide thickness for an oxidation, dopant quantity for a predeposition. A process is in control if the output variables are consistently in the range of values consistent with optimum device performance and yield. This is attained by controlling the related input variables: those quantities such as temperature or diffusion time that can be directly controlled. There are four steps involved in maintaining a controlled process:

1. Determining the output variables that affect device yield, along with their desired values and permissible ranges
2. Determining the relationship between the output variables and the input variables
3. Measuring and controlling the input variables sufficiently to ensure an appropriate range of output variables
4. Making sufficient measurements of the output variables to ensure that control is in fact maintained

The device design process usually determines the acceptable range of output variables which the process engineer must attain. Understanding how these output variables are related to input variables depends on a knowledge of the underlying physics and chemistry, which we attempt to convey throughout this book. In our process control sections, we will stress the application of this knowledge to the measurement and control of input and output variables. For diffusion and related processes, the outputs of greatest interest are diffused dopant quantity and distribution, oxide thickness, and oxide electrical properties. The inputs are furnace temperature, chemical ambient, and the time that substrates are exposed to heat and to source.

The input variables in diffusion and oxidation are conceptually straightforward, and the modern furnace is well designed to control them. Chemical ambient will be reproducible if furnace and jungle function properly, if systems are kept clean and leak-free, and if material purity is ensured. Temperature consistency across the furnace flat zone can be confirmed by *profiling* the furnace: measuring temperature with a thermocouple at numerous points. Similar measurements taken during furnace ramping can confirm the accuracy of furnace ramp rate. Modern furnaces perform these functions automatically. The time duration of furnace cycles and source periods is software controlled and is seldom an issue.

Process problems related to input variables often arise from peripheral causes. For example, in a short predeposition cycle, minor variations in operator technique can result in significant changes in the time-temperature-ambient history of substrates. Whether an operator pushes a timing button before or after attaching a pull rod can cause detectable differences in, for example, a bipolar emitter diffusion. Disastrous contamination of a furnace can result from using the same profiling thermocouple in both p- and n-dopant tubes. Therefore, even the smallest variations should be carefully examined for their influence on the basic input variables. For an oxidation, outputs include electrical properties and film thickness. We have already examined at length the capacitance-voltage (C-V) technique for measuring electrical properties. Film thickness is most conveniently measured using the interference properties of visible light.

An oxide film is a thin, transparent material overlying a reflective semiconductor surface. Therefore, incident light waves must have a node where they reflect at the silicon surface. Figure 3-22 illustrates the concept. Destructive interference of standing waves results when the film thickness x_o is given by

$$x_o = \frac{(i - \frac{1}{2})\lambda}{n} \tag{3-68}$$

where i is an integer, n the refractive index of the film, and λ the wavelength of the light. For layers of a few wavelengths thickness, quite distinct colors result from the selective removal of some wavelengths from incident white light, and these colors can be used to gauge film thickness. Due to the integer factor i above, this method is ambiguous: several thicknesses will produce the same color. Nevertheless, it is useful for quick estimates of oxide thickness, or identification of various process layers. Table 3-4 is a chart giving the observed colors for various film thicknesses. Every process engineer should be at least qualitatively familiar with it.[42]

Example An oxidation takes place in steam for 2 h at 1100°C. After oxidation, the oxide color is observed as red-violet. What is the actual oxide thickness?

According to Fig. 3-17, the expected oxide thickness is 0.32 μm. This gives us an approximate range. According to Table 3-4, the actual oxide thickness is around 0.275 μm.

By using a scanning spectrometer to illuminate the oxide, and mathematically analyzing the resulting absorption spectrum, this color technique can be made precise and automatic. Equipment that will analyze the thickness of a small area of film identified with a microscope and produce a printout of thickness within seconds is now common. These units are referred to as *film thickness analyzers*. A more sophisticated technique, particularly useful for oxides less than 100 nm in thickness, is *ellipsometry*. This method measures phase shifts in reflected polarized light, and determines both thickness and refractive index. Automated ellipsometers are also available.

The output variables for a dopant diffusion are those that measure the distribution of the dopant. As we saw in Chapter 2, the resistivity of the doped semiconductor is directly related to total diffused dopant and is also a direct parameter of circuit performance. The resistivity ρ of a uniform slab or material of length l, width w, and thickness x is related to the applied voltage V and the resulting current I by

$$\frac{V}{I} = \rho \frac{l}{wx} \tag{3-69}$$

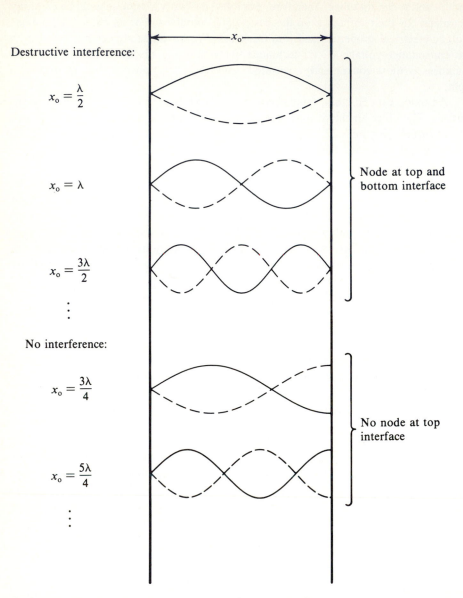

Destructive interference:

$$x_o = \frac{\lambda}{2}$$

$$x_o = \lambda$$

$$x_o = \frac{3\lambda}{2}$$

⋮

No interference:

$$x_o = \frac{3\lambda}{4}$$

$$x_o = \frac{5\lambda}{4}$$

⋮

Node at top and
bottom interface

No node at top
interface

Figure 3-22 Varying patterns of interference in thin transparent films as a function of wavelength.

This equation defines resistivity, which has units of Ω cm. Most diffused layers are thin low-resistivity layers lying atop the thicker, high-resistivity substrate. Under these conditions, almost all current flows in the surface layer, and it is convenient to speak of *sheet resistivity* ρ_s, defined by

$$\rho_s = \frac{\rho}{x} \tag{3-70}$$

Table 3-4 Apparent colors of oxide layers of various thicknesses

Thickness	Order (545 nm)	Color
0.05		Tan
0.07		Brown
0.10		Dark violet to red-violet
0.12		Royal blue
0.15		Light blue to metallic blue
0.17	I	Metallic to very light yellow-green
0.20		Light gold or yellow, slightly metallic
0.225		Gold with slight yellow-orange
0.27		Red-violet
0.30		Blue to violet-blue
0.31		Blue
0.325		Blue to blue-green
0.34		Light green
0.35		Green to yellow-green
0.36	II	Yellow-green
0.37		Green-yellow
0.39		Yellow
0.41		Light orange
0.42		Carnation pink
0.44		Violet-red
0.46		Red-violet
0.47		Violet
0.48		Blue-violet
0.49		Blue
0.50		Blue-green
0.52		Green (broad)
0.54		Yellow-green
0.56	III	Green-yellow
0.57		Yellow to "yellowish" (Not yellow, but is in the position where yellow is to be expected. At times it appears to be light creamy gray or metallic.)
0.58		Light orange or yellow to pink borderline
0.60		Carnation pink
0.63		Violet-red
0.68		"Bluish" (Not blue, but borderline between violet and blue-green. It appears more like a mixture between violet-red and blue-green and overall looks grayish.)
0.72	IV	Blue-green to green (quite broad)
0.77		"Yellowish"
0.80		Orange (rather broad for orange)
0.82		Salmon
0.85		Dull, light red-violet
0.86		Violet
0.87		Blue-violet
0.89		Blue
0.92	V	Blue-green
0.95		Dull yellow-green
0.97		Yellow to "yellowish"

Table 3-4 (continued)

Thickness	Order (545 nm)	Color
0.99		Orange
1.00		Carnation pink
1.02		Violet-red
1.05		Red-violet
1.06		Violet
1.07		Blue-violet
1.10	VI	Green
1.11		Yellow-green
1.12		Green
1.18		Violet
1.19		Red-violet
1.21		Violet-red
1.24		Carnation pink to salmon
1.25		Orange
1.28		"Yellowish"
1.32	VII	Sky blue to green-blue
1.40		Orange
1.45		Violet
1.46		Blue-violet
1.50	VIII	Blue
1.54		Dull yellow-green

Source: Reference 42. Copyright 1964 by International Business Machines Corporation; reprinted with permission.

Note that sheet resistivity has units of ohms and that

$$\frac{V}{I} = \rho_s \frac{l}{w}$$

(3-71)

Equation (3-71) implies that, whenever l and w are equal, regardless of magnitude, ρ_s will be given by V/I. This means that all square sheets have resistance ρ_s and so the units of ρ_s are frequently given as "ohms per square."

Sheet resistivity is measured using a four-point probe, as shown in Fig. 3-23. A current I is forced through the outer two probes, and the resulting voltage drop V is measured between the inner two. For an infinite sheet, it can be shown that[43]

$$\rho_s = 4.5324 \frac{V}{I}$$

(3-72)

Somewhat different values apply if the size of the sheet is not much larger than the interprobe spacing. In everyday practice, the conversion process is frequently bypassed and the V/I ratio is quoted directly in terms of "V-over-I" units.

Although resistivity is related to doping, the precise resistivity of a real diffusion depends on the detailed distribution of the dopant with depth. Even if we assume a

Gaussian distribution, at least two measurements are required to characterize the result. A convenient second measurement of doping is *junction depth* x_j. The junction depth is the distance below the semiconductor surface at which the predominant doping changes from *p*-type to *n*-type or vice versa. This change can be directly observed using a "stain," a chemical solution that darkens *p*-type, but not *n*-type, silicon. Concentrated hydrofluoric acid with an addition of 0.1 to 0.5% nitric acid has this property.[44] Junction depth can thus be measured by a "groove and stain" or "lap and stain" technique, as shown in Fig. 3-24. A sloping groove or bevel is machined into the silicon surface. Stain is applied, and the location of the junction becomes visible. Its depth can be determined by two techniques. If the bevel is precisely lapped, a knowledge of the lapping angle allows depth to be determined from lateral distance. Otherwise, a glass slide can be placed over the groove and illuminated with monochromatic light. Interference bands are formed between the groove and the slide, and can be counted under a microscope. With $n = 1$ for air, and $\lambda = 589$ nm for a sodium lamp, each band or "fringe" observed indicates 0.29 μm of depth.[45]

> **Example** A circuit design requires a structure to have a sheet resistivity of 200 Ω per square. After diffusion, the V/I for this layer is 44.0. The x_j is found by a groove-and-stain technique to be "two fringes." Find (a) ρ_s, (b) ρ, (c) estimate the dopant concentration.
>
> By Eq. (3-72), $\rho_s = 199.4$ Ω per square, quite close to the design value. "Two fringes" is about 0.6 μm, which we take for the junction depth. Then, from Eq.

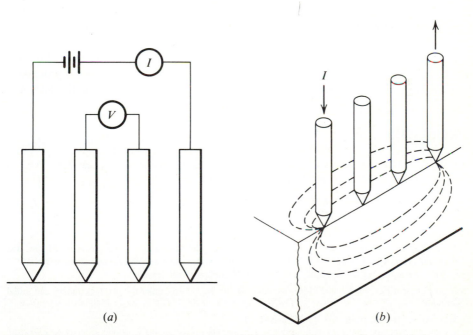

(a)   (b)

Figure 3-23 Four-point probe for resistivity measurement. (*a*) Schematic: outer probes force a current; inner probes measure a voltage. (*b*) Cutaway view of current flow during substrate measurement.

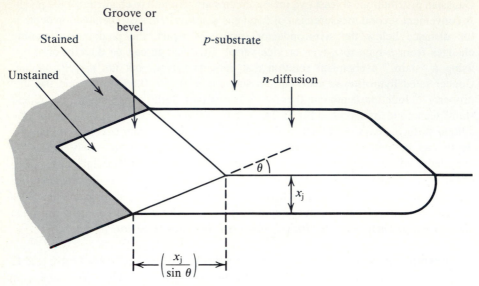

Figure 3-24 Lap- or groove-and-stain technique for junction depth determination.

(3-70), $\rho = 1.15 \times 10^{-2}$ Ω cm. If the doping distribution is uniform, this resistivity corresponds to a dopant concentration of around 10^{19} cm^{-3} according to Fig. 2-6. This is only a rough estimate, because the doping distribution in a diffused structure is more complicated.

More sophisticated methods of determining doping profile are available, but they are slower and more cumbersome. For the everyday control of diffusion processes, knowledge of oxide thickness, resistivity, and junction depth will usually suffice.

PROBLEMS

3-1 Use Eqs. (3-28) and (3-30) to estimate (a) the jump frequency for an atom, ν_{subst}, and (b) the diffusion coefficient D for the following dopants in silicon, at 1200°C:

i. Phosphorus: $E_A + E_V = 3.69$ eV
ii. Boron: $\qquad E_A + E_V = 3.69$ eV
iii. Arsenic $\qquad E_A + E_V = 3.56$ eV

Take 10^{14} as an estimate for the vibrational frequency ν_l. The interatomic spacing d is 0.235 nm, and Boltzmann's constant k is 8.62×10^{-5} eV °K^{-1}. Compare your values with the actual values of 3×10^{-14} cm^2 sec^{-1} for boron and phosphorus, and 4×10^{-15} cm^2 sec^{-1} for arsenic.

3-2 Confirm that Eqs. (3-32) and (3-39) are solutions for Eq. (3-16) under the given boundary and initial conditions.

3-3 Derive Eq. (3-38) from Eqs. (3-37) and (3-32).

3-4 Determine the validity of the approximation expressed in Eq. (3-36) for the following typical diffusions: (a) Phosphorus, 950°C, 30 min. (b) Boron, 1050°C, 60 min. (c) Arsenic, 1200°C, 10 h.

3-5 A substrate has a background doping density N_B. A drive-in diffusion results in a Gaussian dopant distribution containing total dopant Q. Derive an expression for junction depth x_j in terms of Q, N_B, D, and t. Comment on the relative importance of the various factors in determining x_j.

3-6 Discussions of diode behavior often assume that the dopant distribution is linear, that is,

$$N(x) = N(\phi) - [N(\phi) - N(x_j)] \frac{x}{xj}$$

For $\sqrt{Dt} = 0.1$ μm^{-1}, $Q = 1 \times 10^{17}$ atoms per cubic centimeter and $N_B = 1 \times 10^{15}$ atoms per cubic centimeter, evaluate the percentage deviation between a Gaussian and linear distribution, at (a) $x = 0$, (b) $x = 0.5\,x_j$, (c) $x = x_j$.

3-7 A bipolar integrated circuit undergoes the following processing:

Postepitaxial oxidation: steam, 1050°C, 60 min
Etch base window in epitaxial oxide
Base predeposition (boron)
Base drive-in and oxidation: steam, 1000°C, 60 min
Etch emitter window in base oxide
Emitter predeposition (phosphorus)
Emitter oxidation: steam, 950°C, 30 min

Predict the oxide thickness in the following areas: (a) Field (all oxidations present). (b) Base (base and emitter oxidations present, epitaxial oxidation etched away). (c) Emitter (only emitter oxidation present).

What will the apparent colors of each area be? Upon inspection, all colors are as predicted except for the emitter oxidation, which is red-violet. Speculate on the reason for this atypical oxidation rate.

3-8 Determine whether the base oxide thickness found in Problem 3-7 is adequate to mask against an emitter predeposition of 30 min at 950°C. If the phosphorus deposition temperature is raised to 1000°C, will additional masking oxide be required? If so, propose an appropriate total base oxidation time in steam at 1000°C.

3-9 In high-pressure oxidation, silicon is oxidized at an elevated pressure to increase the etch rate for a given temperature. The most obvious effect of increased pressure is to increase the oxidizer density in the gas, N_G, proportional to the pressure. (a) Predict the effect on the parabolic and linear oxidation rates of a change in pressure assuming increased density is the only effect. (b) The predicted pressure dependence is observed for the parabolic and linear rates using steam oxidation and for the parabolic rate using oxygen. However, the linear oxidation rate in oxygen varies as pressure to the 0.7 power.[46] Comment on a possible reason. (Remember that $h \gg k_s$.)

3-10 Express the linear rate constant for oxidation, B/A, in terms of N_G, k_s, h, x_o, and D.

3-11 A silicon substrate is oxidized to a thickness of 100 nm. An aluminum film is deposited on top of the oxide to form an MOS capacitor, which is C-V plotted with the following results:

$$V_{fb} = -3.04 \text{ V}$$
$$\text{Minimum capacitance (in inversion)} = 0.3C_o$$

Given that

$$\epsilon_{ox} = 3.46 \times 10^{-13} \text{ F cm}^{-1}$$
$$\epsilon_S = 10.38 \times 10^{-13} \text{ F cm}^{-1}$$
$$\Phi_{MS} = -0.90 \text{ V}$$

and assuming that $Q_{ss} = 3 \times 10^{10}$ g cm^{-2}, find (a) $x_{d,max}$, (b) $\Delta\phi(\rho)$.

3-12 The MOS structure of Problem 3-10 is subjected to high-temperature bias (HTB) stress, and V_{fb} after stress is found to be -2.2 V. Assume that all oxide charge is mobile and is distributed uniformly through the oxide before stress. After stress, assume all charge to be concentrated at the silicon surface. Find the oxide charge density in q cm^{-2}.

3-13 A silicon substrate is measured with a four-point probe and V/I is found to be 0.01. After an arsenic diffusion, the V/I is remeasured and found to be 10.0. The substrate is grooved and stained, and x_j is measured as 3.0 μm. Find (a) ρ_s, (b) ρ.

Make the simplifying assumptions that the dopant distribution varies linearly from the surface to x_j and that the average doping concentration in the slice determines the resistivity according to Fig. 2-6 in Chapter 2. Now find (c) The average impurity concentration in the slice. (d) The total doping per unit area Q.

3-14 An identical substrate to that in Problem 3-13 is rediffused until $x_j = 5$ μm. Make the same simplifying assumptions, and predict the value of V/I.

REFERENCES

1. Sorab K. Ghandi, *The Theory and Practice of Microelectronics*, Wiley, New York, 1968, p. 66.
2. *Ibid.*, p. 62.
3. Richard B. Fair, "Recent Advances in Implantation and Diffusion Modeling for the Design and Process Control of Bipolar ICs," *Semiconductor Silicon 1977*, H. R. Huff and E. Sirtl (eds.), The Electrochemical Society, Princeton, NJ, 1977, p. 968.
4. A. S. Grove, *Physics and Chemistry of Semiconductor Devices*, Wiley, New York, 1967, pp. 38–39.
5. L. Mei and R. W. Dutton, *Solid State Technol.* **26**:139 (June 1983).
6. See also the July 1981 issue of *IEEE Transactions on Circuits and Systems* and the October 1980 issue of *IEEE Journal of Solid State Circuits*, which are devoted to device modeling.
7. F. A. Trumbore, *Bell System Tech. J.* **39**:205 (1960).
8. H. S. Carlslaw and J. C. Jaeger, *Conduction of Heat in Solids*, 2nd ed., Oxford University Press, London, 1959.
9. D. P. Kennedy and R. R. O'Brien, *IBM Journal of Research and Development* **9**:3 (1965).
10. Ghandi, *op. cit.*, p. 6.
11. S. N. Ghosh Dastidar, *Solid State Technol.* **18**:37 (November 1975).
12. Ghandi, *op. cit.*, p. 95.
13. N. McLouski, *Diffusion of Impurities into Silicon from Gaseous Sources*, NASA Contractor Report CR524, National Aeronautics and Space Administration, Washington, DC, 1966, p. 11.
14. N. Goldsmith, J. Olmstead, and J. Scott Ur., *RCA Rev.* **28**:344 (1967).
15. *Handbook of Chemistry and Physics*, 56th ed., CRC Press, 1975.
16. R. F. Lever and H. M. Demsky, *IBM J. Res. Dev.* **18**:40 (1974).
17. J. Stach and J. Kruest, *Solid State Technol.* **19**:60 (October 1976).
18. McLouski, *op. cit.*, quoting *Federal Register vol. 36, #105*.
19. Fair, *op. cit.*, p. 984.
20. Ghandi, *op. cit.*, pp. 150–51.
21. J. S. Kesperis, *J. Electrochem. Soc.* **117**:554 (1970).
22. D. S. Perloff *et al.*, *Solid State Technol.* **20**:31 (February 1977).
23. D. K. Garg, R. S. Rao, and D. Singh, *Ind. J. Tech.* **8**:172 (1970).
24. P. Negrini, D. Nobili, and S. Solmi, *J. Electrochem. Soc.* **122**:1254 (1975).
25. J. Monkowski and J. Stach, *Insulation/Circuits* **22**:21 (April 1976).
26. Juri Kato and Yoshiteru Ono, *J. Electrochem. Soc.* **132**:1730 (1984).
27. A. S. Grove, *Physics and Technology of Semiconductor Devices*, Wiley, New York, 1967, pp. 102–103.
28. *Ibid.*, p. 24.
29. *Ibid.*, p. 26.
30. B. E. Deal and A. S. Grove, *J. Appl. Phys.* **36**:3770 (1965).
31. R. B. Fair, *J. Electrochem. Soc.* **128**:1360 (1981).
32. G. Camera Roda, F. Santarelli, and G. C. Sarti, *J. Electrochem. Soc.* **132**:1909 (1985).
33. Hisham Z. Massoud and James D. Plummer, *J. Electrochem. Soc.* **132**:1745 (1985).
34. R. M. Burger and R. P. Donovan, *Fundamentals of Silicon Integrated Device Technology*, Vol. 1, Prentice-Hall, Englewood Cliffs, NJ, 1967. Figure reprinted from pp. 41 and 49.

35. B. E. Deal, A. S. Grove, E. H. Snow, and C. T. Sah, *J. Electrochem. Soc.* **112**:308 (1965).
36. C. W. Pearce and G. A. Rozgonyi, "Sources of Oxidation Induced Stacking Faults in Czochralski Silicon Wafers: II. The Influence of Oxygen Content," *Semiconductor Silicon 1977*, H. R. Huff and E. Sirtl, eds., The Electrochemical Society, Princeton, NJ, 1977, p. 606.
37. Cor L. Claeys, Edgard E. Laes, Gilbert J. Declerck and Roger J. Van Overstraeten, "Elimination of Stacking Faults for Charge-Coupled Device Processing," *Semiconductor Silicon 1977*, H. R. Huff and E. Sirtl, eds., The Electrochemical Society, Princeton, NJ, 1977, p. 773.
38. H. L. Tsai, S. R. Butler, and D. B. Williams, *J. Electrochem. Soc.* **131**:411 (1984).
39. M. B. Das, J. Stach, and R. E. Tressler, *J. Electrochem. Soc.* **131**:389 (1984).
40. Stephen C. Su, *Solid State Technol.* **24**:72 (March 1981).
41. K. H. Zaininger and F. P. Heiman, *Solid State Technol.* **13**:49 (May 1970); and p. 46 (June 1970).
42. W. A. Pliskin and E. E. Conrad, *IBM J. Res. Dev.* **8**:43 (1964).
43. Roy A. Colclaser, *Microelectronics: Processing and Device Design*, Wiley, New York, 1980, p. 156.
44. *Ibid.*, p. 157.
45. Ghandi, *op. cit.*, p. 110.
46. Liang N. Lie, Reda R. Razouk, and Bruce E. Deal, *J. Electrochem. Soc.* **129**:2829 (1982).

FOUR

PHOTOLITHOGRAPHY

In Chapters 2 and 3 we discussed how the selective diffusion of dopants can modify the electrical properties of semiconducting substrates. We also described how a patterned film of a masking oxide or nitride is used to achieve this selective doping. In the next three chapters, we will study how the necessary patterns are created.

In this chapter, we describe the basic process of forming a pattern in a photosensitive etch-resistant film, using an optical image. The next chapter examines in more detail the imaging process itself. Chapter 6 describes the transfer of the pattern from the photosensitive film to the substrate, by etching.

4-1 PHOTOLITHOGRAPHY: AN OVERVIEW

A modern integrated circuit contains millions of individual elements, typically the size of a few micrometers. No physical tool is adequate for fabrication on this scale. Instead, microelectronic patterning is performed by radiation: primarily ultraviolet light, but increasingly x-rays or a beam of electrons. The process of using an optical image and a photosensitive film to produce a pattern on a substrate is *photolithography*.

Photolithography depends on a photosensitive film called a *photoresist*. A photoresist, or simply *resist,* has two principal properties:

Exposure to appropriate radiation will cause a change in the solubility of the resist
The resist will *resist* attack (hence the name) by an etchant capable of removing the substrate material

114

In photolithography, a film of the photoresist is first applied to the substrate. Radiation is shone through a transparent *mask plate,* on which has been imprinted a copy of the desired pattern in an opaque material. The resulting image is focused onto the resist-coated substrate, producing areas of light and shadow corresponding to the image on the mask plate. In those regions where light was transmitted through the plate, the resist solubility is altered by a photochemical reaction. Shadowed areas remain unaffected in solubility. This step, by analogy to photography, is termed *exposure.*

Following exposure, the substrate is washed with a solvent that preferentially removes the resist areas of higher solubility. This step is called *development,* again by analogy to photography. Depending on the type of resist, the washed-away areas may be either the illuminated or shadowed regions of the coating. If exposure increases resist solubility, resist is washed away in the areas corresponding to the transparent zones of the mask plate. The resist image is identical to the opaque image on the plate, and the pattern is a photographic positive. Therefore, the resist is called *positive-acting* or just *positive.* A resist that loses solubility when illuminated forms a negative image of the plate and is called a *negative* resist.

After development, the substrate, bearing the patterned resist, is exposed to an etchant. The etchant removes those portions of the substrate unprotected by the resist while the covered areas remain unetched. Finally, the resist coating is removed and discarded, leaving a duplicate of the mask plate pattern etched into the substrate film. Figure 4-1 shows the key steps of the photolithography process.

If photolithography is the reproduction of a pattern on a mask plate, how is the mask plate pattern generated? Actually, mask plates themselves are usually produced lithographically, but eventually they trace their ancestry to a primordial master optical image. In the formative years of microelectronics, the master image was photographically reduced from a macroscopic original. The desired pattern would be cut by hand into a colored plastic sheet, and room-sized reduction cameras would reduce the image to the desired size. This method has been replaced by the *pattern generator,* an apparatus that accepts a computer-generated description of the device (a *design tape*) and analyzes it into individual picture elements. The pattern generator then scans the master plate, "writing" the pattern on the plate using a directed beam of light to expose each element in the pattern. Apertures in the apparatus are varied to produce rectangles of various sizes that are used to construct the overall pattern. Increasingly, the optical pattern generator is being replaced by electron beam techniques. The electron beam has a finer resolution, and can be scanned and modulated electronically without mechanical intervention.

Pattern generation, requiring point-by-point creation of a complex pattern, is inherently slow and painstaking. It is therefore usual to pattern-generate only one of the multiple device images to be printed on the substrate. The single pattern is then reproduced repeatedly to form the array of devices that covers the substrate. The single device image is called a *reticle,* and the equipment used to replicate it into an array is a *step-and-repeat camera* or *stepper.* (This is because it exposes an image of the reticle, takes a *step,* and *repeats* the exposure.) The term *mask plate* is usually reserved for the

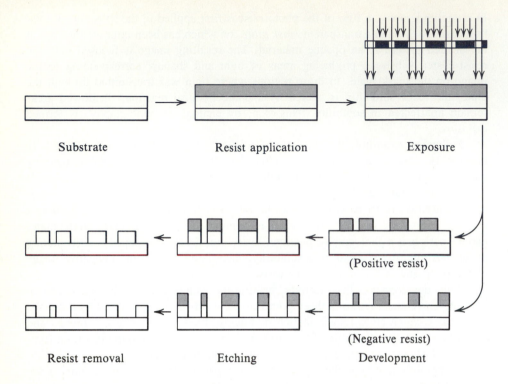

Figure 4-1 Schematic illustration of a photolithography process sequence. Resist is applied to the substrate, and a solubility change is induced by exposure to a pattern of light. In positive resist, exposure increases the solubility; in negative resist, it is decreased. Thus a different pattern results after development. Development is followed by etching to transfer the pattern into the film on the substrate and by resist removal.

plate bearing the entire array. At one time, steppers were used solely to produce mask plates, and mask plates were always used to pattern substrates; there are now steppers that operate on substrates directly without the intervention of a substrate-sized mask. Figure 4-2 illustrates some of these mask-making concepts.

Mask plates must be made of a material transparent to the exposing radiation. In addition, they should have a low coefficient of thermal expansion and be mechanically rigid. Otherwise either temperature fluctuations or mechanical flexing can change the size of the projected image. Quartz or special borosilicate glasses are usually used for ultraviolet lithography. The opaque material comprising the pattern must be highly reflective or absorbent. Metallic chrome is a frequent choice; iron oxide has also been used.

These then are the essential tools of microelectronic patterning: the photosensitive resist and the pattern image. The remainder of this chapter deals with the properties and use of the resist film. Chapter 5 looks in more detail at the nature and exploitation of the pattern image.

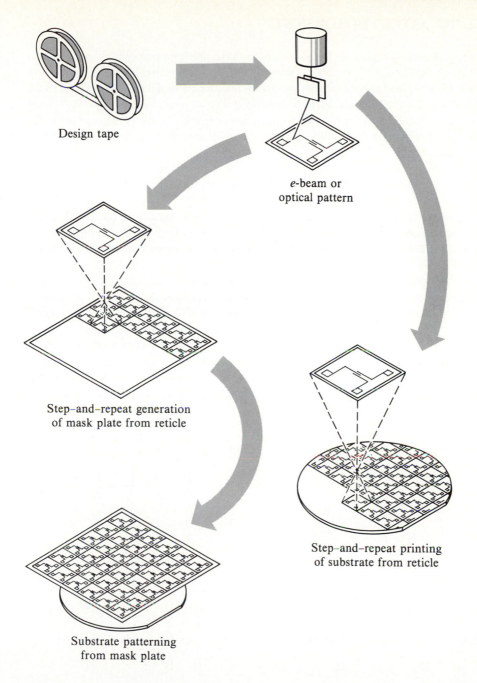

Design tape

e-beam or
optical pattern

Step–and–repeat generation
of mask plate from reticle

Step–and–repeat printing
of substrate from reticle

Substrate patterning
from mask plate

Figure 4-2 Masks, reticles, and mask making. A tape representing desired pattern is used to generate a reticle containing one cell of the pattern. The reticle pattern is repetitively generated on the substrate, either directly or through intermediate generation of a mask plate.

4-2 NEGATIVE PHOTORESIST

Several different materials have proved to have the photosensitivity and etchant resistance required for a successful microelectronic photoresist. Of these, the first to achieve widespread use was a negative-acting photoresist based on polyisoprene in the form of cyclized rubber. (Henceforth, when we refer to *negative resist,* we shall mean cyclized-rubber negative resist.) Two reactions of organic chemistry are relevant to the behavior of this resist: free-radical polymerization of alkenes and the photodissociation of azides.

A review of some chemical terms may be in order. Organic chemistry is the chemistry of carbon compounds. The carbon atom typically bonds to four other atoms and so is called tetravalent. An important class of carbon compounds are the *hydrocarbons,* which consist solely of hydrogen and carbon atoms. Hydrogen bonds only once, to carbon; carbon atoms make their four bonds either to hydrogen or to other carbon atoms. An *alkene* is a hydrocarbon containing one or more carbon-carbon *double bonds,* in which adjacent carbons bond "twice" to each other and additionally to only two other atoms. A *free radical* is any chemical species containing a single electron (rather than an electron pair) available for bonding. Free radicals are usually so reactive that they exist only as short-lived intermediates in chemical reactions. In organic chemistry, two effective ways of creating free radicals are exposure to ultraviolet light or the addition of peroxides. A *peroxide* is a compound containing two oxygen atoms singly bound to each other (H_2O_2, hydrogen peroxide, is the simplest example); these compounds easily decompose to form free radicals, as shown in Fig. 4-3. In the figure, the letters represent atoms of carbon, hydrogen, or oxygen, and the connecting lines represent chemical bonds.

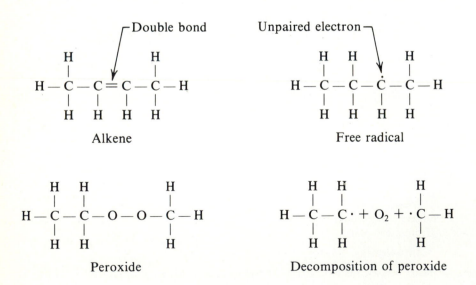

Figure 4-3 Chemical concepts pertinent to negative resist.

The free-radical polymerization of alkenes is a classic example of a chain reaction. Free radicals react with the double bond of an alkene, capturing one of the two electrons of the double bond and leaving the other unbonded. This, by definition, converts the alkene into another, longer free radical, which can then attach to another alkene. By this mechanism, the introduction of a few free radicals into an alkene can initiate a chain reaction in which alkenes become "radicalized" and react with others to form increasingly longer chains. The chains, made up of numerous units based on the original alkene, are called *polymers* of the alkene. Chain propagation continues indefinitely, until the odd electron is deactivated, either by reaction with the reactor wall, an impurity, or another free radical. Figure 4-4 illustrates these concepts using the dialkene *isoprene,* the building block of natural and synthetic rubber.

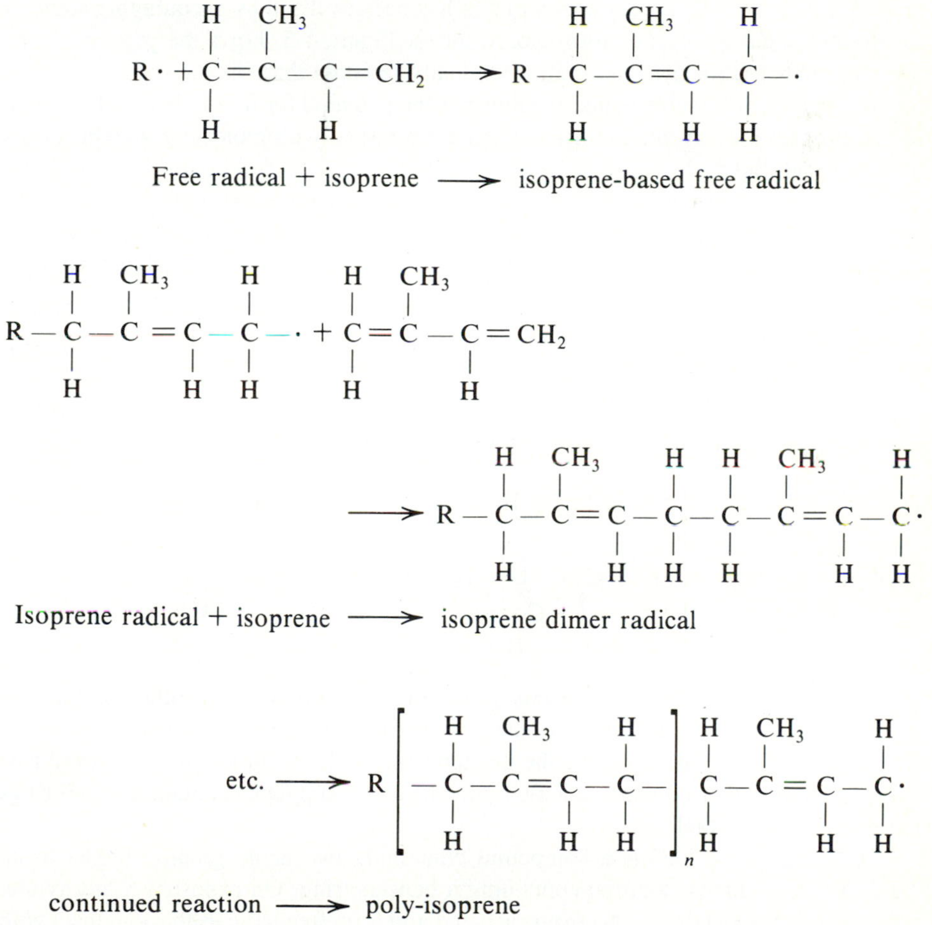

$$
R\cdot + \underset{\underset{H}{|}}{\overset{\overset{H}{|}}{C}}=\underset{\underset{H}{|}}{\overset{\overset{CH_3}{|}}{C}} - C = CH_2 \longrightarrow R - \underset{\underset{H}{|}}{\overset{\overset{H}{|}}{C}} - \underset{}{\overset{\overset{CH_3}{|}}{C}} = C - \underset{\underset{H}{|}}{\overset{\overset{H}{|}}{C}} - \cdot
$$

Free radical + isoprene $\longrightarrow$ isoprene-based free radical

$$
R - \underset{\underset{H}{|}}{\overset{\overset{H}{|}}{C}} - \overset{\overset{CH_3}{|}}{C} = C - \underset{\underset{H}{|}}{\overset{\overset{H}{|}}{C}} - \cdot + \overset{\overset{H}{|}}{C} = \overset{\overset{CH_3}{|}}{C} - C = CH_2
$$

$$
\longrightarrow R - \underset{\underset{H}{|}}{\overset{\overset{H}{|}}{C}} - \overset{\overset{CH_3}{|}}{C} = C - \underset{\underset{H}{|}}{\overset{\overset{H}{|}}{C}} - \underset{\underset{H}{|}}{\overset{\overset{H}{|}}{C}} - \overset{\overset{CH_3}{|}}{C} = C - \underset{\underset{H}{|}}{\overset{\overset{H}{|}}{C}} \cdot
$$

Isoprene radical + isoprene $\longrightarrow$ isoprene dimer radical

$$
\text{etc.} \longrightarrow R \left[- \underset{\underset{H}{|}}{\overset{\overset{H}{|}}{C}} - \overset{\overset{CH_3}{|}}{C} = C - \underset{\underset{H}{|}}{\overset{\overset{H}{|}}{C}} - \right]_n \underset{\underset{H}{|}}{\overset{\overset{H}{|}}{C}} - \overset{\overset{CH_3}{|}}{C} = C - \underset{\underset{H}{|}}{\overset{\overset{H}{|}}{C}} \cdot
$$

continued reaction $\longrightarrow$ poly-isoprene

Figure 4-4 Free-radical polymerization of isoprene. Note the movement of the double bond each time the chain lengthens.

An *azide* is an organic molecule containing the $-N_3$ group, whose bonding can be diagrammed as

$$R-N^--N^+\equiv N \tag{4-1}$$

Here R represents the remainder of the molecule, which is usually some hydrocarbon. Since the reactions of the azide group are usually not much affected by the details of the rest of the molecule, we have no need to consider the detailed structure of R. The azide group is highly unstable and, in the presence of heat or light, liberates nitrogen gas, N_2. The remaining singly bonded nitrogen atom possesses an unbonded electron pair as well as a lone unbonded electron and is thus a highly reactive free radical.[1]

These two chemical reactions are combined to make negative photoresist photo-sensitive. The basic structural material of the resist is called the *resin*. Negative resist resin is a form of rubber (polyisoprene) that has been cyclized by encouraging some of the double bonds to react to form closed loops. Figure 4-5 shows the process.[2] This cyclized rubber is tougher and more chemically resistant than the uncyclized form. It retains a number of double bonds and thus has the potential for free-radical polymeriza-tion. Commercially available formulations are reported with molecular weight ranges of 60 to 150,000.[3]

The resist resin probably retains some degree of photosensitivity, but rapid and complete polymerization upon exposure requires a *sensitizer*, specially designed to be photoactive. A common choice is a bis-aryl diazide such as 4-4'-diazidostilbene:

$$N_3-\bigbenzene-\underset{\underset{H}{|}}{C}=\underset{\underset{H}{|}}{C}-\bigbenzene-N_3 \tag{4-2a}$$

where the hexagon with the circle represents a *benzene ring:*

$$R-\bigbenzene-R'= \tag{4-2b}$$

The bonding in this six-sided carbon ring can be represented by either of the two alternate forms shown. In actuality, the bonding is a quantum-mechanical mix of the two alternate forms. As a result, the benzene ring is lower in energy than would be expected from either bonding scheme by itself, and compounds containing such rings are exceptionally stable.

A "bis-aryl diazide" is a compound containing two azide groups, linked to an "aryl" group—that is, a group containing a benzene ring. On exposure to ultraviolet light, these compounds lose N_2 from each end and form free radicals that can react with an alkene site in the resin. (The benzene rings tend to stabilize the free radical, giving it sufficient lifetime to react.) This results in *cross-linking*, in which the already large resin molecules are now linked to each other by several sensitizer molecules. There is

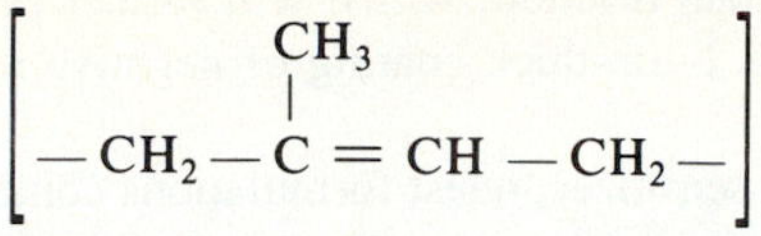

$$\left[-CH_2 - \overset{\displaystyle CH_3}{\underset{\displaystyle |}{C}} = CH - CH_2 - \right]$$

Basic repeating unit of rubber

(a)

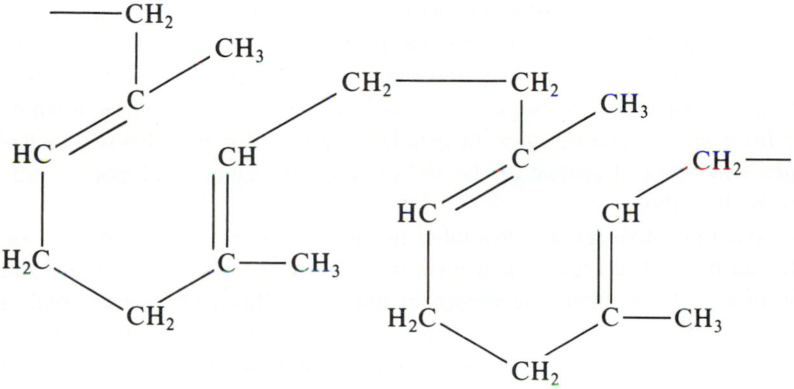

Rubber chain in folded configuration

(b)

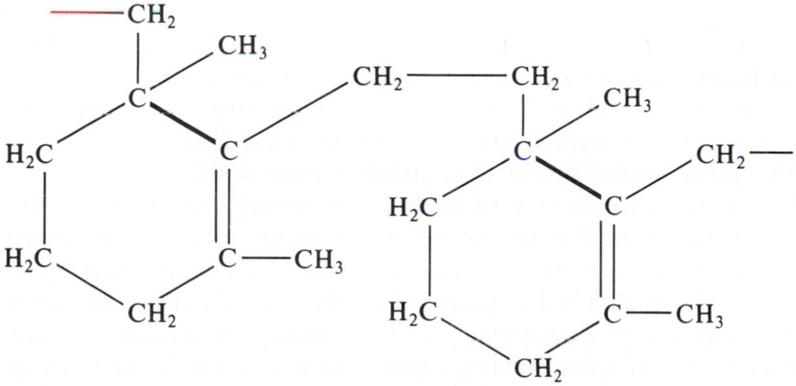

Cyclized rubber—note cyclizing bond
shown extra-dark

(c)

Figure 4-5 Cyclized rubber: (a) The basic monomer, or repeating unit, of rubber. (b) Part of a polymer chain, made up of linked monomers, and folded in a pattern similar to the cyclized form. (c) The molecule after cyclization: the new, cyclizing bonds are shown extra-dark. (*After Ref. 2, used by permission. Copyright 1975, McGraw-Hill Book Company.*)

also an obvious potential for chain reaction, which will further polymerize the resin. Typical exposure energies for a 1-μm-thick coating of negative resist are 10–20 mJ cm^{-2}.

In addition to the resin and sensitizer, resist formulations contain a solvent, which keeps the resist fluid for ease of application. Negative resist solvents are usually mixtures of aliphatic and aryl hydrocarbons.

Development

The cross-linking of negative resist under exposure to ultraviolet light produces a highly insoluble film. A pattern can now be formed by washing away the un-polymerized resist with a solvent. The obvious choice for such a *developer* is a hydrocarbon mixture similar to the original resist solvent. Various proprietary mixtures are available from resist manufacturers. In general, the need for finer linewidths has led to mixtures with reduced content of "harsh" solvents like xylene and more "mild" straight-chain hydrocarbons.

While exposed negative resist is insoluble in the developing solvent, it is by no means unaffected by it. The cross-linked resin will absorb the solvent, swelling and distorting the image. Thus solvent development must be followed by a *rinse* with a fluid that removes solvent from the resin. This compound should have an affinity for the developer while not absorbing into the polymerized resin: *n*-butyl acetate is a common choice. Typical spray develop and rinse times for 1-μm resist layers are 5–15 sec.

Cyclized-rubber negative resist is in many ways an ideal photoresist. A major advantage is the all-or-nothing nature of negative resist development. Polymerized resist is essentially immune to dissolution; unpolymerized resist dissolves so quickly that rates are rarely measured. Therefore the development process requires limited attention. A wide range of solvents can be used, and temperature control is un-necessary. The process engineer can indulge in substantial overdevelopment as a safety factor with no damage to the remaining resist. In addition, the long-chain nature of the resist resin, coupled with its intrinsic chemistry, results in excellent adhesion to the substrate. Other resist systems are not as forgiving in these areas.

However, negative resist has a crucial weakness: its limited ability to resolve fine lines. Since printing of smaller geometries is *the* major means of advancing microelectronic capability, a resist of limited resolution is simply unacceptable in many applications. Two factors limit negative resist resolution. The first is the size of the molecules, coupled with the potential for chain reaction. Resolution is clearly limited to at best one molecular building block, which is a significant factor as resolutions approach 1 μm. Any chain reaction will degrade resolution further. The resolution of a resist is related to its *aspect ratio,* which is the ratio of the thickness of the resist film to the narrowest line which can be printed. Negative resists typically display gently sloping profiles at the edges of geometries, and aspect ratios are limited to less than unity.

The second limitation results from the swelling of polymerized resist during development. When narrow spaces are being printed, the swelling resist on either side may extend into the space and touch. After rinsing, the gap will be bridged by chewing-gum-like streamers of resist. These two factors combine to limit practical negative resist resolution to about 2–3 μm.

4-3 POSITIVE PHOTORESIST

The limitation of negative cyclized-rubber resist have encouraged the use of resists based on another chemistry. These resists are positive-acting and do not rely on polymerization. They are capable of aspect ratios well above unity, which allows the printing of finer lines as well as the use of thicker films. Figure 4-6 shows typical line profiles for positive and negative photoresist.

Chemistry of Positive Photoresist

The photosensitivity of the positive resist system is in the sensitizer. Again, the technology exploits the instability of an N_2 group bonded to an organic molecule. In

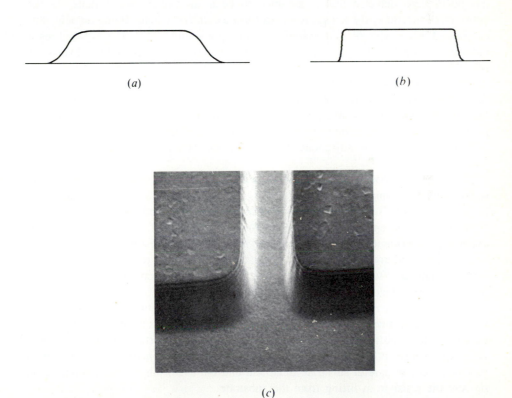

(a) (b)

(c)

Figure 4-6 Profiles of resist images. (a) Negative resist. (b) Positive resist. (c) Scanning electron micrograph of a positive resist image: note the sharper wall and the layered effect due to standing wave patterns. (*Photograph courtesy of Phillip A. Hunt Chemical Corporation. Used by permission; all rights reserved.*)

this case, the N_2 is bonded to a hydrocarbon ring, adjacent to a double-bonded oxygen (ketone) group. The resulting compounds can be referred to as "diazo oxides" or "orthoquinone diazides."[4] Such compounds are constructed using a six-sided carbon ring (with a tendency to assume the stable benzene-ring configuration), a doubly bonding oxygen atom, and a diazide, or N_2 group, as follows:

$$
\begin{array}{c}
\text{O} \\
\| \\
\text{C} \\
\diagup \quad \diagdown \\
\text{C} \qquad \text{C}=\text{N}_2 \\
| \qquad\qquad | \\
\text{C} \qquad \text{C} \\
\diagdown \quad \diagup \quad \diagdown \\
\text{C} \qquad \text{H} \\
|
\end{array}
\tag{4-3}
$$

The three loose bonds can be connected in various ways to the rest of the molecule.

Exposure to ultraviolet light leads to dissociation of the N_2 group with the postulated formation of a five-carbon ring and a *ketene:* a $=C=O$ group. The ketene has never been detected and is assumed to be a short-lived intermediate. In the presence of moisture, the ketene reacts to form a carboxylic acid. Being an acid, this compound reacts with alkaline materials, forming a soluble ester. The exposed resist can thus be washed away in an alkaline solution such as dilute sodium hydroxide (NaOH).

To obtain good results, the resist film must contain sufficient water to react with the ketene. In the absence of water, the reactive ketene will undergo some other reaction, including cross-linking to some other molecule in the resist. For this reason, positive resist processes must be designed to ensure sufficient moisture. Usually, normal air humidity is sufficient. Figure 4-7 shows the positive resist exposure reaction.[5]

The resin in a positive resist usually has no photochemical sensitivity. Instead, the resin provides chemical resistance, viscosity, stability, and other desired characteristics. Resin must be somewhat soluble in alkalis to allow exposed resist to dissolve, but must be insoluble enough to prevent excessive dissolution of unexposed resist. For example, one commercial resin is removed by a developer at a rate of about 15 nm sec^{-1}. Combined roughly 1:1 with a sensitizer, it forms a resist that develops at $100–200$ nm sec^{-1} when exposed, and only $1–2$ nm sec^{-1} when unexposed.[6] Typical positive resist develop times are $30–90$ sec.

Adhesion is a major concern in positive resist formulation. The polymerized network of a negative resist requires only a few attachment sites to adhere tightly to the substrate. But almost all of the individual molecules at the interface of a positive resist must adhere well if the resist is to form an acceptable bond. Therefore, positive resist processes require careful substrate treatment and the use of an adhesion promoter to prevent the resist from lifting from the substrate.

In addition to resin and sensitizer, positive resists contain a solvent. Usually used are ethylene glycol monoethyl ether (EGME or Cellusolve), the analogous monomethyl ether (methyl Cellusolve), or the acetate of one of these compounds. Common

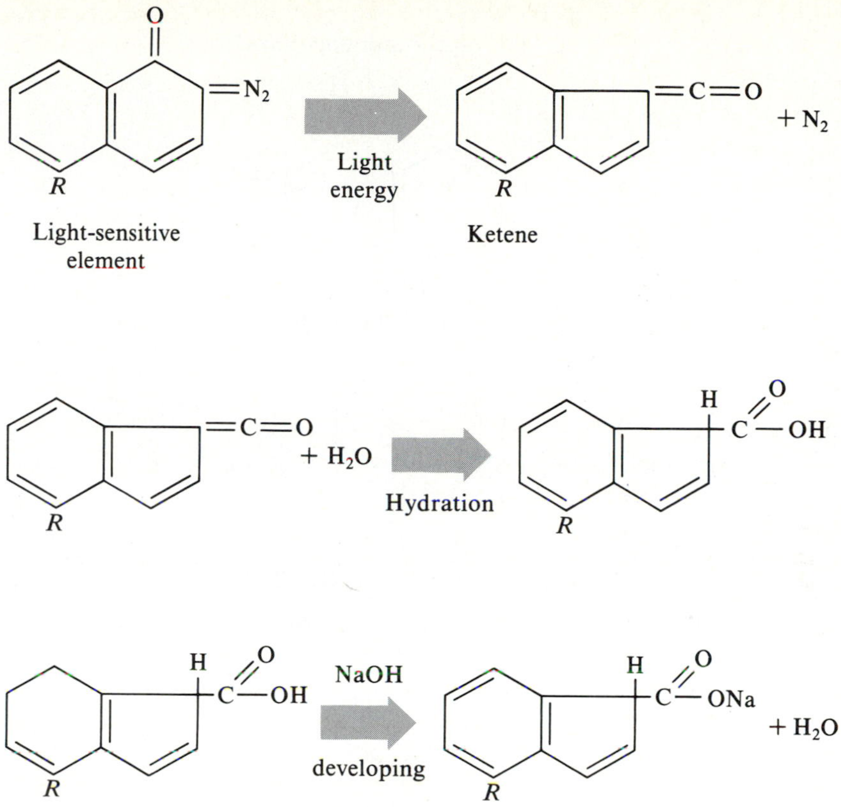

Figure 4-7 The positive resist exposure reaction. (*Reference 5, used by permission. Copyright 1975, McGraw-Hill Book Company.*)

solvents such as xylene or acetone tend to form gel slugs in positive resist and should be avoided, even for minor cleaning tasks.

Figure 4-8 shows some positive resist sensitizers in current use.[7] Table 4-1 lists some commercially available resists.[8]

Development

The nature of the positive resist reaction makes development conditions more crucial. Unlike negative resist, both the "soluble" and "insoluble" portions of a positive resist are removed at observable rates by the developer. Also, the developing speed tends to be a much smoother function of the absorbed light energy. Developer concentration also directly influences development speed. Figure 4-9 shows typical curves of developer speed versus exposure energy, for several developer concentrations.[9] Note that

Napthoquinone diazides

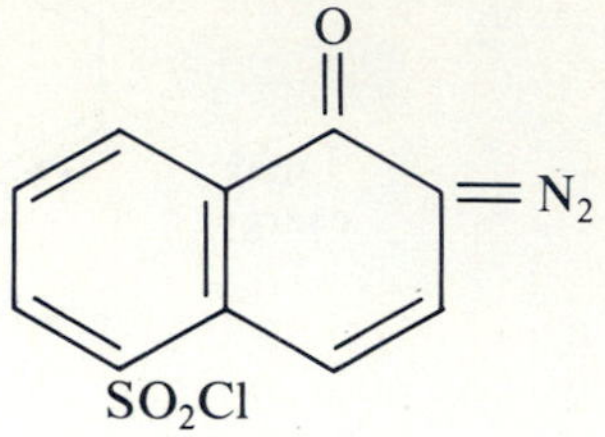

Napthoquinone-1,2-diazide-5-sulfochloride

Benzoquinone diazides

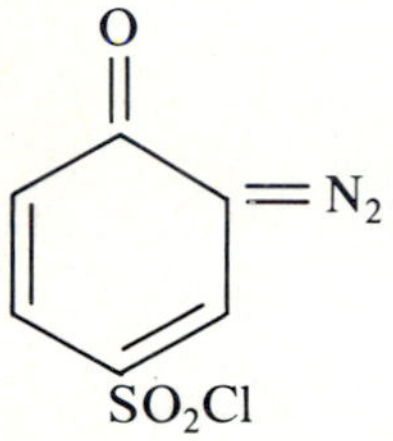

Benzoquinone 1,2-
diazide-4-sulfochloride

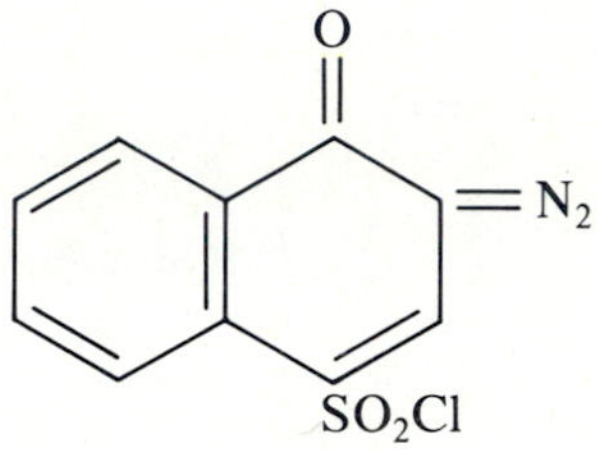

Napthoquinone-1,2-diazide-4-sulfochloride

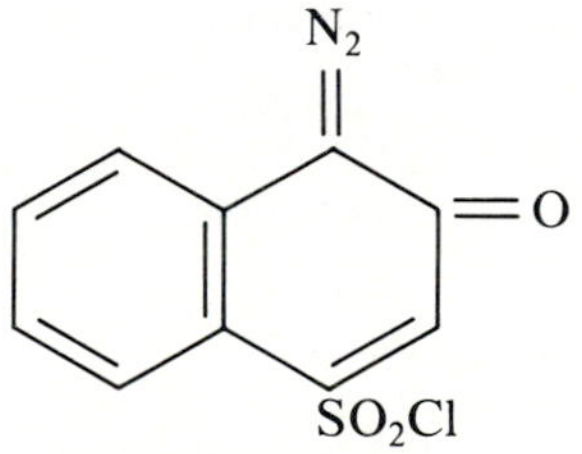

Napthoquinone-2,1-diazide-4-sulfochloride

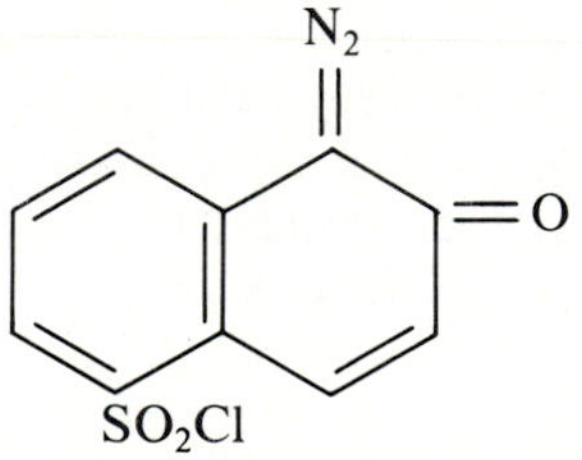

Napthoquinone-2,1-diazide-5-sulfochloride

Figure 4-8 Some positive resist sensitizers. (*Reference 7, used by permission. Copyright 1975, McGraw-Hill Book Company.*)

Table 4-1 Commercially available photoresists

Resist	Solids content (%)	Viscosity	Specific gravity	Index of refraction	Flash point (°C)
Kodak Microresist 809	32 ± 1	23	1.045	1.560	58
Hunt Waycoat HPR 204	28	17.5	1.036	1.469	110
Hunt Waycoat HPR 206	33	41	1.055	1.482	110
MIT Superfine IC 528	28.5 ± 0.5	5 ± 0.5	1.010	1.484	—
Tokyo OKHA OFPR-800	Three available	20 ± 1.5	—	—	—
		30 ± 1.5			
		50 ± 1.5			
Shipley AZ-1370†	27	17 ± 1.5	1.025 ± 0.015	1.64 ± 0.01	41
Shipley AZ-1350J	31	30.5 ± 2.0	1.040 ± 0.010	1.64 ± 0.01	41
Shipley AZ-1470	27	15.7–18.3	1.025 ± 0.015	1.64 ± 0.01	41
Shipley AZ-1450J	31	28.0–33.1	1.040 ± 0.010	1.64 ± 0.01	41
Shipley AZ-1115	20	24.5 ± 1.5	0.990 ± 0.015	1.555	34
Shipley AZ-111H	25	70 ± 5	1.017 ± 0.015	1.555	39
Shipley AZ-2400	26	16.7–20.0	1.000–1.030	—	44

† AZ is a trademark of the Azoplate Division of American Hoechst Corporation.
Note: Resist properties are subject to change by the manufacturer.
Source: Reference 8, used by permission. Copyright 1982, McGraw-Hill.

developer speed is shown in terms of resist thickness removed per unit time. Since resist film thickness is easily measured, and positive development proceeds smoothly down from the resist surface, these units are more meaningful than, say, grams or moles of resist reacting per second.

This more sensitive development reaction adds both difficulty and flexibility to positive resist processes. On the one hand, development must be carefully controlled to obtain desired results. Both concentration and temperature of developer must be carefully controlled, and care taken to prevent the atmospheric carbon dioxide from slowly neutralizing the alkaline solution.

On the other hand, the development process may be used to manipulate the dimensions of the printed pattern. To see why, consider Fig. 4-10. A band of positive resist, of width w, has been exposed. Lines in the figure shows the resist profile after increasing development times. Development begins at the surface and proceeds downward in both exposed and unexposed resist. However, as walls form bounding the exposed area, development also occurs laterally, widening the walls. After resist is completely cleared in the exposed area, the walls continue to recede. This effect is

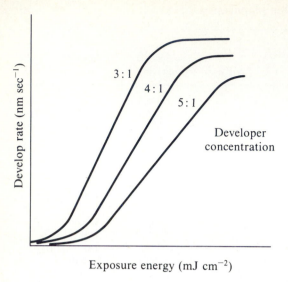

Figure 4-9 A typical set of curves show-ing development speed versus exposure energy for positive resist as a function of concentration of the developer. (*Figure courtesy of, and reprinted by permission of, Intel Corporation. All rights re-served.*)

enhanced in practice, since the edge of the exposed region is not truly sharp. In this way, exposure and development times can be manipulated to print wider or narrower geometries from a given mask plate. This allows "skewing" a dimension to gain slightly higher performance or to experiment with various dimensions without making a new plate.

In summary, positive resists of the "azo" variety sacrifice adhesion and simplicity of development to achieve higher resolution than negative resists. This strategy has made them the resists of choice for present and foreseeable optical lithography processes. Further limitations on resolution lie not in the resist, but in the topography of the device and the wavelength of the light itself. These limitations, and responses to them, will be examined in Chapter 5.

4-4 RESIST APPLICATION

The first step in using any resist is to apply it to the substrate. Three requirements must be met: the resist must adhere well to the surface, the thickness must be uniform across the substrate, and the thickness must be predictable from substrate to substrate.

All imaginable coating techniques have been applied to resists. Spraying and rollercoating are sometimes used in undemanding microelectronic applications, such as coating substrate backs to retain a backside oxide. However, whirl coating, or *spin-ning,* has emerged as the premier technique for coating circular substrates.

In spinning, a small puddle of resist is first dispensed onto the center of the substrate, which is attached to a spindle using a vacuum chuck. The spindle is then spun rapidly, rotating the substrate at several thousand revolutions per minute for some 10 to 60 sec. At the end of this time, the wafer will be coated with a dry resist film that is uniform to several nanometers. Figure 4-11 illustrates the process.

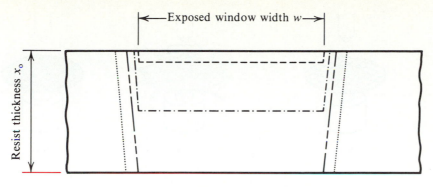

Resist profile at (time $\times$ develop rate) =

$$------ 0.1x_o$$
$$-\cdot-\cdot-\cdot- 0.5x_o$$
$$--\ --\ 1.0x_o$$
$$\cdots\cdots\cdots\cdots 2.0x_o$$

Figure 4-10 The evolution of positive resist development with time. Note that with time the pattern widens as well as deepens.

What happens during this conceptually simple process seems to be complicated. High-speed photography studies have shown the sequence of events. When resist is applied to a stationary substrate, it spreads out into a puddle whose size is limited by surface tension. As spinning begins, centrifugal force throws most of the resist off the edge of the wafer. Various stages in this process have been called the "wave," "corona," and "spiral" stages, after the shape of the waves of moving resist flowing over the substrate.[10] Most of the bulk resist has been thrown from the wafer after the first couple of seconds of spinning. In designing the coater or *spinner,* care must be taken that these flying resist droplets are captured and drained from the spin chamber. Otherwise, droplets redeposit on the wafer, destroying the desired uniformity.

The remaining resist film forms into a smooth layer covering most of the substrate, with an *edge bead,* several times the film thickness, at the substrate edge. The edge bead is presumably formed in response to surface tension, and the remaining film has been described as "squeezed" against the edge bead. Figure 4-12 shows measured edge bead profiles under two different conditions.[11]

The principal determinant of resist thickness is the spinner rotational speed during the 5 to 15 sec following the start of spinning. Although most spinners allow varying a number of other factors, the effect is usually small. However, numerous empirical recipes have arisen in pursuit of a slightly more uniform coating. Many people use a *spread cycle,* spinning the resist at a few hundred rpm to spread it across the wafer, before beginning the final spin at several thousand rpm. Others go further, using a *dynamic dispense,* in which the resist is dispensed on an already rotating wafer. The rotational acceleration of the spinner can be varied. The times devoted to postdispense equilibration, spread cycle, and spinning can be adjusted, and, of course, substrate

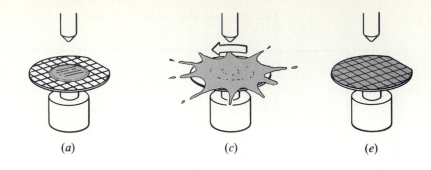

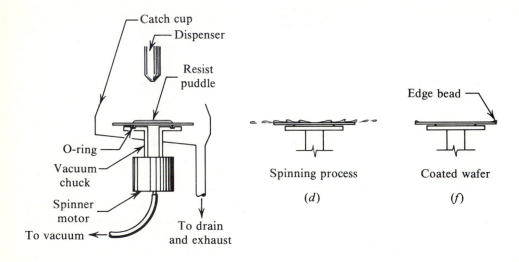

Figure 4-11 Spin coating of resist. (*a*) Resist puddle applied to substrate. (*b*) Profile view of this step, showing some details of the spin coating equipment. (*c*) Spinning begins, throwing off most of resist. (*d*) Profile view of this step, showing waves in resist and function of catch cup. (*e*) Spinning complete, substrate coated. (*f*) Profile view of coated substrate, showing edge bead. Resist thickness is greatly exaggerated.

diameters have increased over the years. Since none of these variables greatly affect the resist thickness, they are usually adjusted empirically to optimize the uniformity of coating.

A commonly used formula relating spin speed to final thickness is

$$z = \frac{kP^2}{\sqrt{\omega}} \tag{4-4}$$

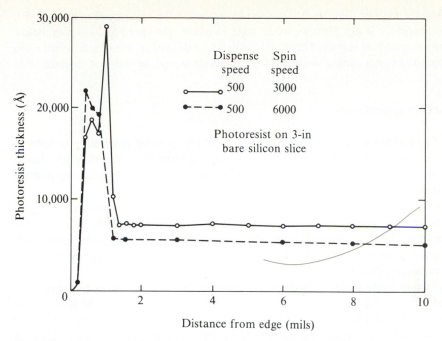

Figure 4-12 Resist edge bead after spin coating. (*Reference 11. Reprinted with permission of Advanced Micro Devices, Inc. All rights reserved.*)

where z is the final resist thickness, P the percentage of solids in the resist, and ω the rotational velocity of the spinner.[12] Here c is an empirical "constant," with units of μm $sec^{-1/2}$ characteristic of the spinner-substrate system. To avoid pinhole formation, negative resist is usually spun at least 500 nm thick (preferably 700 nm); positive resists are usually at least 1 μm thick. Much thicker layers are used when fine resolution is unnecessary. Resist viscosity is usually adjusted (by using more or less solvent) to achieve the target thickness with spin speeds of 2500–5000 rpm. The selection of spin speed does somewhat effect thickness uniformity: higher speeds promote a slight thickening towards the edge ("up-dish") while lower speeds produce a thicker center ("down-dish").[13]

Example With a given spinner and photoresist formulation, a spin speed of 4000 rpm gives a resist thickness of 0.7 μm. Suggest how to increase the thickness to 0.8 μm. To 1.0 μm?

The proper spin speed for a 0.8-μm layer of resist is given, by Eq. (4-4), by the proportionality

$$\left(\frac{\omega}{4000 \text{ rpm}}\right)^{1/2} = \frac{0.7}{0.8}$$

so that

$$\omega = 3062 \text{ rpm}$$

Thus the thicker film can be obtained by adjusting the spin speed to about 3100. To obtain a 1.0-μm film, we would have to adjust spin speed to 2040 rpm, below the range of best results. Therefore, for this film thickness, use of a different resist formulation containing about 20% more solids would be the best method.

Surface Preparation

Several operations usually precede and follow the actual resist application in order to achieve optimum adhesion between the substrate and resist film. The steps are similar for positive and negative resist, although the intrinsically poorer adhesion of positive resist makes them more essential.

Resist must adhere to a variety of substrates. However, silicon dioxide (SiO_2) is perhaps the most common of them, and adhesion techniques are often designed with SiO_2 in mind. Luckily, the techniques seem to work well with other films, even when the chemistry appears different. Therefore, we will explain adhesion promoters in terms of silicon dioxide.

A silicon oxide surface adsorbs water from the atmosphere and becomes covered with both molecular wafer and *silanol* groups (Si—OH). These must be removed for good adhesion. Molecular water is usually reduced by a *prebake* or *dehydration bake* before coating. Typical conditions are 30 minutes at 200–250°C. Longer times or higher temperatures do not appear any more effective. Bakes usually lose effectiveness after 30–60 min due to reabsorption of water, so baked wafers must be promptly coated.

After baking, substrates may be treated with an adhesion promoter. For silicon dioxide, hexamethyl disilazane (HMDS) is frequently used. Figure 4-13 illustrates the SiO_2 surface, the HMDS molecule, and its role in promoting adhesion. The silazane end of the molecule reacts with silanol groups on the oxide surface, removing them and bonding to the oxide. The hydrocarbon end extends into the resist, tying it to the substrate. HMDS is quite reactive with water, yielding *siloxane*. This reaction can be shown in some detail as

$$
H_2O + H_3C - \underset{\underset{CH_3}{|}}{\overset{\overset{CH_3}{|}}{C}} - Si - \underset{}{\overset{\overset{H}{|}}{N}} - Si - \underset{\underset{CH_3}{|}}{\overset{\overset{CH_3}{|}}{C}} - CH_3 \longrightarrow
$$

$$
H_3C - \underset{\underset{CH_3}{|}}{\overset{\overset{CH_3}{|}}{C}} - Si - O - Si + \underset{\underset{CH_3}{|}}{\overset{\overset{CH_3}{|}}{C}} - CH_3 + NH_3 \quad (4\text{-}5a)
$$

or more compactly as

$$
H_2O + [(CH_3)_3Si]_2 - NH \longrightarrow [(CH_3)_3Si]_2 - O + NH_3 \quad (4\text{-}5b)
$$

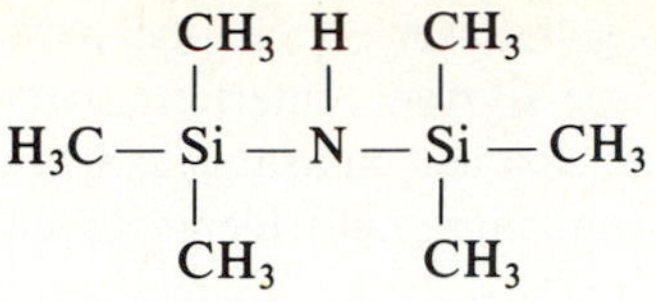

$$
\begin{array}{ccc}
CH_3 & H & CH_3 \\
| & | & | \\
H_3C - Si - N - Si - CH_3 \\
| & & | \\
CH_3 & & CH_3
\end{array}
$$

(a) Hexamethyldisilazane (HMDS)

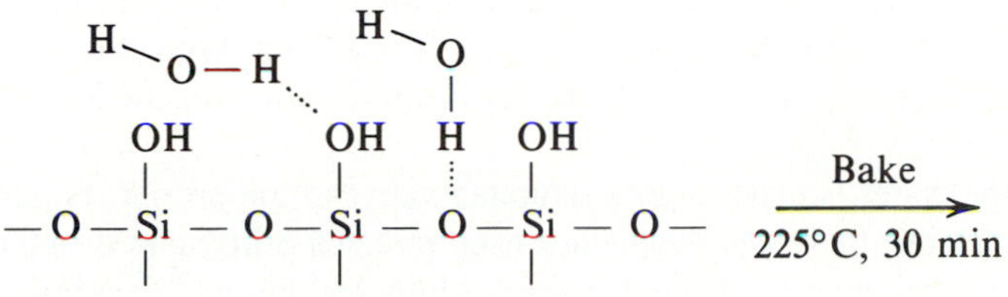

$$
\begin{array}{ccccc}
H & & H & & \\
\diagdown & & \diagdown & & \\
O - H & & O & & \\
& \cdots & | & & \\
OH & OH & H & OH & \\
| & | & \vdots & | & \\
-O - Si - O - Si - O - Si - O - & & & & \\
| & | & | & &
\end{array}
\qquad
\begin{array}{c}
\text{Bake} \\
\xrightarrow{\hspace{2cm}} \\
225°C, \ 30 \ \text{min}
\end{array}
$$

(b) Silicon oxide surface with adsorbed
     water and silanol groups

$$
\begin{array}{ccc}
OH & OH & OH \\
| & | & | \\
-O - Si - O - Si - O - Si - O - \\
| & | & |
\end{array}
\qquad
\begin{array}{c}
\text{HMDS} \\
\xrightarrow{\hspace{2cm}}
\end{array}
$$

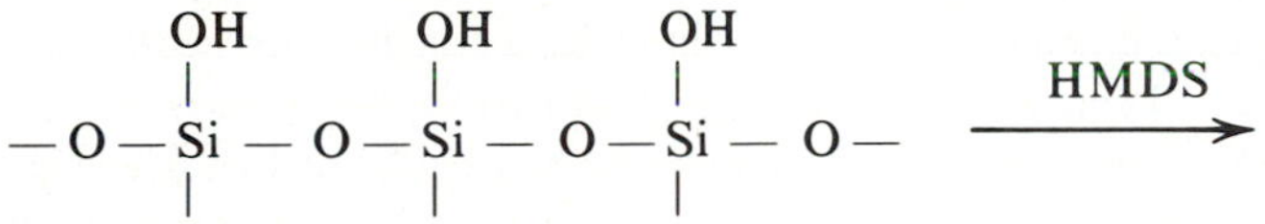

(c) Silicon oxide surface free of molecular
     water, still with silanol groups

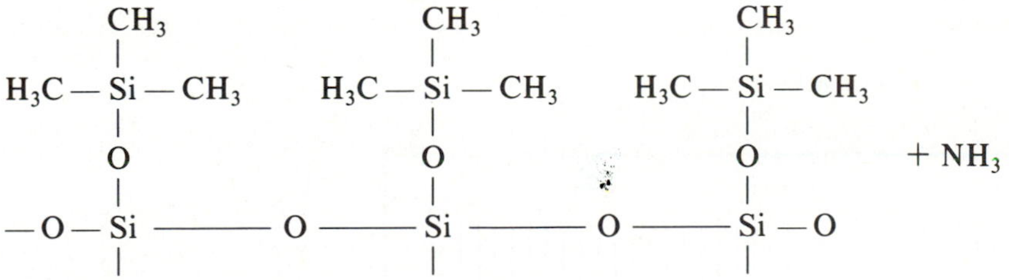

$$
\begin{array}{ccc}
CH_3 & CH_3 & CH_3 \\
| & | & | \\
H_3C - Si - CH_3 & H_3C - Si - CH_3 & H_3C - Si - CH_3 \\
| & | & | \\
O & O & O \qquad + NH_3 \\
| & | & | \\
-O - Si - O - Si - O - Si - O \\
| & | & |
\end{array}
$$

(d) Surface covered with hydrocarbon-rich
     Si(CH$_3$)$_3$ groups for adhesion promotion

Figure 4-13 Hexamethyldisilazane and its reaction with a silicon oxide surface. (a) The HMDS molecule. (b) A silicon oxide surface with absorbed water and silanol groups. (c) Surface after bake, free of molecular water but still retaining silanol groups. (d) Replacement of silanol groups with HMDS to leave a surface coated with hydrocarbon groups that will adhere to the resist.

This reaction can help to remove any remaining molecular water from the oxide surface. However, excessive water will produce excessive siloxane, interfering with the action of the HMDS. Also, HMDS will react with ambient moisture and lose effectiveness unless adequately protected. A strong ammonia smell identifies decomposing HMDS.

Two methods of primer application are in wide use: spray and vapor. In spray priming, a dilute solution of HMDS in a solvent is sprayed onto each substrate. This technique is conveniently combined with spin coating: a primer nozzle is placed alongside the resist dispense line, and the wafer is spray primed and spun dry immediately before resist application. In this method, the spray may also help mechanically scrub stray dust particles from the substrate surface. The solvent used is usually the resist solvent: xylene for negative resist, EGME for positive. Solutions of 5–10% are typical.

In vapor priming, the wafer is exposed to a saturated vapor of the primer. Figure 4-14 shows a typical apparatus. This technique lends itself to batch priming in which a number of substrates are exposed to the vapor for several minutes, but in-line priming combined with coating can also be done. Vapor priming is harder to automate than spraying, but appears to give better results. The use of vapor prevents wafer contamination by liquid or solid impurities in the primer, and the longer time encourages complete reaction with the substrate.

Following resist application, the coated wafer is given a *soft-bake* or *postbake* to remove remaining solvents. Typical temperatures are 90 to 100°C for negative and 80 to 90°C for positive resist. Excessive retained solvent makes a sticky resist surface and will interfere with positive resist development. On the other hand, the photosensitive resist can be exposed not only by light, but by heat, so excessive baking must be avoided.

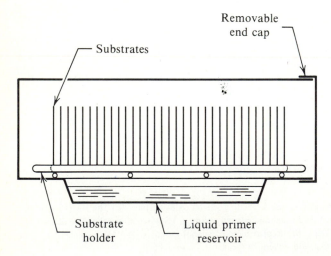

Figure 4-14 A chamber for vapor priming of substrates.

Practical Coating Equipment

Industrial equipment for the baking, priming, and coating of resist must have several attributes. First, it must reliably control and reproduce the important variables of the process. Second, since absolute cleanliness is essential, it must avoid both chemical and particulate contamination of the substrate. Third, it should process the maximum number of wafers per hour with a minimum amount of operator involvement. Fourth, it must be mechanically reliable.

The simplest method of baking is convection baking. Figure 4-15 shows a convection oven, which is simply an insulated chamber, a thermostatically controlled heating element, and a fan to circulate the heated air. Substrates are placed in cassettes on racks in the chamber. Since fans generate particles, the air may be filtered or the fans placed remotely from the substrates.

Convection baking allows large numbers of wafers to be processed. Operator involvement is limited to loading and unloaded at specified times. The steady-state temperature of the oven is easily measured, and regulated to ±1°C at a given location, and ±3°C throughout the chamber. However, in practice the heat delivered to a particular wafer is less reproducible than these figures would suggest. Batch loading

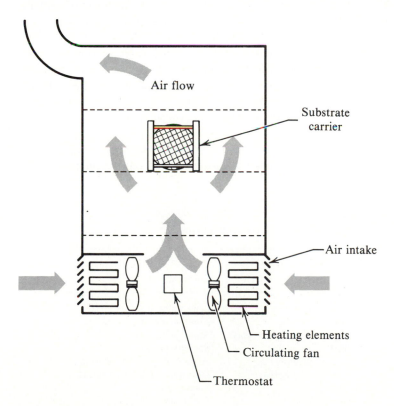

Figure 4-15 A convection oven.

and unloading means that some substrates (the last in, the first out) receive less baking than others. Since batches are loaded in cassettes, different substrates experience different environments; end substrates, for example, are more exposed to the air flow than center substrates. For soft-bake applications, convection baking has another disadvantage. Heated air flowing past the coated substrates heats the surface first, producing a dried crust that inhibits solvent removal from the still moist interior.

In-line baking is an attempt to avoid these drawbacks. In this technique, wafers are moved one at a time down a conveyor, so that all wafers receive identical treatment. In one method, the conveyor passes under infrared radiating elements, and infrared radiation passes through the resist and is absorbed by the silicon substrates. The heated substrate then bakes the resist from the bottom up, avoiding crusting. In another scheme, wafers are deposited one at a time on a hot plate that heats the substrate and thus the resist by conduction. Direct irradiation of the wafer with microwaves while on the spinner chuck has also been utilized. While these methods achieve wafer-to-wafer reproducibility and "bottom-up" heating, they also have their disadvantages. During infrared or microwave heating, no thermal equilibrium is achieved, so it is hard to quantify the temperature attained by the resist. This makes process characterization difficult. Hot plates produce a quantifiable steady-state temperature, but are cooled by each new wafer and thus display a recovery-time cycle like a convection oven.

Resist spinning requires a spindle and motor to rotate the wafer, dispense lines for resist and other fluids, and a cup or shield to capture the waste resist thrown from the wafer. The wafer is secured to the spindle during spinning by a vacuum chuck. Fluids to be dispensed include resist, primer, and a gas blow-off or fluid wash to remove particles prior to spin. Table 4-2 is an example of a typical resist coating cycle that might be programmed into a coater.

Most industrial coating is now done on automated, in-line equipment. An operator is required only to place cassettes of uncoated wafers at the input location, remove

Table 4-2 A typical automated resist coating cycle

Start time (sec)	Duration (sec)	rpm	Dispense	Comments
0	0	0	—	Load wafer
0	2	1000	N_2	Blow off wafer
2	10	500	HMDS spray	Apply primer
12	10	3000	—	Spin-dry primer
22	1	0	—	Allow wafer to stop spinning
23	3	0	Resist	Dispense resist and let spread
26	5	500	—	Spread resist
31	20	3000	—	Spin resist
51	0	0	—	Unload wafer

them from the output, and enter any required changes in process cycle. Therefore, these units are referred to as *cassette-to-cassette* coaters.

Figure 4-16 shows a typical configuration for an automated resist coater. A substrate transfer mechanism removes substrates from the input cassette and transports them to the first bake module. This module, which may be of the infrared or hot-plate variety, prebakes the wafer to remove adsorbed water. Following the bake, the wafer moves to a cool module, thus avoiding dispensing the resist on a hot surface. After cooling, the wafer is conveyed to the spin module and placed on the spinner chuck. Using a program similar to that in Table 4-2, the wafer is primed and spin-coated. Another wafer-handling module transfers the coated wafer to a second bake module for soft-baking. Finally, the wafer is delivered to an output cassette. Such units can process roughly one wafer per minute under highly reproducible conditions and without operator manipulation of substrates. All temperatures, times, and process parameters can be software controlled, allowing both control and flexibility.

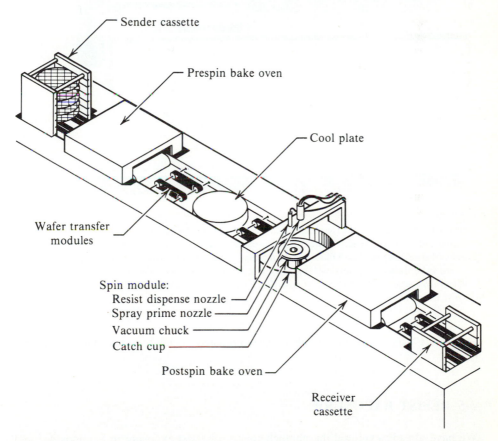

Figure 4-16 An automated spin-coater for photoresist. This sketch is conceptual in nature and is not meant to accurately represent any specific piece of equipment; in fact, it combines features from several different manufacturers.

Figure 4-17 A photograph of an automated spin-coater. There are three parallel tracks for processing. At the left-hand end are three "sender" wafer cassettes, partially raised; at the right are three empty "receiver" cassettes that are lowered almost flush with the machine surface. Wafers move from left to right, through the spin and bake modules. A keyboard and CRT screen allow entry and monitoring of process variables. (*Photograph courtesy of GCA Corporation, Bedford, Massachussetts. Wafertrac® is a registered trademark of GCA Corporation.*)

Figure 4-17 shows a photograph of an automated coater (courtesy GCA Corporation).

4-5 RESIST EXPOSURE

We have already explored the photochemistry of resist exposure in Sections 4-2 and 4-3. This section deals with the execution of the exposure step in practical fabrication.

Light Sources

Optical resists are sensitive to light of 300–500 nm in wavelength, a range that includes visible violet and blue light. Therefore, resist processes occur under "yellow light," from which wavelengths below 500 nm are filtered. Mask-making operations that use photographic emulsion operate under red light only.

Light for exposure is usually obtained from mercury vapor lamps, which produce a line spectrum with maxima near 310, 365, 405, and 440 nm. Of these, the 405-nm line most closely matches the sensitivity maximum of negative resist, and the 365-nm line that of positive. Commercially available meters for calibrating light intensity usually operate at one of these wavelengths. However, all of the mercury lines can contribute to resist exposure, as shown in Fig. 4-18.[14] As lamps and optical surfaces age, there may be preferential absorption of one mercury line, changing the relationship of the measured intensity to the overall exposure energy. This phenomenon, inaccurately called *"spectral shift,"* must be remembered when using light meters for process control.

For negative resist, the degree of polymerization and thus of exposure are major determinants of resist adhesion. Adhesion thus sets a lower limit on exposure light energy. However, as we shall see in Chapter 5, image reproduction suffers as exposure increases. Thus the lowest acceptable exposure energy should be used. Negative resist begins to show a corrugated appearance called *orange peel* as the lower exposure limit is approached. A good general rule for setting exposure for negative resist is to locate the value where orange peel begins and add perhaps 10% more energy.

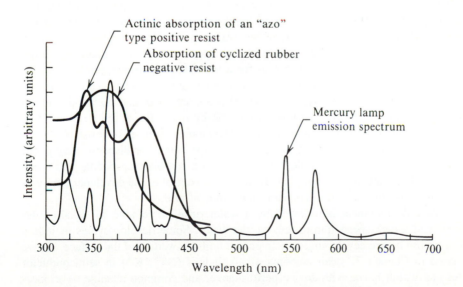

Figure 4-18 Resist absorption spectra together with emission intensity spectrum of a mercury lamp (*See Ref. 14 for sources used.*)

Chemical Issues

Negative resist is also sensitive to ozone and peroxides, which are sources of free radicals and can cross-link the resist. Photochemical smog is rich in peroxides, and therefore charcoal filters are used to purify the air supply for resist processing areas. Ozone can be formed in oxygen by ultraviolet light and is thus present when exposing radiation passes through air. This "oxygen effect" can be prevented by protecting the substrate with a nitrogen blanket during exposure.

Two features of positive resist chemistry should be remembered when designing an exposure process. First, water is required to complete the exposure reaction. When the ambient humidity is very low, or the time between soft bake and exposure quite short, the reaction may be incomplete and the resist may not develop cleanly. Such conditions may require a rehumidification step in the process. Second, positive resist evolves nitrogen gas during exposure, which must escape from the resist. Use of excessive exposure energies or heating wafers too soon after exposure can result in nitrogen bubbles that disrupt the resist. Nitrogen will also accumulate and expand in any small voids left at the resist-substrate interface during coating. Some equipment configurations require special precautions for the escape of nitrogen.

4-6 RESIST DEVELOPMENT

For negative resist, the development process is basically a washing process. The substrate is usually spun on a chuck similar to a spin-coating chuck, while a mist of developer is sprayed over it. Centrifugal force carries the developer with its load of dissolved and suspended resist over the edge of the substrate. After development is complete, a similar mist of rinse is applied. It is conventional to overlap the application of developer and rinse, since allowing the wafer to dry before rinse may leave a residue. After rinsing, the wafer is spun at high speed to expel all remaining fluids. Figure 4-19 shows the cup configuration for a typical spray developer.

Positive resist may either be developed in batches in a tank, or by a spray method similar to negative resist. Positive development is a more controlled chemical reaction than a wash: developer and resist react, molecule by molecule, at the resist surface. Therefore, turbulent flow is avoided in spray developing of positive resist. A steady stream rather than a mist may be projected onto the wafer, or dispense may occur without rotation to form a "puddle." There is also a combination puddle and spray technique called "spuddle" developing. In any case, control of the developer temperature is important to maintain a constant development rate.

Positive development is followed by a water rinse, to remove the alkaline developer, and a high-speed spin to remove the water. Some develop cycles include a water prerinse before developing, to help ensure uniform wetting of the surface. As we discussed in Chapter 3, small amounts of alkali ions (Na^+, K^+) in semiconductor oxides can wreak havoc with device performance, and common alkaline developers are simply dilute solutions of the alkali hydroxides (NaOH, KOH). Therefore, extremely thorough rinsing after positive development is mandatory. To avoid the

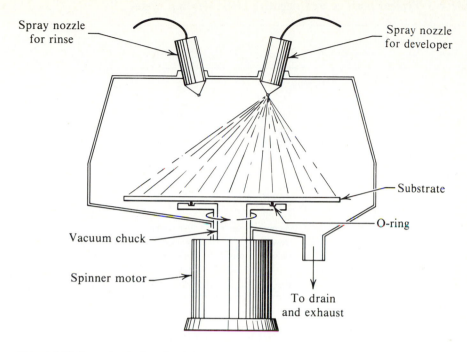

Spray nozzle for rinse

Spray nozzle for developer

Substrate

O-ring

Vacuum chuck

Spinner motor

To drain and exhaust

Figure 4-19 Spray developing of resist.

risk of alkali contamination, various proprietary "metal-ion-free" developers are now available.

Developing lends itself to the same kind of cassette-to-cassette, in-line automated processing described for resist application. In fact, in-line equipment is usually designed in a modular fashion for use either as a coater or developer. The wafer transfer mechanisms and general configuration are identical: the two machines differ only in details of the cup assembly and the associated supply plumbing. Table 4-3 shows typical automated cycles for both negative and positive resist development.

Resist development is usually followed by an inspection, often called *develop inspect* or *develop check*, in which the pattern is inspected for accuracy. If flaws are found, it is fairly easy at this stage to remove the resist and repeat the patterning process. After etch, of course, the flawed pattern becomes a permanent feature of the substrate.

After inspection, resist patterns destined for liquid etch operations are usually *hard baked* in an oven to improve adhesion. Typical hard-bake temperatures are 150°C for negative resist and 130°C for positive resist. Hard bake can be done in a convection oven, or an in-line bake oven can be added to a cassette-to-cassette developer. Obviously, if in-line baking is used, develop inspection must be delayed until after baking, which can make it harder to remove and redo a defective resist pattern.

Table 4-3 Typical positive and negative resist develop cycles

Start time (sec)	Duration (sec)	rpm	Dispense	Comments
			Negative resist	
0	0	0	—	Load wafer
0	10	500	Developer	Develop resist
10	2	500	Developer, rinse	Develop-rinse overlap
12	10	500	Rinse	Rinse wafer
22	15	3000	—	Spin-dry
37	0	0	—	Unload wafer
			Positive resist	
0	0	0	—	Load wafer
0	5	1000	Deionized water	Wet wafer
5	30	1000	Developer	Develop resist
35	15	1000	Deionized water	Rinse wafer
50	15	3000	—	Spin-dry
65	0	0	—	Unload wafer

4-7 PHOTOLITHOGRAPHIC PROCESS CONTROL

While diffusion operations depend on a few major variables, success in lithography requires accuracy in a host of individually minor details. The variables in question are straightforward and their measurement frequently trivial, but each must be executed properly to maintain an effective process. The output "variable" in photolithography is a perfectly printed image. The develop check inspection routinely given to printed substrates is the engineer's primary source of information about the output quality of his process. If substrates fail this inspection, the photoresist must be stripped off and the printing process repeated. The percent of substrates failing, which is carefully monitored, is called the *redo* or *rework* rate and is the key indicator of the engineer's success in maintaining process control. The engineer should be constantly aware of this redo rate and the failure modes of the rejected wafers, and able to recognize process failures by the visual observation of substrates.

However, the inspection of printed patterns is both subjective and complex, and therefore most quantitative efforts in process control must focus on the input variables, which are more directly measurable. Fanatic devotion to the control of all of these "minor" details is the key to a reproducible photoresist process.

One output variable that can be measured directly is the resist thickness delivered by a resist coater. On a routine basis, unpatterned sample substrates should be coated and the thickness measured at a number of points. A spectroscopic film thickness analyzer, such as those discussed in Section 3-5, can be used for this purpose.

The input parameters affecting bake are oven temperature and residence time. Time can be checked with a stopwatch. Temperature measurement is straightforward for convection and hot-plate baking. For radiative baking, temperature can be estimated by affixing a thermocouple to a sample substrate. The temperature obtained is probably not the same as that experienced by a coated substrate with no thermocouple, but at least process consistency can be confirmed.

Coaters and developers require control of cycle times, rotational velocities, and fluid flows. Times can be clocked, and rotation speed checked with a stroboscope. Fluid delivery systems should be fitted with flowmeters so that flows can be monitored. Especially for negative spray development, the shape and consistency of the spray pattern is crucial. Nozzle misadjustment, excessive or insufficient fluid flow, or the entrainment of bubbles in supply lines will affect the spray. Both coaters and developers eject waste material from the substrates, which must be efficiently captured and removed. Proper volume and "head" of exhaust is crucial for this purpose.

Environmental control is vital in photoresist processing. Resist-coated wafers should never be exposed to light with a wavelength below 500 nm to avoid inadvertent exposure. Humidity must be controlled. Relative humidities below about 45% cause loss of adhesion in negative resist; humidities below 30%, incomplete exposure of positive resist. Changes in room temperature can affect resist viscosity and development rates as well as distorting the optics of exposure equipment.

Above all, particulate control is vital, since any piece of dust will destroy the geometry on which it settles. By gloving and gowning all personnel and using highly filtered, laminar-flow air supplies, particulate levels can be kept below 100 particles per cubic foot in general and below 10 ft^{-3} in work stations. Resist processing areas should be checked frequently with a particle counter to see that these levels are maintained. Particulate sources such as paper, boxes, and makeup must be rigorously excluded. Particle generation from equipment is controlled by regular and thorough cleaning of coaters, ovens, cassettes, etc.

Table 4-4 summarizes common resist process failures and their most frequent causes.

Table 4-4 Common resist process failures with possible causes

Observable defect	Possible underlying deviation
Poor resist adhesion	
All resists	Improper or omitted prespin bake
	Improper or omitted prime
	Excessive time between prebake and prime
	Substrate surface, e.g., overly phosphorus-rich, contaminated
Negative resist	Underexposure
	Improper or omitted pre-etch bake
	Excessive time between bake and etch
	Exposure to excessive humidity
Positive resist	Improper or incomplete postspin bake
	Inadvertent exposure to white light

Table 4-4 (continued)

Observable defect	Possible underlying deviation
Incomplete development (scumming)	
Negative resist	Dispense on hot wafers
	Excessive postspin bake
	Grossly excessive resist thickness
	Insufficient developer time/flow
	Exposure to smog or ozone before/during expose
	Inadvertent exposure to white light
Positive resist	Excessive postspin bake
	Underexposure
	Lack of sufficient moisture during exposure
	Excessive resist thickness
	Insufficient developer time/flow
	Developer: too cold, too dilute, too used, too old
Distortion of patterns or attack on resist pattern by developer	
Negative resist	Underexposure
	Incomplete rinse and dry after develop
	Adhesion failure (see above)
	Focus and overexposure problems (Chapter 5)
Positive resist	Improperly regulated soft bake
	Excessive develop
	Developer: too hot, too concentrated
	Resist thickness variations
	Focus and exposure problems (Chapter 5)

	Possible causes of process deviations
Resist thickness	Resist viscosity or temperature
	Spinner rpm
	Resist dispense: excessive, insufficient, off-center, wafer not flat, wafer rotating during spin, etc.
	Resist droplets flying about spinner cup
	Other spinner parameters (ramp, spread, etc.)
	Incomplete soft bake (too thick)
Prime	Time, temperature not as specified
	Primer deteriorated due to ambient moisture
	Excessive time delay between bake, prime, spin
Bakes (pre- and postspin)	Time, temperature not as specified
	Loading/unloading errors (especially convection)
	Gas ambient incorrect
	Insufficient gas flow/venting for solvent removal
	Temperature calibration incorrect (especially radiative baking)
Exposure	Imager mechanical/optical failure
	Incorrect exposure time
	Lamp calibration incorrect (meter error or "spectral shift")
Develop	Incorrect developer concentration or temperature
	Times, flows, cycle settings incorrect
	Spray: sputtering, misoriented, incomplete
	Missing overlap cycle (negative resist only)
	Incomplete spin/dry

PROBLEMS

4-1 The linear coefficient of thermal expansion or quartz used for a mask plate is 5.5×10^{-7} °C^{-1}. Alignment accuracy across a 6-in (15 cm) silicon substrate must be maintained from one layer to the next within 0.5 μm. What is the allowable range of mask plate temperature during alignment?

4-2 The resolution of a negative resist is limited by the length of the resist molecule. A negative resist has a molecular weight of 100,000 daltons, or amu. Assume that the molecule is a straight chain of CH$_2$ units, with a carbon-carbon bond distance of 0.154 nm. Estimate the resolution of the resist by finding the length of one molecule. (The molecular weights of carbon and hydrogen are 12 and 1 daltons, respectively. First find the number of CH$_2$ units in a 100,000-dalton molecule, then compute the length using the bond distance.)

4-3 The resist described in Problem 4-2 has a density of 1.1 g cm^{-3}. Complete exposure of a 1-μm-thick layer of this resist is obtained with an exposure energy of 30 mJ cm^{-2} at a wavelength of 405 nm. Compute the number of photons per molecule required for full exposure. (Avogadro's law implies that if the molecular weight is 100,000 daltons, 6.02×10^{23} molecules will weigh 100,000 g. Use this relation to find the number of molecules in some resist volume and Planck's law to find the energy applied to that volume.)

4-4 A positive resist is composed of 50% napthaquinone-1,2-diazide-5-sulfachloride as a sensitizer and 50% of a resin of molecular weight 200. What proportion of water in the film, by weight, is required to achieve complete reaction of the sensitizer? (You will need the following atomic weights: hydrogen: 1 dalton; carbon: 12; oxygen: 16; nitrogen: 14; sulfur: 32; chlorine: 35.5.)

4-5 A positive resist layer, 1 μm thick, has the following exposure-development relationship:

Unexposed	Develop rate = 0.2 nm sec^{-1}
Exposure 75 mJ cm^{-2}	Develop rate = 10.0 nm sec^{-1}
Exposure 150 mJ cm^{-2}	Develop rate = 20.0 nm sec^{-1}

Consider the situation of Fig. 4-10, in which a window has been exposed in the resist. Let the width of the window be 5 μm. Calculate:

    (a) The thickness remaining in the unexposed area
    (b) The thickness remaining in the exposed area (if development is incomplete)
    (c) The width of the window at the substrate surface (if development is complete)
    (d) The width of the window at the resist surface
for each of the following cases:

1. Exposure 75 mJ cm^{-2}, develop time 90 sec
2. Exposure 75, develop time 150 sec
3. Exposure 150, develop time 90 sec
4. Exposure 150, develop time 150 sec

4-6 The initial puddle of resist dispensed on a 4-in (10-cm) silicon substrate by a spin-coater is 2.5 cm in radius and 0.3 cm high. After spinning at 3500 rpm, the entire wafer is coated 1 μm thick.

    (a) How much of the original resist has been wasted?
    (b) What is the direction of the force vector (gravitational plus centrifugal) at the wafer edge?

4-7 At 4500 rpm, a given resist coater produces a layer of resist 400 nm thick. The solids content of the resist is 35%. A layer of at least 500 nm is desired for pinhole resistance. Recommend ways to achieve this result by:

    (a) Adjusting coater rotational velocity only
    (b) Adjusting solids content only

4-8 In Problem 4-5 we assumed that the exposure energy fell abruptly to zero outside of the window being exposed. Due to diffraction effects, this assumption is unrealistic. Assume that exposure energy falls linearly from the given value at the window edge to the unexposed value at a distance of 1 μm from the window and repeat Problem 4-5.

REFERENCES

1. Robert Thornton Morrison and Robert Neilson Boyd, *Organic Chemistry,* 2nd ed., Allyn and Bacon, Boston, 1966.
2. William S. deForest, *Photoresist: Materials and Processes,* McGraw-Hill, New York, 1975, p. 31.
3. David J. Elliott, *Integrated Circuit Fabrication Technology,* McGraw-Hill, New York, 1982, p. 167.
4. deForest, *op. cit.,* pp. 48–49.
5. deForest, *op. cit.,* p. 52.
6. Elliott, *op. cit.,* p. 168.
7. deForest, *op. cit.,* p. 40.
8. Elliott, *op. cit.,* p. 71.
9. Curves courtesy of Intel Corporation.
10. Elliott, *op. cit.,* p. 126.
11. M. W. Chan, "Another Look at Edge Bead," *Kodak Interface Seminar Proceedings* (1975), p. 16.
12. G. F. Damon, *Collected Papers from Kodak Seminars on Microminiaturization 1965–66,* Kodak Publication (1969), p. 195.
13. Roy A. Colclaser, *Microelectronics: Processing and Device Design,* Wiley, New York, 1980, p. 40.
14. Compiled from Optical Associates, Inc. Publication 7–81 (Sunnyvale, CA); KTI Chemicals, Inc. Publication NR100 8-82 (Sunnyvale, CA); Ref. 2, p. 175.

FIVE

IMAGE CREATION

In Chapter 4, we saw how photoresist is used to convert an optical image (a pattern of light and shadow) into a physical pattern that can be etched into a substrate. In this chapter, we look at the theoretical and practical aspects of creating the optical image that exposes the photoresist and the quantitative aspects of how that image is reproduced in the resist.

Images are created and transmitted using optical equipment. In Section 5-1 we present the fundamentals of optical imaging and the ways in which the physical laws of refraction and reflection determine image reproduction. In Section 5-2 we develop the laws of diffraction and the resulting limits on image accuracy. In applying an image to the photolithography of microelectronic devices, special attention must be given to two key parameters: the reproduction of dimensions and the alignment of the new pattern to the pattern already on the substrate. In Sections 5-3 and 5-4 we deal explicitly with dimensional control and with pattern alignment. In Section 5-5 we present practical systems used for photolithographic imaging, including contact, proximity, and projection printing and the resolution limits of each. In Section 5-6 we discuss new methods of printing that may allow resolution of dimensions below 1 μm.

5-1 REFLECTION AND REFRACTION: THE FORMATION OF IMAGES

In microelectronic processing a designer's mental image of an electronic device is reproduced in physical form on a substrate. The optical equipment used to image the pattern determines the accuracy of this reproduction. Process engineers are seldom involved in the design of optical equipment, and a detailed discussion of optical design is outside the scope of this book. But the engineer will have to obtain maximum performance from the printing equipment and to understand its inherent limitations. Thus the engineer should be familiar with the basic laws of light transmission, the nature of image formation, and the physical limits on image reproduction.

Left to itself, light propagates indefinitely in straight lines. In classical physics, light is a wave phenomenon in which a spherical wave front expands from the light source at a speed c. Far from the source, a small portion of this spherical front can be approximated by a plane. In quantum physics, light propagates by the movement of light quanta, or *photons,* with both wavelike and particlelike properties, in straight lines from the source with speed c. These quanta have an energy given by $E = h\nu$, where h is Planck's constant and ν is the frequency of the light. In both classical and quantum mechanics the wavelength of the light is given by λ, where $\lambda = c/\nu$. An "element" of light—quantum mechanically a photon, classically a normal to the wave front—will move in a straight line; such lines are called *rays.* Optics involves the manipulation of rays in order to create an image.

Reflection and Refraction

The smooth propagation of light is disturbed when the light enters some physical medium in which its speed v is less than the vacuum velocity c. We can characterize such a medium by its *refractive index n* such that

$$n = \frac{v}{c} \tag{5-1}$$

When light crosses an interface between two media of differing refractive index, *refraction* and *reflection* take place. Figure 5-1 shows what happens. Part of the light is reflected from the interface. The law of reflection is simple. The angle of reflection, θ_r, is equal to the angle of incidence, θ_1:

$$\theta_r = \theta_1 \tag{5-2}$$

How much light is reflected depends on the characteristics of the medium. We will not discuss this dependence, except to note that by a proper choice of medium, the reflection can be made essentially complete. In this case, we have a mirror, and the refraction process is of no interest.

In general, however, some of the light energy is transmitted into the interior of the medium, to form a refracted ray. The refracted ray forms an angle with the interface given by

$$n_1 \sin \theta_1 = n_2 \sin \theta_2 \tag{5-3}$$

where n_1, n_2 are the refractive indices of the two media, and the angles are given in Fig. 5-1.

Example The refractive indices of crown glass and of air, for light of wavelength 0.365 μm are, respectively, 1.539 and 1.036. A light ray is incident on a plane

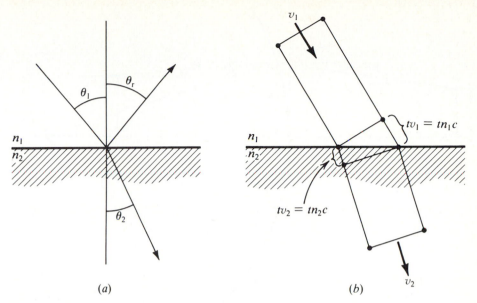

Figure 5-1 Reflection and refraction. (a) Incident, reflected, and refracted rays. (b) Snell's law: the change in propagation direction of a plane wave is a result of a change in propagation speed.

sheet of glass at an angle of 40° to the normal to the surface. What are the angles made with the normal by the reflected and refracted rays?

From Eq. (5-2), the angle of reflection is 40°. Following Eq. (5-3), the angle of the refracted ray is

$$\sin \theta = \frac{1.030}{1.539} \sin (40°)$$

$$= 0.43$$

So $\theta = 28.2°$.

Figure 5-1b shows a simple conceptual derivation of Eq. (5-3) known as *Snell's law*. First we picture a plane wave front, impinging on the interface with speed v_1. As part of the front crosses the interface, its speed is reduced to v_2. The remainder of the front moves with the original speed until it reaches the interface, "catching up" with the other part of the front and changing its overall direction of propagation. At a time t after our first look, the plane wave front is propagating within the second medium with a new direction, which is consistent with Snell's law.

As a specific example, consider the refraction of light rays crossing a spherical interface, as shown in Fig. 5-2. An axis is drawn from the center of the sphere, and a perpendicular is drawn to the axis from an arbitrary point. The length of the perpendicular is x, and its intersection is a distance D along the axis from the surface of the sphere. The sphere has radius R. On the other side of the interface, we construct

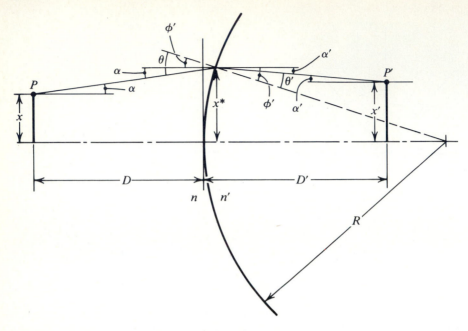

Figure 5-2 Refraction of a ray at a spherical interface.

another perpendicular, at a distance D' along the axis from the interface. The refracted ray will cross this perpendicular at some distance x'.

We now make some simplifying assumptions. We will assume all angles made by rays with surfaces are small, so that

$$\sin \theta \sim \theta \tag{5-4}$$

and that all points are close to the axis, so that

$$x, x' \ll R \tag{5-5}$$

These assumptions, which are quite good for simple systems of lenses, are known in classical optics as the *paraxial approximation*.

By the construction of Fig. 5-2, we can write (in the paraxial approximation)

$$x^* = x + D\alpha \tag{5-6a}$$
$$x^* = x' - D'\alpha' \tag{5-6b}$$

(Note that the angle α' is shown in the negative sense.) The paraxial approximation allows us to write Snell's law as

$$n\theta = n'\theta' \tag{5-7}$$

and to write

$$\phi' = \frac{x^*}{R} \tag{5-8}$$

Also, the construction shows that

$$\theta = \phi' + \alpha \qquad (5\text{-}9a)$$
$$\theta' = \phi' + \alpha' \qquad (5\text{-}9b)$$

Substituting for the angles in Eq. (5-7) gives

$$n\left(\frac{x^*}{R} + \alpha\right) = n'\left(\frac{x^*}{R} + \alpha'\right) \qquad (5\text{-}10)$$

or

$$\alpha' = \frac{n}{n'}\alpha + \frac{x^*}{Rn'}(n - n') \qquad (5\text{-}11)$$

Combining Eqs. (5-6) and (5-11) gives

$$x' = (x + D\alpha) + D'\left(\frac{n}{n'}\alpha + \frac{x^*}{Rn'}(n - n')\right) \qquad (5\text{-}12)$$

which can be rewritten using Eq. (5-6a) to give

$$x' = x\left(\frac{(n - n')D}{Rn'} + 1\right) + \alpha\left(D + \frac{n}{n'}D' + \frac{n - n'}{n'}\frac{DD'}{R}\right) \qquad (5\text{-}13)$$

and similarly, from Eq. (5-11),

$$\alpha' = x\left(\frac{n - n'}{Rn'}\right) + \alpha\left(\frac{n}{n'} + \frac{n - n'}{n'}\frac{D}{R}\right) \qquad (5\text{-}14)$$

These equations give the path of the refracted ray [which is the locus of all points (D', x')] and its angle of propagation, in terms of the refractive indices, the location of the source point, and the radius of curvature of the interface.

We can use these results to discuss the formation of images. The eye perceives an image of an object when light rays diverge as if from a point. The image is perceived to be at the point of divergence. Optical systems use refraction or reflection to manipulate the apparent point of divergence and thus to change the size and location of the perceived image. The most direct way to do this is to bring the original set of diverging light rays back together at a point, from which they will again diverge, as shown in Fig. 5-3a. This is called a *real image*. A real image results when all of the rays leaving some point (within some finite solid angle) are brought together at another point. It is also possible to form an image by causing light rays to diverge *as if* they came from a point, as in Fig. 5-3b. Such an image is a *virtual* image. In any case, the first point is the *object*, lies in *object space*, and is described with unprimed parameters. The second point is the *image*, lies in *image space*, and is described with primed parameters.

Now Eq. (5-13) describes the paths of rays leaving a point. If all of the rays leaving the point P in Fig. 5-2 can be brought together at some point P', regardless of the angle α at which they diverged, we will have formed an image of P. This can only happen at those values of x' and D' that are independent of α. Therefore, imaging of P will result at P' when the coefficient of α in Eq. (5-13) is zero. Thus, for imaging,

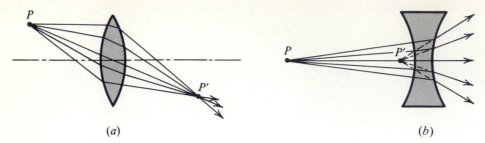

(a)   (b)

Figure 5-3 Imaging. (*a*) A real image: rays converge to and diverge from the point P'. (*b*) A virtual image: rays diverge as if from the point P'.

$$D + \frac{n}{n'}D' + \frac{n-n'}{n'}\frac{DD'}{R} = 0 \qquad (5\text{-}15)$$

Rearranging, imaging occurs when

$$\frac{n-n'}{R} = \frac{n}{D} + \frac{n'}{D'} \qquad (5\text{-}16)$$

If we define a focal length f by

$$\frac{n-n'}{R} \equiv \frac{1}{f} \qquad \text{one spherical surface} \qquad (5\text{-}17)$$

then

$$\frac{n}{D} + \frac{n'}{D'} = \frac{1}{f} \qquad (5\text{-}18)$$

We have now derived a condition for image formation by a single spherical surface in terms of image and object distances and the refractive index and radius of the refracting medium.

Equations of this form are typical of imaging systems. Note that if the object is a very large distance away (e.g., a star viewed through a telescope), the image will be formed at a distance f behind the interface. Conversely, if the object (say, a light source) is placed a distance f before the interface, the "image" will form at infinity, so that the rays emerging will be perfectly collimated—the ideal situation for a searchlight or projector.

Our formula is easily extended to the case of two oppositely oriented spherical surfaces, as shown in Fig. 5-4. It can be shown (see Problem 5-2) that, if the distance between the surfaces is small, we again obtain the form of Eq. (5-18), but with

$$\frac{1}{f} = \left(\frac{n'}{n} - 1\right)\left(\frac{1}{R_1} + \frac{1}{R_2}\right) \qquad \text{two spherical surfaces} \qquad (5\text{-}19)$$

If we let the medium "outside" the spherical surfaces be air, with a refractive index very close to 1 and let the intermediate medium be glass, we have a simple lens.

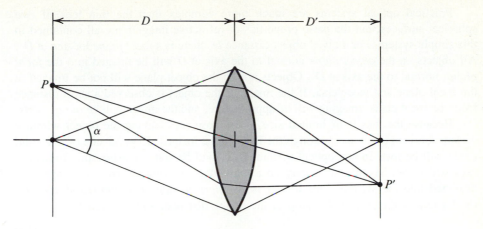

Figure 5-4 A simple lens: two spherical surfaces back to back. Also shown are the focal planes at distances D and D' and the maximum solid angle for ray collection α.

Therefore, the approximation that the surfaces are close is called the *thin lens approximation,* and Eq. (5-19) is called the *lens-makers' equation.* Equation (5-18), with $n = n' = 1$, becomes

$$\frac{1}{D} + \frac{1}{D'} = \frac{1}{f} \qquad (5\text{-}20)$$

Note that, if the object is ½ f from the lens, the image will be formed symmetrically, ½ f behind the lens.

Example A lens is made of crown glass, with two convex spherical surfaces of radii of curvature of 1 m. At what distance will they focus rays from a very distant object ($D \sim \infty$)?

With $n' = 1.539$ and $n \doteq 1.036$ (previous example), and with $R_1 = -R_2 = 1$ m, $1/f$ is

$$\left[\left(\frac{1.539}{1.036} - 1 \right) \left(\frac{1}{1} + \frac{1}{1} \right) \right]^{-1} = 0.97 \text{ m}^{-1}$$

Thus $f = 1.03$ m. Note that R_2 is negative, being measured in object space. With $D = \infty$, Eq. (5-20) becomes

$$\frac{1}{D} = \frac{1}{f}$$

and thus rays from a distance object are focused at $f = 1.03$ m. Similarly, if a light source is located at f, the lens will collimate all its rays into a parallel beam ("focused at infinity").

Practical optical systems are much more complex than the thin lens of two spherical surfaces, but the basic properties of refractive imaging are all contained in this simple system. For a given object distance D, there is a *focal plane* located at D'. All objects on the *object plane* normal to the axis at D will be imaged into the *focal plane*, normal to the axis at D'. Objects not on the object plane will not be imaged in the focal plane and vice versa. If the image of the object is observed a short distance from the focal plane, imaging will be approximate, and the image will be *out of focus*.

Because the lens is of limited size, only those rays leaving the object within a certain solid angle can pass through the lens and be imaged. Rays outside of that solid angle will be lost. In any optical system, there will be some *aperture* that limits the rays which can be imaged: it may be a lens edge, or some aperture or diaphragm designed into the system. This limiter is an *aperture stop*. The *numerical aperture* (*N.A.*) relates the size of this stop to the image distance, and is given by

$$\text{N.A.} \equiv \frac{2r_0}{D'} \tag{5-21}$$

where r_0 is the radius of the aperture. The aperture stop will have its own images in object and image space, which are termed the *entrance* and *exit pupils*. They are not physical objects.

If the image and object are not symmetrically distant from the lens, *magnification* occurs. Linear magnification is the ratio of the off-axis image distance x' to the corresponding object distance x and is given by

$$m_x = \frac{x'}{x} = \frac{nD}{n'D'} \tag{5-22}$$

Angular magnification is the ratio of the angle between two rays approaching the image point and the angle between the rays as they left the object point. It is given by

$$m_\alpha = \frac{D}{D'} \tag{5-23}$$

We note that these quantities cannot be varied independently.

We have derived the behavior of an imaging system under the paraxial and thin-lens assumptions, and for spherical surfaces. To the extent that real systems diverge from these approximations, they fail to image objects exactly. These failures become more obvious as the object point moves farther from the axis, and they are known as *aberrations*. These aberrations can be eliminated by the use of nonspherical lens surfaces.

Another type of aberration is *chromatic aberration*. The refractive index of the lens material varies with wavelength, and so the focal length of a refracting imaging system also varies with wavelength. As a result, differing colors will be differently imaged, and the image of a white object will have colored edges. Chromatic aberration can be made to vanish for some small set of wavelengths by the use of additional lenses (about one for each wavelength to be corrected) and can be minimized over a range of wavelengths by careful design. However, chromatic aberration is an inherent feature of

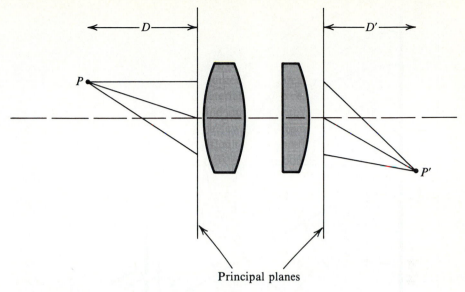

Figure 5-5 The use of principal planes for complex optical systems.

all refracting (but not reflecting) optical systems. It is the effort to eliminate or control aberrations that leads complexity in optical instruments.

Complex systems can be made to fit the form of Eq. (5-18) by the use of *principal planes*. In this formalism, the optical system is treated as a black box. We follow the rays as they diverge from the object point until they cross the *first principal plane*. We then ignore their detailed paths through the system, but pick them up again as they cross the *second principal plane*. The plane is positioned so that the rays cross it at the same distance from the axis as the first plane. From this point, they converge to the image. If we measure the object and image distances from the principal planes, Eqs. (5-20)–(5-23) remain valid independent of the details of the system. Figure 5-5 illustrates this concept.

5-2 DIFFRACTION: LIMITS ON IMAGE REPRODUCTION

Reflection and refraction are phenomena that occur as light moves *through* media: they can be understood in terms of changes in the propagation velocity of light without consideration of its wave nature. By contrast, diffraction is a purely wave phenomena and is associated with the movement of light *past* barriers. While reflection and refraction are actively exploited in optical systems, diffraction is most often an undesirable limiter on system performance.

Perhaps the simplest case of diffraction is shown in Fig. 5-6. Here a plane light wave impinges on two very narrow slits cut in a barrier. The lines in the diagram show the crest of the wave at successive times. Where the wave penetrates the barrier, two

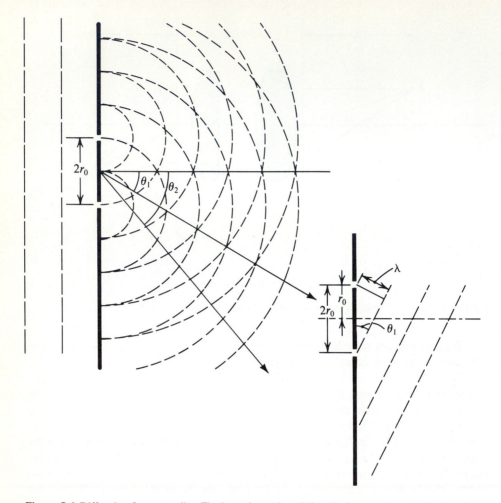

Figure 5-6 Diffraction from two slits. The inset shows the relationship between θ_i and λ for $i = 1$.

new wave fronts are formed that propagate spherically from their points of origin. When the two fronts cross, interference—either constructive or destructive—will occur. The interference will be constructive where crests meet, which is where

$$2r_0 \cos \theta_i = i\lambda \qquad (5\text{-}24)$$

Here $2r_0$ is the slit spacing, θ is the angle made with the original propagation direction, λ is the wavelength, and i is an integer. [The inset to the figure shows how Eq. (5-24) applies in the case of $i = 1$.]

As a result of diffraction, the original plane wave has been scattered in all directions. An interference pattern results, with the intensity maxima in directions given by θ_i. We refer to these maxima in terms of an *order,* equal to the absolute value of i in Eq. (5-24). The zero-order maximum is always in the original propagation direction and represents the undiffracted beam. Other orders are of decreasing intensity and diverge from the original beam. Note that if the slit separation $2r_0$ is more than a few wavelengths λ, only very high order maxima (of negligible intensity) will diverge noticeably from the original direction. Therefore, diffraction is noticeable only at dimensions approaching the wavelength.

As situations become less idealized, diffraction phenomena quickly become more complicated. If we let the incident beam of Fig. 5-6 vary arbitrarily in phase and wavelength, like real light, the criterion for constructive interference would be less simple. Altering the angle of incidence and the shape of the apertures would add additional complications. To express even a fairly simple real case requires specialized mathematical functions. But all diffraction patterns share general characteristics: a maximum intensity in the original propagation direction, periodic maxima in other directions, decreasing amplitude as the diffraction angle increases, and a strong dependence on the ratio of aperture dimension to wavelength.

As we discussed earlier, light passing though an optical system is limited either by a lens edge or by some other aperture stop. This aperture will create a diffraction pattern that will divert some of the light from its desired path, thereby decreasing the quality of the image. Most often, this aperture is circular. The diffraction pattern for a point imaged with monochromatic light through a circular aperture of radius r_0 is known: the flux density I is given by

$$I = I(0) \left(\frac{2\pi}{2D'} \right)^2 r_0^4 \left(\frac{J_1(u)}{u} \right)^2 \tag{5-25}$$

where $I(0)$ is the flux density incident on the aperture, D' is the distance from the aperture, and $u = (2\pi/\lambda)r_0 \sin \theta$. $J_1(u)$ is called the first Bessel function and is tabulated in standard mathematical tables.

The Bessel function term of Eq. (5-25) is plotted in Fig. 5-7. The first minimum occurs at $u = 3.83$ or $0.61 \lambda D'/r_0$. A geometrical point will thus have a circular image of this radius, known as an *Airy disk,* surrounded by fainter rings. This smudging of the image is an inescapable result of diffraction and is a function only of the aperture size and light wavelength. A conventional criterion for the maximum resolution of an instrument, known as the *Rayleigh limit* after Lord Rayleigh, holds that two points are distinguishable only if their Airy disks do not overlap. This criterion can be expressed for angular or linear separation by

$$\Delta\theta = \frac{\Delta x'}{D'} = \frac{0.61\lambda}{r_0} \tag{5-26}$$

The Rayleigh criterion defines the *diffraction-limited resolution* of an instrument. No improvement in optical quality, magnification, etc., can increase resolution past this limit.

Example The numerical aperture of an optical system is 0.3. It uses light from the 365-nm line of the mercury arc. What is the resolution $\Delta x'$ using the Rayleigh criterion?

From Eq. (5-26)

$$\Delta x' = \frac{0.61\lambda}{r_0/D'}$$

$$= \frac{1.22\lambda}{\text{N.A.}}$$

$$= \frac{(1.22)(0.365\ \mu m)}{0.3}$$

$$= 1.48\ \mu m$$

The Modulation Transfer Function

Optics was first applied to astronomical telescopes. The Rayleigh criterion for resolution is excellent for astronomical applications, which involve the observation of self-luminous points. However, microelectronic photolithography requires the imaging of gridlike patterns rather than points. In addition, photolithographic patterns are imaged from a mask plate, which is illuminated by some external light source with its own optical characteristics. For this reason, the performance of photolithography equipment is usually measured using criteria other than the Rayleigh limit. To use these criteria, we must first look at the concept of *coherence*.

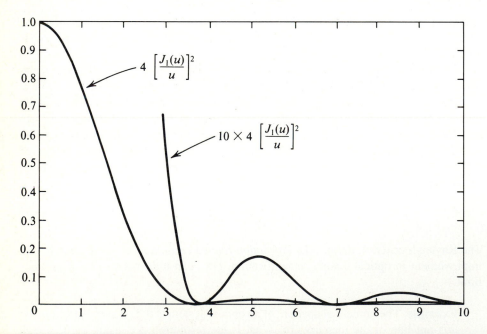

Figure 5-7 Diffraction from a circular aperture: the term $4[J_1(u)/u]^2$.

Monochromatic light waves all have the same frequency, but may differ in phase. Similarity in phase is termed coherence. Laser light is the premier example of coherent light: a laser beam is perfectly monochromatic and perfectly in phase. Other types of light sources are usually *incoherent:* there is no predictability to the phase of their light.

Despite this inherent incoherence, conventional light sources can be used to provide coherent illumination to optical systems. This is possible because light from a given point within the source *must* be coherent: a point can radiate only at one intensity at one time. Incoherence results from the uncoordinated radiation of multiple source points. Coherence is thus obtained by illuminating the object with light from only a small part of the source.

Figure 5-8 shows an optical system consisting of a light source, an object to be illuminated (such as a mask plate), an objective lens that images the object at the desired point P', and a *condenser* lens whose purpose is to direct light from the source onto the object. In Fig. 5-8*a*, the system is configured so that the image of a point in the source is formed at the objective lens. This design guarantees that an image of the

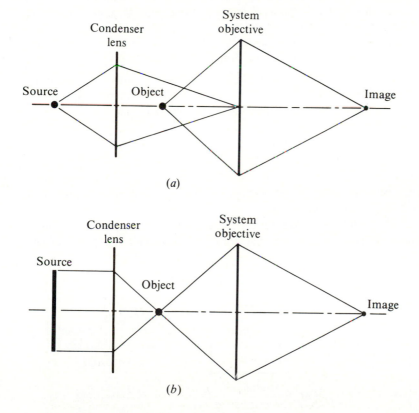

Figure 5-8 Coherent and incoherent imaging. (*a*) A single source point is imaged into the system objective; $S = 0$. (*b*) Entire source imaged into the objective; $S = 1$.

source is *not* formed either in the object or image focal planes. As a result, only a very narrow cone of light from the source passes both through the object and through the imaging system, and the image is formed with coherent light. In Fig. 5-8b, the condenser images the entire source onto the object. Rays from the extremities of the source can pass through the object and the objective to contribute to the image, and the illumination is incoherent. The degree of coherence S in such a system can be expressed by

$$S = \frac{r_s}{r_0} \tag{5-27}$$

where r_s is the radius of the image of the source at the objective and r_0 is the radius of the objective. In Fig. 5-8a, $S = 0$; in Fig. 5-8b, $S = 1$.

For a given degree of coherence and numerical aperture, the performance of an optical system can be quantified using the *modulation transfer function (MTF)*. Figure 5-9 illustrates the concept. Let the object be a grid or grating of parallel lines, spaced some distance x apart. The light transmission of this grating varies sinusoidally, from a minimum transmitted energy I_{min} at the densest part of the "line" to a maximum I_{max} at the brightest part of the "space." The *modulation* of the object, M, is then given by

$$M = \frac{I_{max} - I_{min}}{I_{max} + I_{min}} \tag{5-28}$$

After passing through the system, an image will be formed with some differing maximum and minimum intensities, I'_{max} and I'_{min}. The modulation of the image, M', is defined analogously to M. The modulation transfer function is then

$$\text{MTF}(x) = \frac{M'}{M} \tag{5-29}$$

For an ideal, diffraction-limited system, the MTF can be calculated as a function of x, λ, S, and N.A. The performance of real systems can be evaluated by measuring their MTF and comparing it to the ideal case.

Figure 5-10 shows the ideal MTF for coherent and incoherent illumination, as a function of *spatial frequency* $1/x$.[1] Note that, for coherent imaging, the MTF is unity

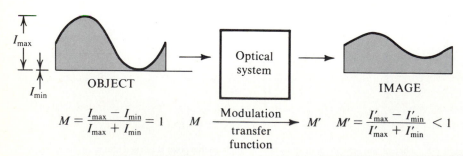

Figure 5-9 The modulation transfer function.

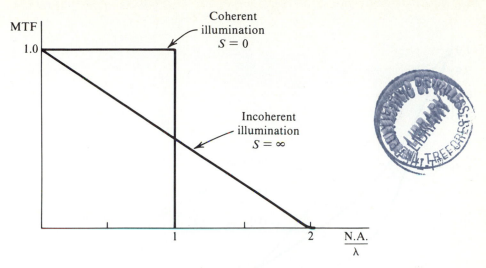

Figure 5-10 Modulation transfer function for an ideal imaging system.

out to a sharp cutoff frequency equal to N.A./λ. The MTF then drops to zero. For fully incoherent illumination, the MTF declines monotonically to zero at a frequency equal to 2N.A./λ. Most real printing systems use partially coherent illumination: The MTF in this case declines somewhat less at low frequencies than the incoherent case and cuts off at higher frequencies than the coherent case.

The MTF is also affected by changes in focus. Figure 5-11 shows MTF as a function of defocus for the incoherent case.[2] If the location of the substrate to be printed varies from the ideal focal plane by one-quarter wavelength, there is perceptible degradation of MTF at intermediate wavelengths; defocus by one-half wavelength cuts the maximum usable spatial frequency to roughly one-third.

Deviation from focus is an important limiter of photolithographic performance and can be conveniently expressed in terms of the *Rayleigh depth, d_R*:

$$d_R = \frac{\lambda}{2(N.A.)^2} \tag{5-30}$$

From Eqs. (5-26) and (5-30), we can draw an important conclusion. The maximum resolvable spatial frequency is inversely proportional to numerical aperture, and thus higher resolution can be obtained by increasing N.A. However, depth of focus is inversely proportional to the square of numerical aperture, so that an increase in N.A. sharply decreases the depth of focus. In any real system, both object and imaging plane deviate somewhat from their ideal locations, and some degree of defocus must be tolerated. If we attempt to improve resolution by increasing aperture, we eventually produce a system so sensitive to focus deviations that it is unusable. Since numerical aperture is related to image distance, which determines magnification, a similar trade-off occurs between depth of focus and magnification. The result is that increasing

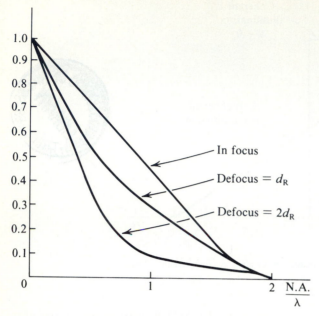

Figure 5-11 Modulation transfer function as a function of defocus. (*Reference 2. Reprinted with permission of* Solid State Technology, *published by Technical Publishing, a company of Dun & Bradstreet.*)

the power of the objective does not and cannot produce an indefinite improvement in the effective resolution of an instrument. As a result of these trade-offs, dramatic improvements in resolution can be had only by decreasing the wavelength of the exposing radiation.

Example An optical system has an N.A. of 0.3 and operates with light of 0.365 μm wavelength. The resist in use requires a MTF of 0.5 or greater for adequate pattern reproduction. The process requires a focus tolerance of ± 2 μm. (a) Estimate the resolution, assuming incoherent illumination. (b) What will be the change in resolution if the N.A. is increased to 0.42?

From Eq. (5-30), the Rayleigh depth for this system is

$$d_R = \frac{0.365 \ \mu m}{2(0.3)^2} = 2.03 \ \mu m$$

Thus the 2-μm focus tolerance corresponds to about 1.0 d_R. Consulting Fig. 5-11, we see that an MTF of 0.5, at a defocus of 1.0 d_R, is exceeded if the spatial frequency is below 0.6N.A./λ. This frequency corresponds to

$$\frac{(0.6)(0.3)}{0.365 \ \mu m} = 0.49 \ \mu m^{-1}$$

The corresponding resolution is the reciprocal of the spatial frequency, or 2.03 μm. This compares with the value of 1.48 obtained using the Rayleigh criterion in the previous example.

If the numerical aperture is increased to 0.42, the Rayleigh depth is decreased to 1.03 μm. A defocus of 2 μm now corresponds to $2d_R$. Consulting Fig. 5-11, we see that a MTF of 0.5 is now obtained only at spatial frequencies below 0.4N.A./λ. This is a reduction of 33% in frequency, which almost balances the increase in N.A. and results in a resolution of 2.17 μm. No benefit is gained by increasing N.A. in this case.

5-3 DIMENSIONAL CONTROL

In microelectronic photolithography, the patterns to be printed fall into two basic classes: windows and lines. We may cut an array of windows into a diffusion mask to allow dopant to enter the substrate, or into an insulating film to allow connections with a conductor. Conductive interconnects and resistors are usually formed into line patterns. In both cases, it is important that the pattern have both the shape and the size intended by the designer. Once the basic process is well established, the photolithography engineer is chiefly concerned with maintaining proper pattern definition and dimensions.

Both shape and size are determined by the same variables, and poor pattern definition will affect both. For this reason, we can effectively control pattern shape by controlling pattern dimension. This is convenient for several reasons. Dimensional variation is easy to quantify; shape is more a matter of judgment. Dimensions appear explicitly in device design equations, and their effects on device characteristics can be calculated. And dimensional control is basically a one-dimensional problem while pattern definition is two-dimensional.

Dimensions are most often measured on line-type patterns. One reason is that it is usually the dimensions of interconnect or isolation lines that determine cost or performance. For another, measurements on repeating arrays of lines are easily calibrated using the line pitch. Thus dimensional control is frequently referred to as "linewidth control." Another common term is "C.D. control" (or just "C.D.s"): this is an acronym for "critical dimensions."

We can thus reduce the complex problem of image accuracy to the simpler question of the width of a single line. Consider, for example, a photomask, transparent except for an opaque line feature of width x, which is to be printed in positive photoresist. Figure 5-12 shows some of the factors that will affect the printed linewidth x'. They include the quality of the image, the image-resist interaction, reflectivity effects, and substrate topography. The image-resist relationship turns out to be the key one, so we begin by finding an expression for the shape of the resist line in terms of the image characteristics. To do this, we turn to Fig. 5-13, which shows the response of the resist, that is, the dependence of development speed on incident light energy.[3]

Near a line edge, incident light energy varies while develop time is constant.

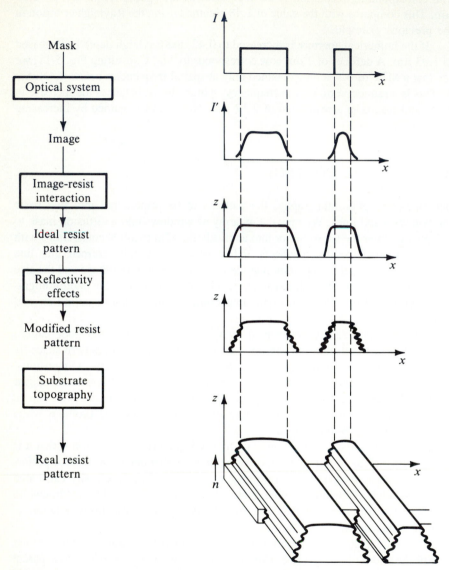

Figure 5-12 Factors affecting linewidth accuracy of resist images.

Therefore, instead of measuring response by develop rate as in Chapter 4, the response curve in Fig. 5-13 shows thickness of resist remaining, z, after a constant develop time, as a function of the logarithm of incident light energy E. The edge of the line is at the point where the resist thickness goes to zero. Being interested in the behavior near the zero-thickness point, we can approximate the resist response curve by its tangent at $z = 0$. Let the exposure at this point be E_0 and let the tangent intersect the original (zero

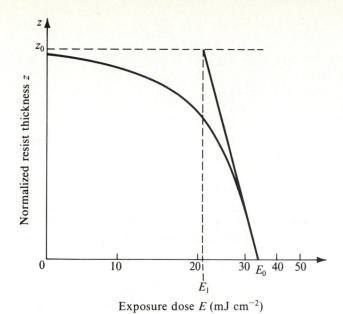

Figure 5-13 Resist pattern thickness as a function of exposure energy, for a typical resist and set of process conditions. (*Reference 3. Reprinted with permission of* Solid State Technology, *published by Technical Publishing, a company of Dun & Bradstreet.*)

exposure) resist thickness, z_0, at E_1. The resist *contrast* γ is defined as the reciprocal of the absolute value of the slope of the tangent. So

$$\gamma = \frac{1}{\ln (E_0/E_1)} \tag{5-31}$$

If the tangent approximates the resist curve, the thickness as a function of exposure is given by

$$z(E) = z_0\gamma[\ln E_0 - \ln E(x, z)] \tag{5-32}$$

since γ is the slope of the tangent. Here $E(x, z)$ is the light energy distribution of the image. The x direction is perpendicular to the axis of the line being printed.

The exposure energy $E(x, z)$ is the product of the incident intensity $I'(x)$ times the exposure time t times an exponential absorption factor involving the resist absorption coefficient α or

$$E(x, z) = tI'(x) \exp [-\alpha(z_0 - z)] \tag{5-33}$$

and thus

$$z(E) = z_0\gamma \left(\ln \frac{E_0}{tI'(x)} + \alpha(z_0 - z) \right) \tag{5-34}$$

Solving for z yields the shape of the printed line, $z(x)$, in the vicinity of the line edge:

$$z(x) = \left(\frac{1}{z_0\gamma} + \alpha\right)^{-1} \left(\alpha z_0 + \ln \frac{E_0}{t I'(x)}\right) \qquad (5\text{-}35)$$

Actually, we are often more interested in the slope of the resist line near the edge; this is given by differentiating $z(x)$:

$$\frac{dz}{dx} = -\left(\frac{1}{z_0\alpha} + \alpha\right)^{-1} \frac{1}{I'(x)} \frac{dI'(x)}{dx} \qquad (5\text{-}36)$$

In deriving this expression, we have ignored the effect of lateral development, which will preferentially narrow the top of the line. Equation (5-36) can thus be regarded as an upper bound on the true line slope in real resist systems.

The ideal printed line has sharp edges (infinite slope) and is identical in dimension to the mask. This will be true only if the image is perfect, so that I' is zero within the designed line area and large elsewhere. For an imperfect image, the line approaches the ideal if (1) the image is near-perfect, (2) the resist contrast is large, and (3) the absorption is small.

The properties of the incident intensity $I'(x)$ are determined by the imaging system. We have seen that, even in the focus plane, the image of the mask is diffraction-limited and thus imperfect. The intensity falloff near the line edge has been approximated[4]:

$$\frac{dI'}{dx} \simeq K \frac{\text{N.A.}}{\lambda} \qquad (5\text{-}37)$$

where K is a factor dependent on the optical system. For a particular piece of equipment and a specific image, the diffraction-limited light intensity I' at the image edge can be calculated. Figure 5-14 shows such a calculation for a representative situation.[3]

The abscissa of the figure shows distance from the geometrical edge of the feature. At the true feature edge, the image intensity is about 40% of maximum, and the maximum value is approached only about 0.3 μm into the illuminated zone. One the other hand, the intensity is still about 10% of maximum 0.2 μm away from the true edge. For a sufficiently long exposure time, even a very small intensity will fully expose the resist. Therefore, an increase in exposure time will widen the exposed region and thus (in positive resist) increase the linewidth. It is less feasible to reduce linewidths by underexposure, however, since the energy soon falls below the exposure limit even at 100% intensity. The process engineer is frequently asked to "push" dimensions by adjusting exposure. For common resists and optical system, dimensions can be pushed by a few tenths of a micrometer in the direction of overexposure, but only by a tenth or two by underexposure. If CD "pushing" is a common part of a process, masks should be ordered with the bright-field dimensions at their minimum expected values. Attempts to push beyond these limits will lead to an irreproducible process.

Incident energy is also affected by poor optical focus. Moving off the true focal plane degrades the image, making the falloff of intensity more shallow and in general reducing the overall intensity. The effect on linewidth will depend on the details of the

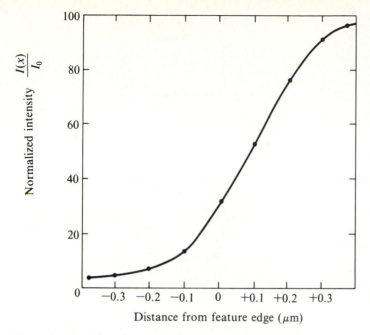

Figure 5-14 Light intensity near the image edge. The geometrical edge of this mask feature is at $x = 0$. N.A. $= 0.42$, $\lambda = 436$ nm, $S = 0.7$, linewidth $=$ spacewidth $= 0.75$ μm. (*After Ref. 3. Reprinted with permission of* Solid State Technology, *published by Technical Publishing, a company of Dun & Bradstreet.*)

resist contrast and exposure energy: Linewidths may either increase or decrease for small variations in focus. At sufficiently poor focus, lines become poorly defined and eventually disappear. Marginal poor focus is one condition which is more apparent in shape than in size: as focus degrades, corners of features show a typical degradation called the Maltese Cross. Figure 5-15 shows the characteristic loss of shape definition caused by poor focus.

Returning to Eqs. (5-36) and (5-37), we find that for a given image intensity $I'(x)$, line definition depends on two other factors: the resist contrast γ and the absorption α. Contrast is a function of the resist and developer system, and in general should be made as high as possible. It should be noted, however, that at sufficiently high contrast, the $1/z_0\gamma$ term will become insignificant in relation to the absorption term α. At this point, improvements in contrast no longer result in sharper definition.

Resist absorption is determined by the resist itself, the reflectivity of the substrate, and by standing wave effects. Some energy must be absorbed by the resist to drive the exposure reaction, so a zero-absorption resist is not attainable. However, interference and reflection effects are usually more important than simple resist absorption.

Figure 5-16 shows light of wavelength λ penetrating a layer of resist on a substrate. There is reflection of the light both at the surface of the resist and at the surface of the substrate. This has two effects. The first is the creation of intensity maxima and minima within the resist, which lead to variations of exposure with

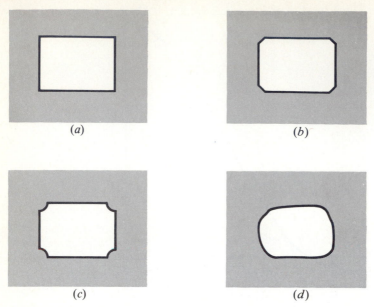

Figure 5-15 Decay of shape definition with increasing poor focus. (*a*) Ideal shape at good focus. (*b*) Slight corner notching. (*c*) Well-developed "Maltese Cross." (*d*) Complete loss of corner definition.

distance from the substrate. The result is an oscillating profile of the line edge, as shown in Fig. 5-12. The second effect is interference between the first and second reflected waves, affecting the overall film reflectivity and hence the amount of incident light actually absorbed by the resist. Figure 5-16*a* shows this effect. In Fig. 5-16*a*, the resist thickness (corrected for refractive index) is an odd number of quarter-wavelengths, and the interference between waves is destructive. Absorption is thus a maximum for this case. In Fig. 5-16*b*, the thickness is an even number of quarter-wavelengths, and absorption is a minimum due to constructive interference. In the first case, the reflectivity is lower, and the effective exposure energy greater, than in the first. The magnitude of this effect depends strongly on the reflectivity of the substrate. A strongly reflecting film like aluminum may well require several times the exposure energy as a weakly reflecting substrate like silicon dioxide. Figure 5-17 shows some calculated reflectivities for differing values of substrate reflectivity, resist thickness, and resist absorption.[3] Note how resist absorption can vary significantly with small changes in resist thickness.

We have now listed the major determinants of linewidth on a flat substrate. But a microelectronic device substrate is rarely flat, so we must also consider the effect of substrate topography. Each oxidation removes silicon from the substrate, and each deposited film adds height. After several patterning steps, the wafer surface contains steps some micrometers in height. When a resist line crosses such steps, its linewidth is affected.

One effect is geometrical. Figure 5-18 shows a resist line crossing a step. Since spun photoresist forms a flat surface across the substrate, the resist thickness is greater

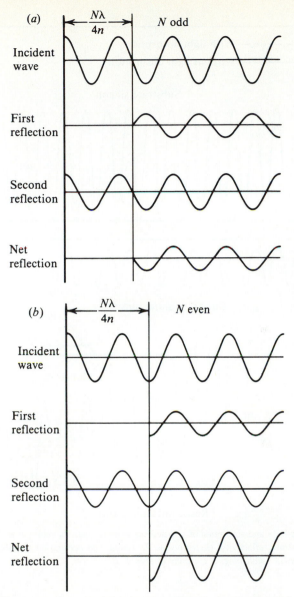

Figure 5-16 Standing wave effects in resist. (*a*) $z_0 = (N/4n)\lambda$, N odd. Destructive interference, low reflectivity, high absorption. (*b*) $z_0 = (N/4n)\lambda$, N even. Constructive interference, high reflectivity, low absorption.

at the bottom of the step than at the top. Since the resist profile is not vertical, the linewidth at the substrate surface must be wider at the bottom of the step. If the step height is h, the increase in width due to the extra resist height Δx_h will be

$$\Delta x_h = w_2 - w_1 = 2h \left(\frac{dz}{dx} \right)^{-1} \tag{5-38}$$

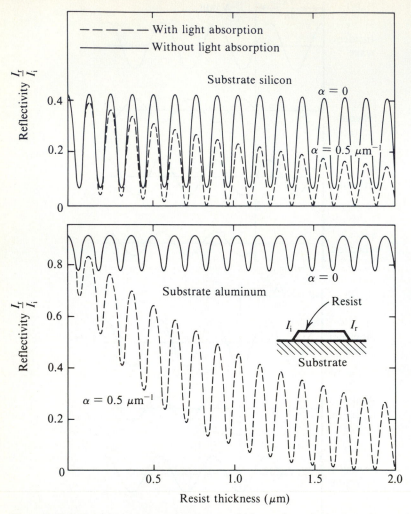

Figure 5-17 Reflectivity and absorption of a coated substrate for various values of resist absorption, resist thickness, and substrate reflectivity. (*Reference 3. Reprinted with permission of* Solid State Technology, *published by Technical Publishing, a company of Dun & Bradstreet.*)

where dz/dx is given by Eq. (5-36).

Interference effects can also vary linewidth at a step. If the step is an odd number of quarter-wavelengths in height, the standing-wave pattern, and hence the energy coupled into the resist, will change significantly as the line is crossed. The result can be a significant and hard-to-control linewidth change.

Topography can have an additional effect on linewidth through the scattering of light. Figure 5-19 shows a case in which a line is printed in close proximity to a topographic feature on a reflective substrate. If the feature is illuminated, there can be significant light scattering into the supposedly dark area of the line, exposing the resist.

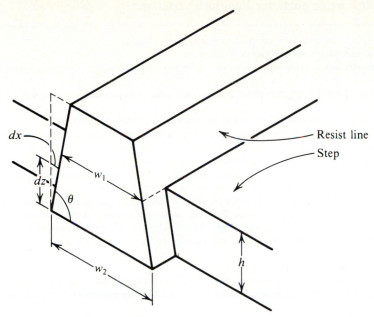

Figure 5-18 Geometrical effects on the linewidth of a resist line crossing a step. (*After Ref. 3. Reprinted with permission of Solid State Technology, published by Technical Publishing, a company of Dun & Bradstreet.*)

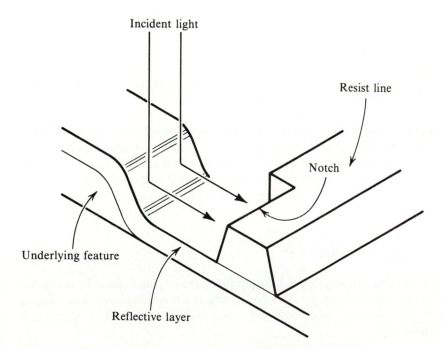

Figure 5-19 Line notching due to reflection of a nearby topological feature.

Table 5-1 Troubleshooting guide for linewidth variation

I. Incorrect dimension; consistent over substrate, distinctly printed edges.
 A. Check mask dimension, applying known process bias.
 B. Exposure may be too low or too high—determine which way for your resist type and run a matrix of exposures.
 C. Is the substrate material highly reflective? If so, a much higher exposure than expected may be required.

II. Incorrect dimension; fairly consistent over substrate, fuzzy edges, notched corners, etc.
 A. Look for the characteristic notched corners of poor focus. If found, check focus as appropriate for your printer.
 B. With negative resist, scumming or gross overexposure may be present.
 C. With positive resist, gross underexposure or underdevelop may be present. Are you trying to push C.D.s more than a few tenths of a micrometer with a diazo-type resist?

III. Incorrect dimension; only in zones of the substrate, usually with fuzzy edges or notched corners.
 A zone of the substrate is out of focus. Frequent causes are substrate warpage, frontside contamination in contact printing, contamination of points where the substrate contacts the chuck of the printer. Look for substrate warpage and examine contact points under microscope for contamination.

IV. Dimension varies widely over steps.
 A. Wider lines at step bottoms using positive resist: underexposure or underdevelop of the thicker resist in the step. Calculate actual resist thickness and double-check expose/develop times; try higher exposure energy.
 B. Standing-wave effects, especially with reflective substrates or thin resist. Calculate actual resist thickness in terms of exposure wavelength; try small changes in resist thickness.
 C. Shallow resist edge profile that aggravates effect of step. May require process recharacterization to steepen profiles.

IV. Notching or deformation of particular features, particularly on reflective substrate.
 Reflected light from adjacent topography. Look for a possible "mirror" in the topography near the feature.

The resulting *line notching* can be a limiting factor on achievable resolution. For example, this situation is quite common on MOS devices, where a metal line is fabricated close to a polysilicon gate.

In summary, linewidth control is affected by optical imaging, by resist chemistry, by exposure and development times, by resist thickness and absorbance, and by substrate topology and reflectivity, in fact, by everything. As a result, linewidth control in an operating environment is a constant care and challenge for the engineer. Table 5-1 summarizes the causes and cures of linewidth variation with an emphasis on diagnosis and treatment.

5-4 PATTERN ALIGNMENT

In addition to being properly defined, a photolithographic pattern must be positioned properly relative to prior patterns. This positioning is termed *alignment*. Any imaging instrument for microelectronic fabrication must include a method of aligning to previous patterns.

The simplest way to align patterns is to physically observe the previous pattern and line up the new one to match it. For primitive devices it may be fairly easy to determine alignment by visual examination of the device patterns, but with more sophisticated devices the proper alignment is less obvious. Therefore, alignment is usually performed using a specialized structure called an *alignment mark*.

Figure 5-20 shows some features of alignment marks. The mark, in this case, is a square inside of a box, as shown in Fig. 5-20*a*. The square is fabricated on the substrate at some previous step and serves as an alignment *target*. The box is part of the mask for the current step and is the alignment *guide*. Alignment is achieved by moving the mask until the box is positioned surrounding the square.

A good alignment mark has several characteristics. First, like the square in the box, it should show in an obvious manner when alignment is achieved. Second, since some alignment error is inevitable, the mark should indicate the tolerance for error. In our case, we could design the mark so that the box and square just touch at the maximum allowable alignment error. Therefore, as long as there is space all around the square, the alignment is acceptable.

Another requirement for the aligner operator is good visibility of the mark. The target must be fabricated in such a way that it is clearly visible through the resist coating at the required alignment step. During alignment, the target must be sighted by looking through the mask in the vicinity of the guide. The design of the guide must thus allow sufficient view of the substrate so that the target may be seen and tracked. Figure 5-20*c* shows a violation of this rule. Here our opaque box has been merged into an opaque region of the mask so that the target can be viewed only through a small window. The aligner operator is thus unable to acquire the target even though it is almost at the correct position. This fault can be corrected by placing the box in a transparent region of the mask.

Alignment marks become part of the pattern fabricated onto the substrate and can thus be used at inspection steps to check alignment tolerances. For this reason, a well-designed mark mimics the dimensional behavior of the device being created. For example, our guide is an opaque box, and the width of the box borders will be reduced by overexposure. This will increase the open area in the center of the box and thus allow more tolerance for the box to move relative to the target. Therefore, overexposure increases the apparent tolerance allowed by our mark.

Assume that the purpose of the mask is to fabricate a set of lines. The design requires that each line not touch a nearby window, fabricated at the previous step along with the alignment target. Figure 5-20*d* shows the structure. For this case, our mark is well designed. Overexposure will decrease the size of the line and increase its distance from the window structure. Tolerance to misalignment has been increased, both in the mark and in the device structure, by this process variation.

On the other hand, assume that the line we are printing must completely *surround* the window on the previous layer, as shown in Fig. 5-20*e*. In this case, our alignment mark should be redesigned. Overexposure will shrink the linewidth, decreasing the tolerance around the window, while overexposure makes our alignment tolerance look wider. The result is an alignment guide that implies increased tolerance despite the fact that the tolerance in the device has become tighter.

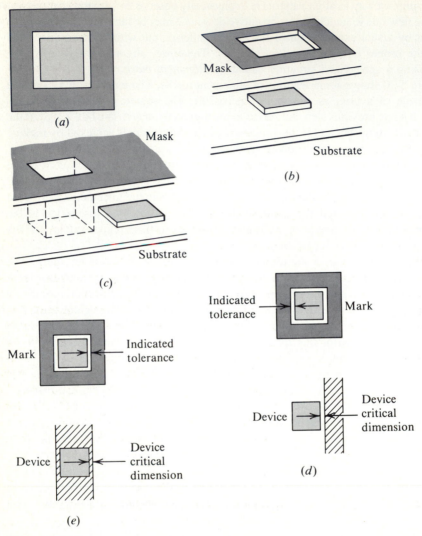

Figure 5-20 Alignment marks. (*a*) Box (target) and square (guide). (*b*) Alignment method: the target is previously fabricated on the substrate, and the box is printed on the mask. (*c*) A poorly designed alignment guide: the target is invisible until almost perfectly aligned. (*d*) Good match between alignment mark and device characteristics: overexposure increases tolerance both of device and mark. (*e*) Poor match between alignment mark and device characteristics: overexposure increases tolerance of mark but decreases tolerance of device.

In addition to square-in-the-box, cross-in-a-box, and cross-in-a-cross alignment masks, there are more intricate designs including various vernier arrangements meant to allow quantitative measurement of alignment error. Many of these designs are quite ingenious. However, it is important to remember that alignment marks are ultimately intended to be used for alignment, either by a human operator or a machine. Machines

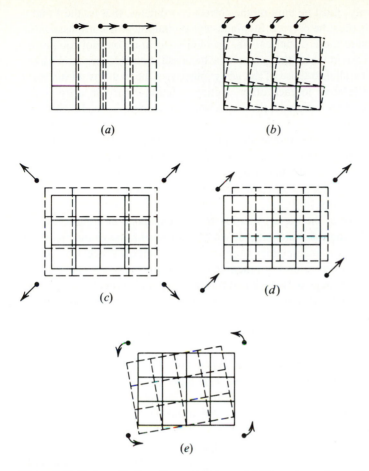

Figure 5-21 Factors affecting alignment accuracy. (*a*) Step-and-repeat error: linear spacing. (*b*) Step-and-repeat error: reticle rotation. (*c*) Run-in/run-out of mask relative to substrate. (*d*) Alignment error: translational. (*e*) Alignment error: rotational.

are currently incapable of dealing with much complexity; human operators can do so—at reduced efficiency. Therefore, the simplest feasible alignment guide is usually the best.

Perfect alignment requires a perfect fit between the patterned substrate and the image to be printed. Such perfect fit is rarely found: Fig. 5-21 demonstrates some of the factors that degrade alignment. These causes are collectively referred to as *overlay error*.

The first two factors involve the mask. Masks are created by a step-and-repeat method, in which a master device pattern, or reticle, is repeatedly imaged on the mask to build up a rectangular array. For two layers to align properly, their masks must have identical arrays. Errors in the step-and-repeat process can thus prevent proper alignment of all devices.

First, the mask arrays must be built on the correct coordinates: that is, the x and y distances between cells must be correct. Step-and-repeat equipment usually uses laser interferometry to measure these distances. The method is accurate but not foolproof. Figure 5-21a shows a case in which the x interval is incorrect. Second, the x and y axes of the reticle must be parallel to those of the array: otherwise the patterns within each device will be rotated relative to each other. Reticle rotation error can arise if the mechanical positioning of the reticle on the step-and-repeat camera is not flawless, as shown in Fig. 5-21b.

Marks can be built into the reticle that allow checking for translational and rotational error in the step-and-repeat process: they are essentially alignment marks, in which (say) the target is in the left-hand upper corner of the reticle and the guide is in the right-hand lower corner. Step-and-repeat error is indicated if the target and guide fail to align.

When the imaged size of the mask pattern is slightly expanded or contracted in a linear manner relative to the substrate pattern, we refer to *run-in* or *run-out*. Run-in/out, shown in Fig. 5-21c, can result either from the step-and-repeat process or during photolithography. Run-out can result from thermal expansion if the temperature varies between successive process steps or if mask and substrate are at different temperatures. For this reason, mask and substrate temperatures must be controlled so that all alignments occur at the same temperature. Run-out can also occur due to magnification error by the imaging equipment or due to mechanical flexing of the substrate or mask.

Finally, if the mask and substrate dimensions are all correct, we come to true alignment error: failure by the aligner (mechanical or human) to achieve the optimum alignment of mask to substrate. As with any physical process, error is inevitable. The odds are zero of perfect alignment, and there will be a Gaussian distribution of error about the ideal location. Alignment error can be translational (x, y) or rotational (θ). In Fig. 5-21d, the array is translated up and to the right; in Fig. 5-21e, it is rotated counterclockwise. In general both types of error can be present, and the observed pattern of deviation can become complex. Therefore, when overlay error is observed, some care should be taken in analyzing it into its components to be sure whether simple error or something more fundamental is present.

Example Overlay error is checked at the top, bottom, left and right edges, and in the center, of a 4-in silicon substrate. The observed errors in x and y, in micrometers, are found to be

	Top	Right	Center	Left	Bottom
x	0.0	0.7	0.5	0.3	1.0
y	0.7	1.0	0.5	0.0	0.3

Analyze the error into run-in/out, rotational and translational misalignment.

The center location cannot exhibit run-in/out or rotation. Therefore, the error at the center is all translational, and we know that the mask is translated $(0.5, 0.5)$

from nominal. Subtracting out the translational error gives the following values for the remaining rotational and run-in/out components:

	Top	Right	Center	Left	Bottom
x	−0.5	0.2	0.0	−0.2	0.5
y	0.2	0.5	0.0	−0.5	−0.2

Rotation will cause a tangential error at all locations (y error at left and right; x at top and bottom). Run-in/out causes radial error (y at top and bottom, x at left and right). By inspection, we can see that the run-out is 0.2 μm and the rotation corresponds to 0.5 μm counterclockwise at the wafer edges. This is a rotation of about 10^{-5} radians and a run-out of about 4 ppm. The student is encouraged to sketch out this example and get a feeling for the combined effects of error.

5-5 PRACTICAL IMAGING EQUIPMENT

An imaging instrument for microelectronic use must have several characteristics. First, it must be capable of transmitting, with high accuracy, an image from a mask plate or reticle to a resist coating on a substrate of varying topography. To do this well, it must not only be of high optical quality, with a spatial resolution of at worst a few micrometers, but must be able to fix the substrate accurately at the focal plane of the image.

The instrument must also allow image alignment. This implies a method, either manual or automated, of observing and adjusting the position of the projected image until alignment is achieved.

A practical printer must have some additional attributes. If used in a profit-making enterprise, it must have a sufficient throughput of patterned substrates to be economical. It must be reliable enough to be operated by semiskilled personnel, without constant adjustment. And, finally, a real instrument must live in a real world of ubiquitous dust and dirt. Since image sizes are small compared to the average dust mote, the instrument should either keep particles from the mask and substrate or have some means of controlling their effect on the image.

Defect avoidance, output, low cost, alignment accuracy, and optical quality are difficult to get in one piece of equipment. As a result, several different types of printing equipment are currently in use, each with different strengths and weaknesses. We will now review the principal types of printers.

We should note that, for each type of printer, there are usually a variety of makes and models available. The individual characteristics of each model of printer are important, and the engineer needs to become intimately familiar with the instrument he uses. However, neither space nor propriety allows us to single out specific machines for our discussion. Therefore, the text and figures that follow deal with generic

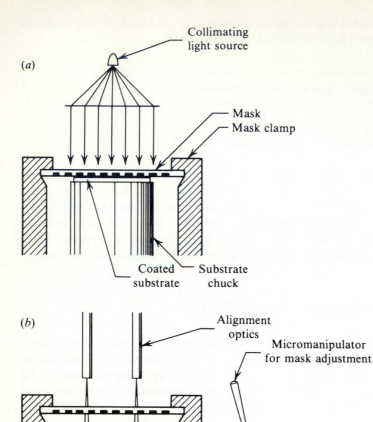

Figure 5-22 Conceptual drawing of a contact printer. (*a*) In contact, during exposure. (*b*) Out of contact, during alignment.

varieties of printing and are specifically intended *not* to represent the product of any particular equipment manufacturer.

Contact Printing

The simplest way to transmit a mask image to a substrate is to clamp the two together and shine light through the mask. This is *contact* printing. Figure 5-22 shows a contact printer. There is no optical processing of the image: the image is used directly from the mask itself. An optical system *is* required for illumination: it is designed to furnish a coherent, highly collimated beam, essentially perpendicular to the mask. (In practice, exactly perpendicular illumination leads to some dramatic diffraction effects, so the beam is decollimated about 3°.) With coherent illumination, the MTF of a contact printer shows no degradation up to the sharp cutoff at N.A./λ.

Alignment can be observed directly in contact printing by looking through the mask to the substrate with a microscope. When the mask and substrate are in contact, they are both in focus at the same point and observation is easy. However, alignment cannot be *adjusted* in contact, because the mask would scrape against the substrate. Therefore, some apparatus must be provided to clamp and unclamp the mask, moving it a few micrometers from the wafer to allow alignment. When this happens, the mask and wafer are no longer simultaneously in focus, so the aligning microscope must have sufficient depth of focus to allow both to be observed. There must also be some type of *micromanipulator,* to allow the alignment to be corrected in micrometer-sized increments.

In contact printing, there are no optics to limit alignment accuracy, but there is one systematic overlay error characteristic of contact printers. Pressing the substrate against the mask can lead to *bowing* of the mask, so that the mask and substrate become arcs rather than planes. In this case, points on the convex surface extend past their corresponding points on the concave surface, leading to run-out. Mask bowing is difficult to eliminate and must therefore be adequately controlled.

The focal plane of a contact printer is at the mask, and therefore, optical depth of focus is not an issue. The resolution of contact printing is limited only by the wavelength of the light used. Where other instruments may be limited to certain wavelengths by the characteristics of their optics, contact printers have no such limitation. They are also conceptually simple and inexpensive.

The main limitation of the contact printer is the number of defects it causes. For one thing, contact printing is highly sensitive to particles. Particles can easily get between the mask and substrate, either from the environment or from contamination of the substrate or the mask. Once there, they keep the substrate out of contact with the mask in the area around the particle. The substrate is thus out of the focal plane, and the entire zone around the particle is poorly defined. The particle is also likely to damage the substrate or mask as it is pressed between them. If the substrate is moved relative to the mask during the alignment process, the particle may tear an extended gouge in the resist. Finally, if the particle becomes embedded into the mask, it will be carried along to each substrate exposed, causing defects in each of them. Particles also shorten the useful mask life, so it is seldom worthwhile to invest in an expensive, low-defect mask plate. Even in the absence of particles, contact printing requires physical contact, and thus abrasion, between the mask and substrate. Therefore, contact printing is inherently defect-prone.

Whether it is economical to accept the defect level associated with contact printing depends on the application. For very small devices, especially those involving fairly large feature sizes, even a large number of defects will have little impact on the device-per-substrate yield. In such applications, cheap and easy contact printing may be quite appropriate. For larger devices and smaller geometries, however, a lower defect level is required.

Proximity Printing

An extension of contact printing is *proximity printing*. Proximity printing uses the direct image of the mask just as contact printing does, but maintains some small space between the mask and substrate. Thus there is no mask-to-substrate abrasion, no embedding of particles into mask or substrate, and no poor definition zone surrounding small particles. (Particles present on mask or substrate do block light and cause defects, but the defect is limited to the size of the particle.) Proximity printing is thus less defect-prone than pure contact.

However, proximity printing is more difficult to do well. It is harder to control a small mask-to-substrate gap with precision than to press two objects tightly together. Also, the mask image degrades as the gap increases. The minimum printable linewidth has been reported to be approximately x where[5]:

$$x \cong \sqrt{g\lambda} \tag{5-39}$$

and g is the mask-to-substrate gap. With gaps in the range of 5 to 50 μm and UV radiation of about 400 nm, linewidths must be at least several micrometers for effective printing.

Alignment is also more complicated since the mask and substrate can no longer be simultaneously imaged by the same microscope. In general, proximity printing is a compromise application: it offers less defect induction than contact printing at moderately higher cost and some loss of resolution, but it is much less expensive than full projection printing.

Whole-Substrate Projection Printing

From the beginning of photolithography, the advantages of projection printing have been obvious. If the mask image is projected onto the substrate using an optical system, the problems of mask-substrate contact (or near contact) are avoided. The surface of the mask can be shielded from particle contamination with a transparent cover called a *pellicle*.[6] The surface of the pellicle is several millimeters from the mask surface, so that any particles falling upon it are well out of focus and are not printed. It is usually not feasible to cover the substrate surface with a pellicle. But since the substrate is physically separated from the mask, it can be flushed with a continual flow of clean air to avoid particle contamination. The mask, untouched by abrasive surfaces, can be used indefinitely, so it is economical to invest both time and money in a defect-free mask.

Projection printing, however, poses a substantial challenge in optical design. Not only must features of 2 to 3 μm size be resolved, but the image must be accurate in size (magnification) to better than 1 ppm. Aberrations, including chromatic aberration, must be similarly controlled. The substrate must be positioned accurately in the image focal plane: errors not only in placement but in parallelism must be tightly controlled. As a result, no practical printer operates by the simultaneous projection of an entire mask image onto a substrate. However, whole-substrate projection printing has been made practical, using the strategy of *scanning projection printing*.

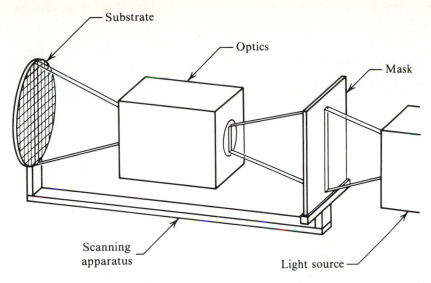

Figure 5-23 Conceptual drawing of a scanning projection printer. Features not shown include the alignment optics and hardware, the details of the optics and of the scanning apparatus.

Figure 5-23 shows the concept behind the scanning projection printer. A narrow beam is used to illuminate a stripe on the mask, and the image of this stripe is projected onto the wafer. By projecting only a narrow region of the pattern, the optical difficulties are reduced. The mask and substrate, held rigidly in relation to each other, are then transported through the beam so that the entire mask surface is scanned. In this manner, the mask image is projected onto the substrate. The scanning technique also allows the mask-to-substrate distance to be rigidly fixed, simplifying the issue of focus control.

A more detailed description of one popular type of practical printer was given by D. A. Markle in 1974.[7] The optics of the instrument are designed to project only a narrow arc-shaped portion of the pattern, and have some novel features. They are reflecting, rather than refracting, so that chromatic aberration is not an issue. The design turns out to be inherently nonmagnifying, so that magnification control is not an issue. The numerical aperture is 0.16. The result has been a highly successful printer. Based on announcements of ongoing improvements, scanning projection printers should continue their usefulness well into the future.[8]

The limitations of scanning projection are the limitations inherent in simultaneous imaging across the entire substrate diameter. A scanning projector must maintain focus at each point on the substrate. This requires not only precise temperature control and constant equipment adjustment, but extremely flat masks and substrates. Also, satisfactory alignment must be maintained over the entire substrate diameter. Local causes of overlay error, such as physical distortion of the substrate or optical errors in the system, cannot be compensated for and thus must be controlled rigidly. As substrate diameters grow and feature sizes shrink, it becomes increasingly difficult to maintain the required whole-substrate alignment and focus.

Step-and-Repeat Projection

Step-and-repeat techniques have long been in use for the fabrication of mask plates, where alignment to an underlying pattern is not an issue. Their use in device fabrication was impractical until the advent of dependable automatic alignment systems. Modern electronics having made the alignment systems feasible, the alignment systems can now be used to fabricate modern electronics.

Figure 5-24 schematically represents a step-and-repeat projection printer, often called a *wafer stepper*. The printer exposes only a small portion (one or a few devices)

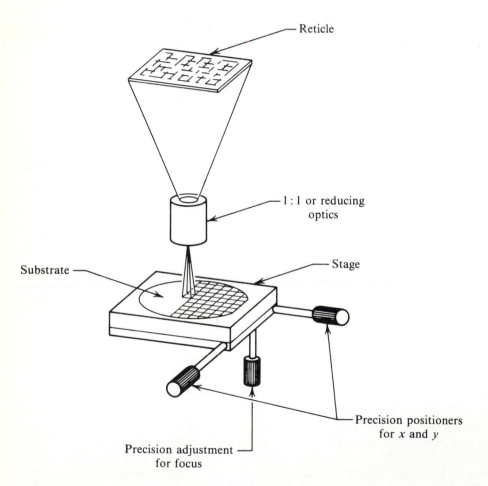

Figure 5-24 Conceptual drawing of a step-and-repeat projection printer. Among the features not shown are the sensors and computation equipment required to determine alignment, focus, and *x-y* position.

of the pattern at any one time. At each exposure, the printer can determine the optimum alignment and focus distance for the area to be exposed. This results in much more tolerance for substrate deformation than occurs with whole-substrate projection.

Further advantage results from step-and-repeat projection if the optics are made reducing, so that the reticle dimensions can be made larger than the device. (This is called $N:1$ projection, where N is the magnification, e.g., a $10:1$ wafer stepper. The technique is termed reduction step-and-repeat photolithography.) This means that, say, a 2-μm feature, which must be accurately sized to ± 0.25 μm, can be fabricated on the reticle at 20 ± 2.5 μm, allowing much more tolerance for error. It also means that a 2-μm defect or particle on the reticle, fatal if reproduced directly on the device, will be shrunk to a harmless 0.2 μm in the projected image. While increased magnification reduces the useful depth of focus of the image, the wafer stepper can refocus itself at each step and therefore can maintain tighter focus. A number of step-and-repeat instruments are currently available with magnifications of from 1 to 10 and numerical apertures from 0.2 to 0.35.

The power of reduction step-and-repeat imaging to produce low-defect, small-dimension, tight-tolerance devices is unparalleled. However, the equipment is expensive, slow, and delicate. Complex automated systems are required to control the x- and y-movement, to determine the alignment and focus, to image the reticle, and to provide the light for exposure. Because each small portion of the substrate is independently exposed, the technique is inherently slow. Acceptable throughput is obtained only by making alignment and exposure times quite short, requiring very fast alignment software and an intense exposing light source. Therefore, although we can expect to see reduction wafer steppers take over the high-technology end of fabrication, there will continue to be applications where scanning projection, proximity, and even contact printing are more economically appropriate.

5-6 FUTURE DIRECTIONS IN PHOTOLITHOGRAPHY

Microelectronic progress continues to demand the fabrication of smaller feature sizes. Currently, dimensions on devices in large-scale production are frequently a few micrometers in size, but it is generally accepted that submicrometer dimensions will be required in production by the late 1980s. There are two principal obstacles to penetrating the 1-μm barrier: (1) The inherent resolution of the imaging equipment is constrained by the wavelength of the imaging light; (2) the empirical limitations of the resist-wafer system, including lateral development, standing waves, topography, and reflection, make the theoretical resolution unattainable. Both of these obstacles are under active attack.

An immovable barrier is the physical one: for a given wavelength of light, indefinitely small dimensions cannot be imaged. Therefore, there is a strong motivation to reduce the light wavelength. Below the near-ultraviolet wavelengths currently used lies the far ultraviolet ($\lambda \sim 200$ nm). Below the far ultraviolet lies a broad spectral range where virtually nothing is transparent; the next usable window lies in the x-ray regime ($\lambda = 0.4-5$ nm). The use of electron beams, as in electron microscopy, offers even smaller wavelengths ($\lambda \sim 0.01$ nm). Each of these types of radiation has been used experimentally for photolithography.

Far-Ultraviolet Photolithography

Moving from near-UV illumination around 400 nm to the far-UV regime offers a factor-of-2 reduction in diffraction-limited feature size. Since systems that can print 2-μm dimensions are widely used in production, this change alone would at least approach 1-μm printing. Far-UV imaging is also attractive because it could be done with minor adaptations of existing equipment and methods. Refractive glass optics are opaque to far-UV radiation, but quartz optics could be used. Reflecting optics would require even less adaptation. A vacuum is required to use UV below 100 nm, but printing in the 200-nm region could occur under atmospheric conditions. Even conventional diazo-type photoresists can still be used, although they become inconveniently absorbent.

Far-UV illumination has been utilized for contact, proximity, and scanning-projection printing. Resolution of geometries below 1 μm for contact and well below 2 μm for proximity and projection have been reported (for reviews and further reading, see Refs. 9 and 10).

Far-ultraviolet is an evolutionary improvement of current techniques. Since it builds on existing techniques, it is fairly easily implemented. However, it does not offer dramatic improvements; in particular, it does not promise significantly submicrometer resolution. For this reason, interest continues in even shorter wavelengths.

X-Ray Photolithography

Below about 100 nm, most materials become opaque to radiation, until another window is reached in the x-ray regime at 5 nm. Research in x-ray photolithography, for a number of technical reasons, concentrates on wavelengths in the 0.4–0.8 nm range, a factor of 100 smaller than conventional ultraviolet radiation. X-ray photolithography thus has the potential to remove the wavelength limitation on feature size indefinitely. However, the practical and widespread use of x-rays requires grappling with some significant technical difficulties.

The problem with x-rays is their well-known penetrating power. Neither reflecting nor refracting materials are available for x-rays. Therefore, x-ray optics are essentially nonexistent, and x-ray printing must be performed without optics. This is an obvious situation for contact or proximity printing. Due to the short wavelength, proximity printing can be done at reasonable gap spacings. The methods used are fairly conventional, except as regards the light source. Conventional x-ray sources cannot be collimated. Thus the radiation is derived directly from a pointlike x-ray source, as shown in Fig. 5-25. Since the source is not truly a point, the image will be smeared. If the source dimension is $2r_s$, and its distance from the mask is D, the result will be a smearing of the image Δx of

$$\Delta x = 2r_s \frac{g}{D} \qquad (5\text{-}40)$$

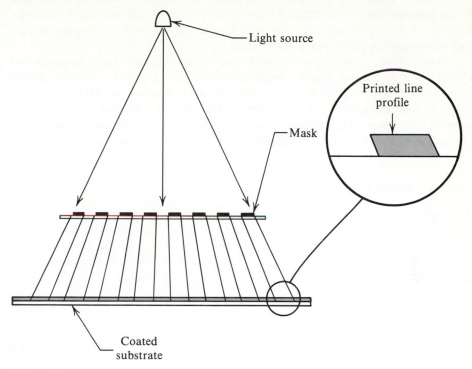

Figure 5-25 An x-ray printer using a point source, illustrating the magnification effect. The inset shows the effect of the point source on line profile. The relative magnitude of both effects is highly exaggerated in the figure.

Also, because the illuminating radiation travels radially from the point source, there will be a slight magnification of the image (due to the gap) of

$$m_x = 1 + \frac{g}{D} \tag{5-41}$$

(It has been found useful to make the gap and hence the magnification adjustable so that minor mask size errors can be smoothed out.) X-ray resolution and overlay from a point source are thus limited by the geometry of the source rather than diffraction. These limitations may be avoidable if practical use can be made of another source of x-rays, synchrotron radiation. This radiation, which results from the acceleration of electrons orbiting in a synchrotron at relativistic speeds, is inherently collimated.

The penetrating power of x-rays also makes mask making difficult. The opaque material is usually gold: the supporting "transparent" substrate can be a membrane of silicon, silicon oxide/nitride, boron nitride, or an organic film. Such masks are neither durable nor very rigid mechanically. It is also difficult to inspect an x-ray mask for defects, both because of the small dimensions involved and the ability of x-rays to penetrate many visible defects.

Effective production also requires reasonable throughputs, which are determined by the exposure time per substrate. X-ray exposure times are still longer than desirable, because of the inefficiency of x-ray sources and the insensitivity of x-ray resists. Active research is under way in these areas.[11]

Nevertheless, reasonably uncomplicated x-ray printers have demonstrated effectiveness in imaging submicrometer geometries. Even in 1981, an x-ray printer using conventional alignment was reported to have achieved an MTF of 0.6 for 0.25-μm lines and spaces, with nonlinear overlay errors of ~0.1 rms. Exposure times were 0.5 to 3 min. The exposure was performed in helium to avoid atmospheric absorption.[12] Such performance is simply not practical using conventional lithography.

Electron-Beam Lithography

The limitations imposed on optical microscopy by the wavelength of light have been overcome using electron microscopy. It is natural, then, to investigate the potential of electron beams in lithography. Certainly wavelength should not be a concern. The wavelength associated with a particle such as an electron is related to its momentum by de Broglie's equation:

$$\lambda = \frac{h}{mv} \tag{5-42}$$

where h is Planck's constant, m is the mass, and v is the velocity of the particle. For a 10-keV electron, the wavelength is 0.012 nm.

Electron beams can be focused and scanned over a substrate using techniques similar to those used in television and cathode-ray tubes. The required device pattern can be built from a design tape, using software to dissect the pattern into a series of picture elements. Indeed, electron beam generation of mask reticles and plates is an established technology. Using electron beam lithography on substrates, devices can be patterned directly from the design without an intervening mask-making step.

Electron beam pattern resolution, like x-ray resolution, is not directly limited by diffraction. Instead, the limiting factor with electron beam (or "e-beam") is electron scattering.[13] Figure 5-26 shows the scattering pattern of electrons entering a fairly thick layer of photoresist: they diverge from the point of entry, forming a pear-shaped zone of exposure that determines the minimum printable spot. Scattering also causes a *proximity effect:* two adjacent spots require less exposure than two separated spots, due to their mutual scattering. Sufficiently sophisticated software can compensate for these effects.

A major limitation of electron-beam lithography is speed. When a mask is used for lithography, vast numbers of picture elements are exposed simultaneously. With an electron beam, each element must be formed independently, putting the *e*-beam method at a huge disadvantage in speed. A great deal of ingenuity has been expended to overcome this handicap by finding more efficient means of scanning, by varying the beam size and shape, by increasing the power of the electron beam, and by developing more sensitive resist systems.[14] Nevertheless, speed appears an inherent limitation of

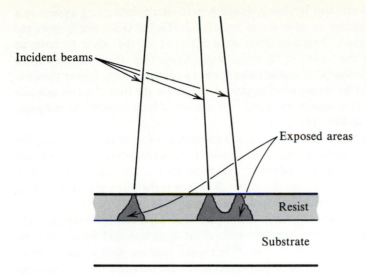

Incident beams

Exposed areas

Resist

Substrate

Figure 5-26 Scattering in electron-beam lithography.

the technique. No resist system can require less than one electron per picture element, and there are theoretical reasons to believe that beam current above 2×10^{-7} A are not achievable. Under these conditions, the exposure time for 0.1-nm features on a 4-in substrate has been reported to be 60 sec, a time considered barely acceptable in many production environments.[15]

The advantages of electron-beam lithography are resolution and flexibility. Its inherent disadvantages are speed and complexity. As a nonoptical technique, it is a fundamentally greater change than competing methods, requiring more education and retraining for successful introduction. These factors suggest that electron-beam lithography will not soon become appropriate for the routine mass production of microelectronic devices. However, because it eliminates the mask-making loop, e-beam is ideal for "one-of-a-kind" and "almost-one-of-a-kind" applications, for example, the fabrication of experimental and prototype devices. Another possible application is for ROM printing, in which a mass-produced device is customized with an instruction set peculiar to a single user.

Photoresists for Shorter Wavelengths

The use of higher-energy radiation requires new photoresists. Although work has been done with the diazo-type resists, other materials are in demand, especially for electron-beam and x-ray applications. A number of replacements have been proposed.

The most popular high-energy resists are derivatives of *polymethyl methacrylate (PMMA)*. These resists are exposed by "depolymerization": scission of molecular chains by the incident radiation. They are thus positive acting and are truly the opposite in mechanism of the cyclized-rubber negative resists.

Inorganic resists are also receiving study. A particularly interesting system is a combination of germanium selenide and silver selenide (GeSe/AgSe) that is sputtered onto the wafer in layers. Exposure leads to a migration of the silver to form an insoluble material in the GeSe, so the resist is positive-acting. However, this resist demonstrates an edge-sharpening effect that leads to contrast much higher than expected, perhaps due to the formation of electron-hole pairs in the film.[16] Silver-arsenic-sulfur materials have also been investigated.[17] A number of both organic and inorganic resists are surveyed in Ref. 18.

In addition to seeking new materials, resist research has focused on controlling the detrimental effects of wafer topography: linewidth widening over steps, standing waves, and reflection from the substrate. All these effects are functions of resist thickness and would vanish at zero thickness. Research is therefore aimed at making the effective resist thickness for image formation very thin.

A thin resist would also alleviate some particular high-energy imaging problems. Electron-beam scattering increases with resist thickness and thus a thin resist will display better *e*-beam resolution. When x-ray printing with a point source, the diverging rays are not in general perpendicular to the substrate. As a result, resist lines "lean" toward the source. A sufficiently thin resist will minimize that effect.

Very thin resists are not compatible with normal wafer topographies, but it is possible to achieve near-zero *effective* resist thicknesses by using *multilayer resist techniques*. Figure 5-27 shows two such methods.

In Fig. 5-27a–c, a two-layer method is illustrated.[19] A thick layer of a deep-UV resist is first applied to the wafer, smoothing out any topographical effects. Such a layer is often called a *planarizing* layer. A thin layer of a near-UV resist, which is highly absorbent in the far-UV, is then applied. Patterning of the upper layer is performed conventionally: the resolution is excellent because the resist is thin and flat. The entire wafer is then subjected to intense far-UV illumination, which penetrates and exposes the second layer only where the upper layer has been removed. Conventional developing of the underlayer with a high-contrast developer achieves excellent transfer of the overlying pattern.

Figure 5-27d–f shows a three-layer resist process.[20] Here an intermediate, inorganic layer is placed between the lower and upper polymers. The upper resist is patterned and the intermediate inorganic layer is etched. The thinness of both layers allows excellent pattern reproduction. The underlying polymer can then be patterned using a method like *reactive ion etching,* which is highly directional and maintains dimensional accuracy. The three-layer system has several advantages over the two-layer system. It is not dependent on the photochemical behavior of the lower layer. It avoids the interactions that can occur between two organic films in contact. Also, antireflective or absorbing agents can be placed either in the inorganic or planarizing film, further reducing standing-wave and reflection effects. On the other hand, this system requires sputtering or spinning a defect-prone inorganic film and is inherently more complex.

Multilayer resist techniques can result in significant improvements in the achievable resolution of current imaging equipment.[21,22] As such, they can postpone the day

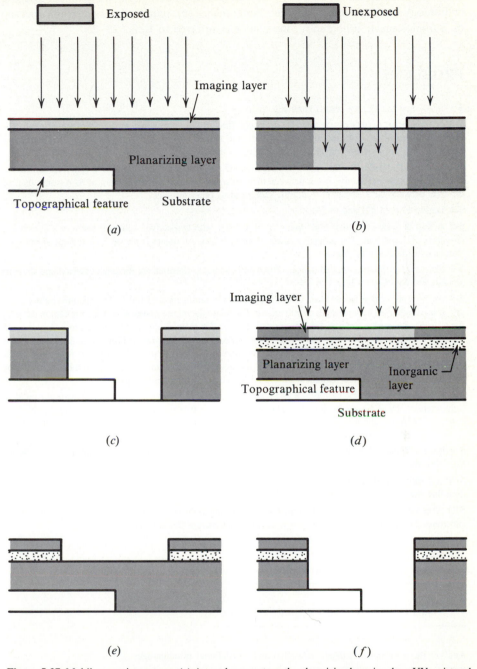

Figure 5-27 Multilayer resist systems. (*a*) A two-layer system: the planarizing layer is a deep-UV resist and the imaging layer is a near-UV resist. (*b*) Deep-UV illumination after patterning the imaging layer. (*c*) Pattern after developing the planarizing layer. (*d*) A tri-layer system with an intermediate inorganic layer. (*e*) Etching of the intermediate layer after patterning the imaging layer. (*f*) Reactive ion etching of the planarizing layer.

when the microelectronic industry is forced to adapt inherently more complex imaging technology, such as x-ray or e-beam. For some recent, practically oriented information on submicrometer lithography, the reader is referred to Refs. 23–25.

PROBLEMS

5-1 Show that Snell's law results from Fig. 5-1b.

5-2 A light beam is incident on a sheet of "heavy flint" glass at an angle of 30° from the normal. The refractive index of the glass (at 365 nm) is 1.705, and that of air is 1.036. Find the angle of the reflected beam and the angle of the refracted beam relative to the normal.

5-3 Derive Eq. (5-19) in the thin lens approximation. That is, assume two back-to-back spherical surfaces of radii R_1, R_2, with a refractive index between them of n' and an index outside them of n. Assume that the distance between the surfaces is negligible compared to other distances.

5-4 Derive Eqs. (5-22) and (5-23).

5-5 A lens is made of "heavy flint" glass ($n' = 1.705$), with two convex spherical surfaces with radii of curvature of 1 and 2 m. The refractive index of air is 1.036. An object is placed 2 m in front of the lens. Where will the image be formed?

5-6 Sketch the diffraction-limited images of two point sources. Illustrate the Rayleigh criterion and show its consistency with Eq. (5-26).

5-7 An imaging system has an N.A. of 0.4 and a linear magnification of 100×. The aperture radius is 0.5 cm. It uses light with a wavelength of 600 nm. (a) Calculate the minimum resolution and maximum depth of focus using Rayleigh's criteria. (b) This system is used for manual alignment on a contact printer, with minimum feature sizes of 3 μm and a mask-to-substrate gap during alignment of 1 μm. Comment on its suitability for this purpose. (c) A well-meaning process engineer attempts to improve the resolution of the instrument by replacing the instrument eyepieces, increasing magnification to 200×. Find the new resolution and the new depth of focus; comment on the advisability of this improvement.

5-8 Calculate the resist line profile of a line feature of a nominal width of 5 μm. Approximate the incident light intensity, by assuming that Eq. (5-37) holds and that $I' = 0.5I'(0)$ at the nominal line edge. Let N.A. = 0.42, $\lambda = 456$ nm, and $K = 1$ for the optical system. Let $z_0 = 1$ μm, $\gamma = 0.4$, and $\alpha = 0.5$ μm^{-1} for the resist. Define the exposure time so that $tI'(0) = 2E_0$. Plot the result.

5-9 Repeat Problem 5-8, using twice the exposure time. How much does the linewidth change? (Define linewidth by the points where z vanishes.)

5-10 Examine the effect on the resist line of Problem 5-6 caused by (a) doubling the absorption α; (b) doubling the contrast γ.

5-11 Overlay error is checked at the top, bottom, left and right edges, and in the center of a 4-in silicon substrate. The observed errors in x and y, in micrometers, are found to be

	Top	Right	Center	Left	Bottom
x	0.0	0.7	0.5	0.3	1.0
y	−0.3	0	−0.5	−1.0	−0.7

Analyze the error into run-in/out, rotational and translational misalignment.

5-12 Estimate the achievable resolution of the following printers. Assume a resist process using 1-μm-thick resist and requiring a MTF of 0.5.

   (a) Scanning projection printer, focus tolerance ±6.0 μm, N.A. = 0.16, operating wavelength, 0.365 μm.

(b) Step-and-repeat projection printer, focus tolerance ±2.0 μm, N.A. = 0.3, operating wavelength, 0.365 μm.

(c) Step-and-repeat projection printer, identical to (b) except operating in the far-UV at 0.200 μm wavelength.

(d) Electron beam writing, limited by electron scatter in the resist to a resolution of 0.25 times the resist thickness.

(e) X-ray printer, limited by geometric effects. Source size 2-cm, mask-to-wafer distance 10 μm, source-to-wafer distance 0.5 m.

REFERENCES

1. M. Lacombat, G. M. Dubroeucq, J. Massin, M. Brevignon, *Solid State Technol*. **23**:115 (August 1980).
2. M. C. King and M. R. Goldrick, *Solid State Technol*. **19**:37 (February 1977).
3. W. Arden, H. Keller, and L. Mader, *Solid State Technol*. **26**:143 (July 1983). The following discussion of linewidth is based on their article.
4. S. Wittekoek, *Proc. Microcircuit Engineering, Amsterdam* (1980), pp. 155–170.
5. R. K. Watts and J. H. Bruning, *Solid State Technol*. **24**:99 (May 1981).
6. T. A. Brunner, C. P. Ausschnitt, and D. L. Duly, *Solid State Technol*. **26**:135 (May 1983).
7. D. A. Markle, *Solid State Technol*. **17**:50 (June 1974).
8. J. D. Buckley, *Solid State Technol*. **25**:77 (May 1982).
9. Daryl Ann Doane, *Solid State Technol*. **23**:101 (August 1980).
10. David J. Elliott, *Integrated Circuit Fabrication Technology*, McGraw-Hill, New York, 1982, p. 369.
11. Gary N. Taylor, *Solid State Technol*. **27**:124 (June 1984).
12. R. K. Watts and J. H. Bruning, *Solid State Technol*. **24**:99 (May 1981).
13. T. S. Chang, D. F. Kyser, and C. H. Ting, *Solid State Technol*. **25**:60 (May 1982).
14. R. K. Watts and J. H. Bruning, *Solid State Technol*. **23**:127 (May 1980).
15. R. E. Howard, *Solid State Technol*. **23**:127 (August 1980).
16. H. Nagai, A. Yoshikawa, Y. Toyashime, O. Ochi, and Y. Mizshima, *Appl. Phys. Lett*. **28**:145 (1976).
17. D. A. Doane, *Solid State Technol*. **23**:101 (August 1980).
18. M. P. C. Watts, *Solid State Technol*. **27**:111 (February 1984).
19. B. J. Lin, "Developments in Semiconductor Microlithography IV," *SPIE* **174**:114 (1979).
20. J. M. Moran and D. J. Maydan, *J. Vac. Sci. Technol*. **16**:1620 (1979).
21. E. Ong and E. Hu, *Solid State Technol*. **27**:155 (June 1984).
22. M. Hazatkis, *Solid State Technol*. **23**:74 (August 1981).
23. D. P. Kern, P. J. Coane, P. J. Houzego, and T. H. P. Change, *Solid State Technol*. **27**:127 (February 1984).
24. V. Miller and H. L. Stover, *Solid State Technol*. **28**:127 (January 1985).
25. See also the May 1983 issue of *Proceedings of the IEEE*, which is a special issue on submicron fabrication.

SIX

ETCHING

The last two chapters have dealt with the use of photolithography to form patterns in photoresist. However, photolithography is only a prelude to the actual pattern-forming step: the etching of the substrate. The photolithographic pattern becomes part of the substrate and assumes a useful physical form, only after etching away all of the areas not protected by the resist.

In this chapter, we study etching using two quite different media: liquid chemicals and reactive gas plasmas. In Section 6-1, we deal with the theory of the most basic method of etching: isotropic etching by immersion in a chemical bath. In Section 6-2, we describe some wet etchants for common microelectronic materials. In Section 6-3, we explore the principles and methods of *plasma etching*, using a reactive, low-pressure gas plasma as the etching medium. In Section 6-4, we look at the chemistry of specific plasma systems and the film that they etch.

After etching, the photoresist film is no longer useful and must be removed. Resist removal, which can itself be viewed as an etching process, is examined in Section 6-5. Also in this section we study *lift-off* techniques, which can be used as an alternative to etching. Finally, in Section 6-6, we review important factors in etch process control, including measurement methods used to characterize etching.

6-1 WET CHEMICAL ETCHING

In etching to reproduce a photoresist pattern, we desire to remove a film or layer from the substrate in those areas not covered with photoresist. The most obvious way to do so is to immerse the substrate in a bath of some chemical substance that attacks the film. Ideally, this substance or *etchant* should have several characteristics. It should react with the film to be etched in a smooth and reproducible manner, producing soluble products that can be carried away from the surface. It should not react with the photoresist, so that the resist mask will remain unaffected by the etching process. It should be chemically selective, so that it reacts only with the film to be etched and not with other microelectronic materials. In particular, an ideal etchant will not attack any layer underneath the film being etched, so that the etch process will be self-limiting.

Figure 6-1a shows some of the quantities of interest in a wet chemical etching process. There are three layers of material: the underlying bulk substrate; a thin film to be etched, of thickness z; and an overlying resist pattern. The resist pattern consists of lines of width x_L, separated by spaces of width x_S. Also shown is the *pitch P*, which is equal to $x_S + x_L$.

To have a specific example in mind, we can let the bulk substrate be silicon and the film be a layer of silicon dioxide ("oxide") grown on it by the methods of Chapter 3. We immerse the substrate in a solution of dilute hydrofluoric acid (HF), which will etch the oxide with a rate r with units of length per time. Under the proper conditions, the HF will attack neither the underlying silicon nor the resist mask.

Figure 6-1b shows the situation some time t_1 after the etch has begun. In the areas between the resist lines, the film has been eroded to a depth R given by

$$R = rt_1 \qquad (6\text{-}1)$$

Next to the resist line, downward etching is not occurring, but etching does occur laterally and obliquely at a uniform rate in all directions. As a result, at each edge there is a quarter-circle profile of radius R centered on the lower corner of the resist line. Etching of this type is called *isotropic*, because it occurs in all directions equally. It is typical of wet chemical etching.

In Fig. 6-1c, etching of the film is complete. This occurs at a time τ given by

$$\tau = \frac{z}{r} \qquad (6\text{-}2)$$

At this time, $R = z$, and the *top* of the etched cut extends under the photoresist by a distance equal to z, a phenomenon referred to as *undercutting*. The linewidth is thus reduced by $2z$. The *bottom* of the etched line is of nominal width.

To stop the etch at time τ requires pinpoint control, since any deviation in film thickness or etch rate can cause incomplete film removal. It is customary, therefore, to *overetch* the film by an amount sufficient to compensate for any process variation. Such a situation is shown in Fig. 6-1d. Note that, since the HF etchant is totally selective and does not etch silicon, downward etching has stopped, but lateral etching continues, and the radius R increases with time. The degree of overetch, given by

$$\% \text{ Overetch} = \left(\frac{t}{\tau} - 1 \right) \times 100\% \qquad (6\text{-}3)$$

is frequently set at 20–50% and can be 100% for noncritical applications. Not only does it result in a decrease in the etched linewidth, but it also causes a steeper slope at the edge of the etched cut. Neither of these results is usually desirable.

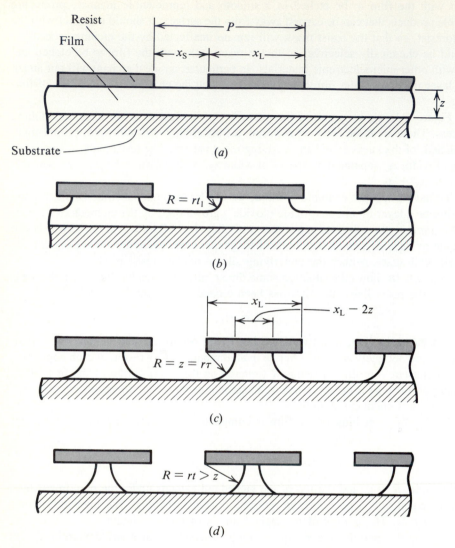

Figure 6-1 Isotropic etching in a wet chemical bath. (*a*) Unetched, masked film, showing parameters to be used. (*b*) Partially etched film, etch time t_1. (*c*) Ideally etched film, etch time $\tau = z/r$. (*d*) Film after overetch, etch time $t > \tau$.

Example A 0.34-μm (3400-Å) film of silicon dioxide is to be etched with a buffered oxide etchant of etch rate 850 Å min^{-1}. Process data shows that both the thickness and the etch rate may vary up to 10%. (*a*) Specify a time for the etch process. (*b*) How much undercut will occur at the top of the film?

The ideal etch time is

$$\tau = \frac{3400 \text{ Å}}{850 \text{ Å min}^{-1}} = 4 \text{ min}$$

With 10% expected variation in two independent variables, at least a 20% overetch should be added, to give 4.8 min. Under these conditions, the undercut will be

$$850 \text{ Å min}^{-1} \times 4.8 \text{ min} = 4080 \text{ Å} = 0.408 \ \mu\text{m}$$

Etching of Multiple Layers

When a film is to be applied over an etched step later in the process, steep slopes can lead to thinning over the step. Therefore, various methods have been developed to produce shallower slopes. One method involves the wet etching of two films of dissimilar etch rates. Figure 6-2 shows this situation. The upper film is thin and fast-etching, with thickness z' and etch rate r'. The lower film etches at a rate r appreciably less than r' and has thickness z.

What will be the edge profile after etching such a bilayer system? We can easily identify two extreme cases. In the open space between lines, the thin film is etched almost immediately, and the thicker film is etched downward almost as if the thin film were not there. Therefore, in the open space and near the resist edge, we have "normal" etching with $R \simeq rt$. At the surface, just under the resist, the upper film will etch laterally just as illustrated in Fig. 6-1, with a quarter-circle profile of radius $R = r't$. Since the film is so thin, its profile rapidly approaches the vertical.

At intermediate locations, the situation is more complicated, as shown in Fig. 6-2b. Areas of the thicker film that would not normally be reached by the etchant become exposed as the thin film is etched from above it. The etchant penetrates the film along a path like S' and gives rise to a gently sloping profile extending between

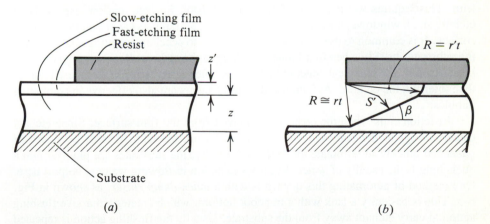

(a) (b)

Figure 6-2 Etching of a bilayer film. (a) Unetched, masked film. (b) Generation of a sloped profile as the fast-etching surface layer recedes. β defines the slope of the edge; S' is a typical path by which etchant reaches the sloped area.

the two end points. In the limit of a thin upper layer, the exact profile can be shown to be a straight line with

$$\sin \beta = \frac{r}{r'} \tag{6-4}$$

with β defined as in Fig. 6-2.

The bilayer system is quite useful when intentionally employed to produce sloped profiles. However, an effect of this kind, unintentionally produced, can be disastrous. For example, consider what happens if resist adhesion is lost during etching. If the etchant attacks the bond between the substrate and the resist, the etchant can penetrate rapidly along the resist-film interface. The result can be treated as a bilayer system, with an upper layer of zero thickness and an etch rate given by the rate of penetration under the resist. The resulting etch profile is undercut deeply under the resist, drastically decreasing the etched linewidth. In extreme cases, the undercut regions on two sides of a line can meet, lifting the resist away completely. In a translucent film like silicon dioxide, a shallow profile is easily recognized in a microscope by a widely spaced pattern of interference fringes. Any such unexpected shallow profile is a strong indication of resist adhesion failure.

Rate and Time

We have now developed several equations for the shape of etched lines, each involving the two variables of time and etch rate. These are in fact the key parameters in etching, and each of them deserves some further discussion.

Time seems quite straightforward. For immersion in an etchant bath, the time presumably starts upon immersion and ends on removal. However, there are some considerations that can alter the true time of etching from the apparent time.

Etching begins when the etchant comes into intimate contact with the substrate film. This requires wetting of the film by the etchant. Surface tension, especially in etching small windows, may retard wetting, as may air bubbles caught on the substrate surface. It is common to use a surfactant in etchants to accelerate wetting and to agitate the substrates briskly when first immersing them in the etchant. Reaction time can also be affected by the chemical generation of gas bubbles at the etch surface; these bubbles then interrupt the etching. Agitation and wetting agents are also effective in combating bubbles.

Reaction ends when the etchant is removed from the film surface. Since etchant usually clings to the substrate as it is removed from the bath, it must be rinsed or *quenched* quickly to terminate the reaction. Usually, the substrates are plunged into a quenching bath, usually of water. Vigorous agitation in flowing water is helpful here. One method of automating this quench is with a *quick-dump rinser,* as shown in Fig. 6-3. This is basically a tank with a trapdoor bottom, which creates a massive flushing action to carry etchant away from the substrate. Usually the flushing action is repeated several times to ensure complete rinsing.

Not only the control of time, but the specification of time, is a problem in wet etching. Given a known etch rate and thickness, it is elementary to select a time to give

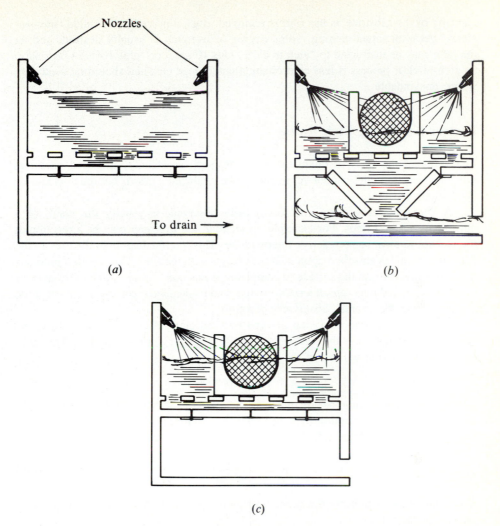

Figure 6-3 A quick-dump rinser for thorough quenching of the etching reaction. (*a*) Full rinser before immersion of substrates. (*b*) Quick-dump action for massive flushing. (*c*) Rinser refilling for repeat of dump cycle.

the desired degree of overetch, using Eq. (6-3). This enables production operators to run an etch operation using a timer, a simple, unambiguous, and thus ideal situation. However, as we shall see, reaction rates are not always sufficiently predictable to allow use of this "etch-to-time" method.

In such cases, the etch must be controlled by observing the occurrence of complete film removal, each time the etch is performed. This point is the *end-point* of the etch, and the technique is *end-point detection (EPD)*. In wet etching, end-point detection is usually performed by the eye of a human operator, who observes a color change or *clearing* of the substrate as the film is removed. Such a procedure is called "etch-to-clear." As with timed etching, some degree of overetch is usually desired, and an operator may be instructed to "etch to clear, plus 10 sec," or "plus 10%." Obviously, an etch-to-clear process is less controllable than a timed etch and should be used only where unavoidable.

Factors Governing Etch Rates

The study of the rates of chemical reactions, such as etching, is called chemical kinetics. Reaction rates cannot yet in general be predicted a priori, but much is known about their behavior.

A chemical reaction, as macroscopically observed, is usually the result of a number of reaction steps. For example, in etching, the first step may be a reaction in the etchant to form some reactive intermediate. This intermediate then diffuses to the film surface and is absorbed upon it. It then reacts with the film. The reaction products must then desorb from the surface to complete the reaction. Usually, one of these steps is the slowest, and thus *rate-determining* step, and its detailed mechanism may be quite different from the overall, observable reaction.

Let the rate-determining step involve two reactants, *A* and *B*. (These reactants could be—or not be—the etchant molecule and the film molecule.) The rate-determining reaction will be of the form

$$iA + jB \rightarrow \text{products} \tag{6-5}$$

where *i* and *j* are the number of molecules involved. The rate of the reaction can be measured by the consumption (negative change with time) of species *A* (or *B*), and so we can write the reaction rate as

$$r = -\frac{1}{i}\frac{d[A]}{dt} \tag{6-6}$$

where the brackets indicate the *concentration* of species *A*. A typical rate expression will be of the form

$$r = -\frac{1}{i}\frac{d[A]}{dt} = k'[A]^i[B]^j \exp\frac{-E_A}{RT} \tag{6-7}$$

Here k' is an elementary rate constant, which depends on the shape and size of the reacting molecules. *T* is the temperature, and *R* is the *gas constant*, with a value of 1.987 calories mole^{-1} °K^{-1}. E_A is the *activation energy*. This is the minimum energy that must be invested to initiate the reaction. If the reaction is a net consumer of energy, then the activation energy must at least equal the energy consumed. But even an energy-producing reaction will usually have an activation energy, because some energy is required to bring the reactants together and overcome chemical or physical barriers to their bonding. Activation energy is essentially constant for a given reaction.

Clearly, temperature is a major determinant of reaction rate, with an effect determined by E_A. For many common reactions at room temperature, E_A is of the order of tens of kilocalories per mole. This leads to the rule of thumb that reaction rates double with every 10°C of added temperature. A ±1°C temperature change can thus change rates on the order of 10%, and temperature control is important in etching reactions. Most etches near room temperature will require control to ±1°C.

The other determinant of reaction rate is concentration. Before discussing the effects, we should describe the measurement of concentrations in solutions. A straightforward measure of concentration is the mass of reactant in a unit volume (e.g., grams per liter). However, chemicals actually react not by weight, but molecule by molecule. It is therefore more fundamentally meaningful to describe concentrations in terms of molecules per volume.

The two systems are interconverted using the concept of *molecular weight*. Each type of atom has an atomic weight, equal to the number of nucleons in its nucleus: hydrogen has one, carbon twelve, etc. The molecular weight of a molecule is the sum of the atomic weights comprising it. The number of molecules in a given mass of reactant can be found by dividing the mass by the molecular weight.

Since the weights of atoms are inconveniently small, it is conventional to work in terms of *moles*. One mole is the quantity of molecules contained in one *gram-molecular-weight*. One gram-molecular-weight is a mass, in grams, equal to the numerical value of the molecular weight. For example, oxygen atoms have eight nucleons, and the O_2 molecule has a molecular weight of 16. Hydrogen has one nucleon, and the H_2 molecule has a molecular weight of 2. Sixteen grams of oxygen, or two grams of hydrogen, equal one gram-molecular-weight and contain 1 mole of molecules. Due to humanity's lack of foresight in defining the gram, one mole contains the inconvenient number of 6.02×10^{23} molecules. When we deal with concentrations in a scientific discussion, we shall use mole concentrations (units of moles per liter) unless otherwise noted.

Both mass and moles are rational units for concentration. However, in common industrial practice, neither of these systems is used. Instead, the concentration of a solution is frequently given in terms of the *volume* ratio of the reagents mixed to form it. (These reagents are frequently solutions themselves, and their concentration also varies, being determined by convenience in preparing and shipping the chemical. Table 6-1 gives the concentrations of common reagents as sold, together with their molecular weights and some other properties.) For example, a "six-to-one" mixture of water and hydrofluoric acid would contain six parts by volume of pure water and one part by volume of commercially available hydrofluoric acid, of 49% concentration (by weight) in water. The resulting mixture would *not* be exactly seven parts of volume, because volume is *not* conserved in mixing of chemicals. This system is convenient for the person who mixes the chemicals, but not for the engineer attempting to understand molecular concentrations in the etchant.

Concentrations in Acids

Most microelectronic etchants are acids. Acids display some interesting concentration effects that are important in the etching process.

Most materials that dissolve in water undergo *dissociation:* the molecule breaks up into various charged portions or ions. An acid is, by definition, a substance that dissociates in water to liberate a hydrogen ion, H^+. A generic formula is

Table 6-1 Properties of common chemical reagents

Name	Formula	Molecular weight	Concentration†	Dissociation constant‡
Hydrofluoric acid	HF	20.0	49%	3.53×10^{-4}
Nitric acid	HNO_3	63.0	69.5%	(Strong acid)
Acetic acid, "Glacial"	$H_4C_2O_2$ CH_3COOH	60.0	99%	1.76×10^{-5}
Sulfuric acid	H_2SO_4	98.1	98%	1.20×10^{-2}
Phosphoric acid	H_3PO_4	98.0	85%	7.52×10^{-3}
Ammonium fluoride	NH_4F	37.0	40%	Salt, dissolves in water
Ammonium hydroxide	NH_4OH	35.05	29%	1.79×10^{-5}

† Concentration by weight, in water, as commonly supplied.
‡ At 25°C. For multibasic acids, the first dissociation constant is given.

$$HA \rightarrow H^+ + A^- \tag{6-8}$$

where A is the negatively charged fragment of the molecule (or *anion*). Examples include

$$HF \rightarrow H^+ + F^- \quad \text{hydrofluoric acid} \tag{6-9}$$

$$HNO_3 \rightarrow H^+ + NO_3^- \quad \text{nitric acid} \tag{6-10}$$

The hydrogen ion is a strong oxidizing agent, which accounts for the reactivity of acids. It usually appears in the rate-determining step of an acid reaction, even when it does not appear in the overall chemical equation, and its concentration $[H^+]$ is of major importance. The negative logarithm of this concentration, $-\ln[H^+]$, has the special name *pH*. The concentrations of chemical species in a reaction are determined by an *equilibrium constant K*, which is fixed for a given reaction at a given temperature. For a reaction of form

$$iA + jB \rightarrow kC \tag{6-11}$$

the equilibrium constant is given by

$$K = \frac{[A]^i[B]^j}{[C]^k} \tag{6-12}$$

Thus for the acid dissociation reaction of Eq. (6-8), the equilibrium constant is

$$K_a = \frac{[H^+][A^-]}{[HA]} \tag{6-13}$$

This equilibrium constant K_a has the special name of the *acid dissociation constant* or *ionization constant*.

Acids can be classified as *strong* or *weak* depending on their ionization constant. For strong acids, the constant is very large, and they are almost entirely dissociated in water. If $[HA]_0$ is the concentration of the acid in the original reagent, then, for a strong acid,

$$[H^+] \simeq [HA]_0 \qquad (6\text{-}14)$$

In a weak acid, the dissociation is very incomplete, and much of the original reagent is undissociated. We can find the hydrogen ion concentration by noting that

$$[H^+] + [HA] = [HA]_0 \qquad (6\text{-}15)$$

because all of the reagent is either dissociated or undissociated. Also, because of Eq. (6-8),

$$[H^+] = [A^-] \qquad (6\text{-}16)$$

Substituting into Eq. (6-13) gives a quadratic equation for $[H^+]$, which can be solved to give the hydrogen ion concentration in an unbuffered solution of a weak acid:

$$[H^+] = \frac{1}{2}(-K_a + \sqrt{K_a^2 + 4K_a\,[HA]_0}) \qquad (6\text{-}17a)$$

$$\simeq -\frac{K_a}{2} + \sqrt{K_a\,[HA]_0} \qquad \text{weak, unbuffered} \qquad (6\text{-}17b)$$

Table 6-1 gives the dissociation constants for common acids. We note that both acetic and hydrofluoric acids are weak, with dissociation constants $\sim 10^{-5}$. The actual amount of hydrogen ion in the solution at any time is thus quite limited.

Because of the low value of $[H^+]$ in weak acids, the pH and thus the etch rate is quite vulnerable to change. Small dilutions, or consumption of the reactant during etching, or other factors can significantly alter pH. Such deviations can be limited by the technique of *buffering* the solution.

Buffering involves the use of a *salt* of the acid. A salt is a substance made up of an acid fragment A and a basic fragment B, which dissociates completely in water as follows:

$$BA \rightarrow B^+ + A^- \qquad (6\text{-}18)$$

If the acid fragment A is the same as that in Eq. (6-13), it will help determine the pH. If we add the salt in approximately the same proportions as the acid, then the contribution to $[A^-]$ from the salt will overwhelm that from the dissociation of the weak acid and

$$[A^-] \simeq [BA]_0 \qquad (6\text{-}19)$$

Then, from Eq. (6-13),

$$[H^+] \simeq K_a\,\frac{[HA]_0}{[BA]_0} \qquad (6\text{-}20)$$

Because this equation contains the *ratio* of two concentrations, dilution will not affect pH. A buffered solution is also more stable against consumption of the reagent, since further dissociation can supply additional H^+ without changing the A^- concentration much. Note that the effect of buffering depends on the roughly equal concentrations of

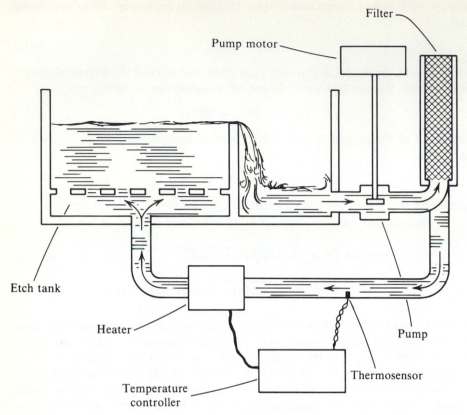

Figure 6-4 Recirculating, filtered, temperature-controlled bath for the etching of microelectronic substrates.

salt and acid. "Buffered" solutions of 50:1 concentration are more of psychological than of chemical benefit.

Even if buffering maintains control of the bulk concentration in the etch bath, there is a tendency for reactant to be depleted in the immediate vicinity of the reaction. Therefore, good etching control is encouraged by circulating or agitating the acid, to provide a constant supply of fresh etchant. The requirements of circulation, temperature control, and also of particulate control can be met by an apparatus such as that in Fig. 6-4. Such equipment is common in microelectronic etching. A pump continually recirculates the etchant, passing it through a filter that removes particles and a thermostatically controlled heater that regulates the temperature. Using such equipment, buffered HF etch baths can be used for weeks before acid replacement is required. When filtering is not used, it is advisable to change etch baths at least daily.

Chemical Safety

Chemical etchants are by their nature corrosive and hence hazardous. The precautions required to use them safely are simple, but they *must* be observed. The microelectronic engineer is obligated, not only morally but legally, to ensure that the processes he establishes protect the safety of those who operate them.

Anyone working with acids should wear safety glasses or goggles, gloves, and arm and body protection (aprons, sleeve guards). Immersion of the gloved hand should never be required: handles must be provided to prevent this requirement. Adequate ventilation must be provided to isolate all acid fumes from the workplace. A barrier to splashing of acid is also advisable.

Most accidents with acids happen while filling or emptying etch baths. Persons pouring acids should always wear a full face shield, in addition to the other protective equipment, to protect against splashing. Mixing of acids should occur in the proper order, and in a sink or station where spills, container breakage, and fumes can be controlled. Glass acid bottles should always be transported in buckets so that breakage does not cause a spill, and all chemical processing areas should be provided with materials for spill control.

Emptying of baths should not require lifting and pouring, but should be possible by operating a remote control. For some acids, *aspiration* is appropriate, in which the acid is sucked from the bath by a water-driven suction device. For acids that react strongly with water, or heated acids, aspiration is inappropriate and a remote-controlled drain should be provided. A qualified professional should be involved in waste etchant disposal to ensure that both safety and environmental regulations are satisfied.

Figure 6-5 shows a typical industrial etch station that incorporates these principles. Note the exhaust vent, which draws air through the perforated tabletop, sweeping fumes away from the workspace. Waste acid discharges into the lower portion of the sink, either from the drain or the aspirator. A *plenum rinse* (not shown) is usually provided to wash out the lower chamber or plenum when required. Also note the drain-exhaust configuration, which helps separate liquid and gaseous effluent.

Special precautions are required with hydrofluoric and with sulfuric acids, which are mentioned in Sections 6-2 and 6-5, respectively. The reader should be thoroughly familiar with these precautions before using these acids.

6-2 CHEMICAL ETCHANTS

There are a large number of materials used in microelectronic processing, and a larger number of chemicals available to etch them. An extremely comprehensive list has been given by Walter Kern and Cheryl A. Deckert.[1] Nonetheless, most microelectronic professionals, most of the time, are engaged in etching only a few films, and we shall concentrate on these. They include silicon dioxide, silicon nitride, silicon in either single-crystal or polycrystalline form, and aluminum.

Silicon oxide can be thermally grown on a silicon crystal, as described in Chapter 3. Both silicon oxide and nitride can also be deposited as a film by chemical vapor deposition, as we shall see in Chapter 8. These films can be used as diffusion masks or as insulating layers to isolate conductors laid on the substrate surface. Aluminum is the primary metal used to fabricate these conductors. It is deposited by the various vacuum

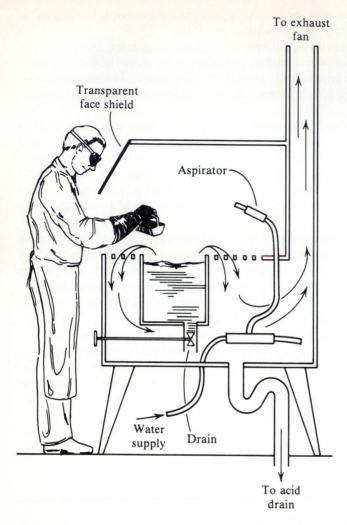

Figure 6-5 A workstation for acid etching. Note the face shield to prevent splattering, the perforated top for fume control, the remote-controlled drain and aspirator for acid removal, and the arrangement of drain and exhaust systems.

methods covered in Chapter 7. Polycrystalline silicon, which is also formed by chemical vapor deposition, can also serve as a conductor.

For each of these materials, there are various possible etchants. Ideally, we desire a chemical that will etch the film smoothly and reproducibly, at a predictable rate, while not attacking photoresist or other common microelectronic materials. It should be chemically stable and safe to handle. Furthermore, we want to avoid chemistry based on the formation of trace or unstable intermediates, which can decay over time or vary with mixing conditions. As we look in turn at each of these films, we shall find (surprise!) that our ideal etchant is not available, and we will discuss the trade-offs required for effective etching.

Etching Thermal Silicon Dioxide with Hydrofluoric Acid

It is the ease of etching thermal silicon dioxide ("oxide") with hydrofluoric acid solutions (HF) that has made silicon the premier microelectronic material. Thermal oxide is oxide grown by the high-temperature oxidation of silicon (Chapter 3) as opposed to deposition from a vapor (Chapter 8). It is an extremely consistent, homogeneous film and, unless heavily doped, etches at a predictable rate in a manner very close to that described by the theory we have just studied.

Oxide is etched using solutions of hydrofluoric acid. The nominal reaction is

$$SiO_2 + 4HF \rightarrow SiF_4 + 2H_2O \tag{6-21}$$

It is customary to buffer the HF with *ammonium fluoride*, NH_4F, a salt that dissociates to fluoride ion (F^-) and ammonium ion (NH_4^+). Volume ratios of six or seven parts NH_4F to one part HF are customary. This mixture is often called buffered oxide etch (BOE). BOE is chemically stable and can be stored for considerable time without degradation. However, ammonium fluoride is of limited solubility and is prone to crystallize out of BOE solutions slightly below room temperature. Therefore, thorough mixing is required, and storage conditions, particularly in cold weather, deserve attention.

BOE is a reasonably selective etch for oxide. It will not etch silicon, so the removal of an oxide film from a silicon crystal is self-limiting. However, HF does attack silicon nitride, and caution is required in etching combined oxide-nitride structures. HF also attacks photoresist to some extent, although the addition of ammonium fluoride reduces this effect. At room temperature, resist adhesion is usually satisfactory up to etch times of perhaps 30 min on undoped oxide.

The etch rate is approximately 1000 Å min^{-1} at room temperature. (In most of this book, we have suppressed the somewhat disreputable angstrom, but the measurement of etch rates in angstroms per minute is so common that we have used it throughout this chapter.) Figure 6-6 shows etch rates as a function of temperature for a 6:1 BOE mixture. It has been shown[2] that the etch rate depends not directly on the HF concentration, but on the concentration of the HF_2^- ion, which is formed in solution by the reaction of HF with the dissociated fluoride ion F^-.

Because thermal oxide etch rates are so predictable, oxide etching can usually be specified by time. When end-point detection is required over silicon, we can make use of the wetting properties of silicon and of oxide. Oxide is *hydrophilic* and is easily wetted by water. Silicon is *hydrophobic* and repels water. Therefore, a completely etched silicon substrate, dipped in water, will shed the water instantly when removed. By contrast, a substrate with even 100 Å of oxide on the surface will remain wet.

Oxide etch rates will increase severalfold when phosphorus is present in the oxide. If the phosphorus is concentrated near the surface, as after a phosphorus diffusion, the result is effectively a bilayer etching process and a sloped profile. This effect is sometimes exploited to produce sloped contacts after a phosphorus diffusion. How-

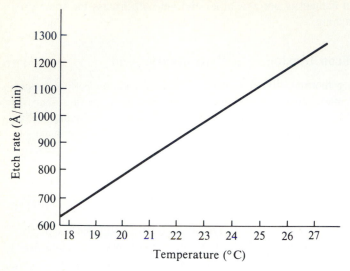

Figure 6-6 Etch rates as a function of temperature for 6 : 1 buffered oxide etch (BOE). (*Figure courtesy of and reproduced by permission of Intel Corporation. All rights reserved.*)

ever, it is difficult to control, due to the unpredictability of the doped etch rate, and the loss of resist adhesion on phosphorus-doped oxides. If the fast surface etch rate is not desired, it is wise to "dip off" the phosphorus layer before applying photoresist. This can be done with a brief HF dip, which strips off the fast-etching doped layer without significantly penetrating the slower-etching, underlying oxide.

Safety Precautions for HF Solutions

Hydrofluoric acid is unusual in that it does *not* cause pain on contact with the skin. However, it will penetrate the body and can cause severe damage to tissue and to bone if not immediately treated. As a result, victims of HF burns can be unaware of their exposure until the pain of bodily damage becomes apparent many hours later.

Therefore, it is vital that persons using HF be extremely cautious (i.e., paranoid) about the possibility of exposure. If there is any suspicion of contact with HF, contact should be *assumed* and appropriate measures taken.

Etching of Chemical-Vapor-Deposited Silicon Dioxide

Silicon dioxide can also be deposited by chemical vapor deposition (CVD), using methods we shall study in Chapter 8. This material goes by many names, including CVD silica, Pyrox, and Vapox. It is frequently doped heavily with phosphorus, and called glass, phosphosilicate glass, p-glass, or Pyroglass. It may also be doped with other materials. By any name, CVD oxides are less chemically stable and predictable than thermal oxides. They tend to be porous and inhomogeneous and of nonuniform thickness. They are also often nonstoichiometric (that is, their chemical formula is not exactly SiO_2 but SiO_x with $x \simeq 2$), and contain impurities from the deposition process such as hydrogen.

For these reasons, although CVD oxide is still etched in HF solutions, its etch rate is much faster under given conditions than thermal oxide. A thermal treatment or *densification* of the oxide, for about 15 min at 1000°C or more will reduce the etch rate and also make it more reproducible. Such a treatment can also reduce the surface phosphorus concentration of a P-doped oxide, improving resist adhesion. Densification can be useful when only high-temperature materials are on the substrate (such as in MOS fabrication), but are not feasible if aluminum or some other low-melting material is present.

Etch rates (in angstroms per minute) of CVD oxides in BOE are about[2]

	Undensified	Densified
Undoped	3000	1000
Phosphorus-doped (8 mole %)	5500	3000

Actual rates vary with the nature of the deposition process and must be determined individually. However, it is usually possible to run a CVD-oxide etch by time once the film has been characterized. One must be alert to any changes in the deposition process or subsequent thermal history, which can change the etch rate. To obtain a reasonable etch time with these high etch rates, CVD oxides are usually etched with a diluted form of BOE. If the oxide is deposited over aluminum, additional proprietary chemicals are often added to reduce acid attack on the metal.

Phosphorus-doped CVD oxides are quite common and deserve some comment. It has been shown that the etching consists of attack on the SiO_2 by the HF component coupled with dissolution of the phosphorus-containing P_2O_5 by water in the etch. As a result, the etch rate is linearly dependent on the phosphorus content.[3] Measuring the etch rate of densified p-glass in a controlled etchant can in fact be used as a quick test of the phosphorus content of the glass.

It is often desirable to produce sloped etch profiles in CVD oxides. One way to do so is to produce a bilayer situation, by increasing the phosphorus doping of the top layer of the oxide. Other ways of accelerating surface etch rates have also been tried, including ion bombardment of the oxide surface.[4] When selective removal of a phosphorus-rich, surface layer is required, an alternate formulation is P-Etch,[5] which is a 2:3:20 mixture of nitric acid, HF, and water. This solution is quite selective for phosphorus oxides over thermal oxide. However, HF-nitric acid solutions will etch silicon, so use of P-Etch is risky when exposed silicon is present on the substrate.

Etching of Silicon Nitride

Silicon nitride (Si_3N_4) is a deposited film that can be formed either under furnace conditions around 1100°C or in plasma-assisted CVD reactors at several hundred degrees. Nitride properties vary greatly with deposition conditions, and are extremely sensitive to the amount of hydrogen or oxygen incorporated into the film. It is possible to etch silicon nitride in HF solutions such as BOE (Ref. 6). Etch rates in BOE are slow, around $5-15$ Å min^{-1}. (In concentrated HF, the rate rises to 15 to 1000 Å min^{-1}.)

HF etching of nitride is complicated by the fact that nitride and oxide are frequently used in alternating layers. Consider the use of BOE to etch a 1000-Å nitride layer lying over an oxide layer. Nitride etching takes roughly 100 min. Once the nitride is removed at any spot on the substrate, oxide etching commences at rates of around 1000 Å min^{-1}, so even a 1% error in end point removes large amounts of oxide. Also, the 100-min etch time may well exceed the resist endurance in BOE. Therefore, BOE etching of nitride over oxide is usually not feasible. Conversely, however, it is quite practical to etch oxide over nitride with BOE.

A preferable wet etchant, from a selectivity point of view, is refluxing phosphoric acid, H_3PO_4. Refluxing is the process of continually boiling the acid while collecting the vapor on a condenser and letting it flow back into the etch tank. The reaction thus takes place at the boiling point of the solution, which depends on the water content in the acid and is around $130-150°C$. This solution etches nitride at around 100 Å min^{-1}, thermal oxide at $10-20$ Å min^{-1}, and silicon at about 2 Å min^{-1} (Ref. 7). The selectivity is thus appropriate for most purposes. However, the temperature and thus the rate are not easy to control, and the end point is difficult to observe directly in the refluxing environment.

Another disadvantage is that boiling phosphoric acid quickly destroys photoresist, requiring the use of an *auxiliary mask*.[7] Figure 6-7 illustrates this technique. A layer of silicon dioxide is deposited over the nitride layer to be patterned, and the oxide is then masked using standard photolithography. Next the oxide is etched to reproduce the masked pattern. Now the resist can be removed, and the oxide serves as an etch mask for the patterning of the nitride. If it is undesirable to leave the oxide in place, it can now be dipped off in BOE.

There are significant disadvantages to both processes for wet etching of nitride. For this reason, nitride was an early candidate for etching with plasma. Today plasma is used for most nitride patterning operations, although wet methods are still useful for stripping of nitride layers and other noncritical applications.

Etching of Aluminum in Acidic Solutions

Metallic films are used to form conducting connections between elements of microelectronic circuits, and aluminum is the most common metal in use. Aluminum is deposited either by sputtering or evaporation. In either case, there is a distinct and observable crystal grain structure to the film. Aluminum etch rates are difficult to control, for a number of reasons.

The first has to do with the nature of metallic etching by acids. When metals are etched, there is a transfer of charge among the reactants. The basic reaction involves the hydrogen ion of the acid and is

$$3H^+ + Al\ (metal) \rightarrow \tfrac{3}{2}H_2 + Al^{3+} \tag{6-22}$$

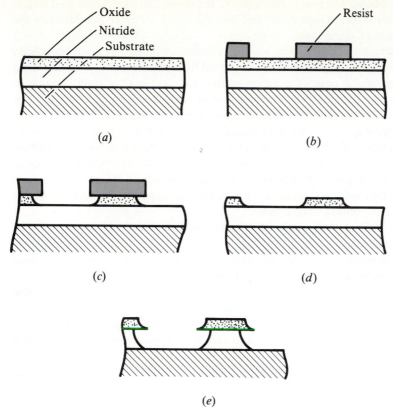

Figure 6-7 Use of an auxiliary mask for etching in resist-destroying etchants. (*a*) Nitride layer with oxide film deposited upon it. (*b*) Photomasking of the oxide layer. (*c*) Etching of the oxide layer. (*d*) Resist removal. (*e*) Etching of the nitride layer using the oxide layer as a mask.

The removal of electrons from the metal to form a soluble positive ion is called *oxidation* and occurs at the site of etching. The addition of electrons to the hydrogen ion to form hydrogen gas (*reduction*) can occur at any site electrically connected to thé etch site. The etch rate is thus affected by the electrical environment of the metal. Etch rates of aluminum on bare silicon, which conducts electricity, can be twice that of aluminum over an insulating oxide layer. The situation can be further complicated by the presence of other metals or by the built-in potential of devices on the substrate.

Such electrochemical effects are common to all metals, but aluminum has some unique complications. Most metals are reasonably unreactive, and the challenge of etching them is to find a powerful enough acid (with a sufficient *oxidation potential*) to convert them to ions. Aluminum is highly reactive, and a bare aluminum surface would in fact react violently with water just as sodium does. Aluminum remains stable because a bare metal surface is immediately covered with a thin, inert, protective layer of aluminum oxide (Al_2O_3), which prevents reaction. To etch aluminum, we thus require a chemical that will continually dissolve the aluminum oxide layer as it forms. Because the oxide layer is much more inert than the metal, its properties dominate the apparent etch rate of aluminum.

The grain structure of aluminum also complicates etching. The grain boundaries differ chemically from the grain interiors, and tend to etch more quickly. Attack along grain boundaries at the edge of a patterned line can "cut out" a grain, leading to *notching* of the metal lead. (Metal notching due to grain boundary etching should be distinguished from notching due to reflection during photolithography. The two effects are similar in appearance and are often confused, but the reflectivity effect requires the presence of a nearby reflecting geometry.) On the other hand, grain boundary etching of a grain sitting atop an insulating layer can electrically isolate it from its neighbors. If this happens, its etch rate will slow and it may become very difficult to etch. Such scattered, refractory unetched grains have a distinctive appearance under the microscope and are descriptively called snow.

Incomplete etching of aluminum can also be caused by the formation of hydrogen bubbles according to Eq. (6-22), which can cling to the surface during etching and stop the reaction. The result will be circular areas of unetched aluminum on an otherwise properly etched substrate, frequently called snowballs.

Aluminum is commonly etched with a mixture of phosphoric acid (H_3PO_4), nitric acid (HNO_3), and acetic acid ($C_2H_4O_2$ or CH_3OOH). The exact composition varies, and most establishments have their own favorite, proprietary recipe that is felt to optimize resist adhesion, line definition, and etch rate. Typical ranges of composition and conditions are

Phosphoric acid	70–80% (by volume)
Deionized water	10–20%
Acetic acid	3–7%
Nitric acid	3–7%
Temperature	40–50°C
Etch rate	3000–10,000 Å min^{-1}

A mixture of this type is totally unreactive with silicon, oxide, and nitride. Not being a strong oxidant, it reacts only slowly with most other metals. However, it will attack gallium arsenide, so hydrochloric acid (HCl) solutions are used to etch aluminum over GaAs.

Because the etch rate of aluminum is so undependable, aluminum etching is almost always done "etch-to-clear." Usually the end-point detector is the eye of a human operator, who is also charged with agitating the etching substrates to prevent bubble formation. As a result, aluminum etching is more an art than a science, greatly dependent on the practiced skill of the etcher.

Etching of Polycrystalline Silicon

Polycrystalline silicon (frequently called poly) is commonly used as an interconnect, a fuse, or to define a gate structure on a MOS device. It is deposited by chemical-vapor-deposition methods and has a distinct grain structure. It thus displays the unpredictable etch rate of a CVD film and the grain-structure complications found with aluminum. It also has the low reactivity typical of silicon, so that it cannot be etched with hydrofluoric acid alone.

Silicon can, however, be etched by adding nitric acid to HF. Nitric acid partially decomposes to form small amounts of nitrogen dioxide (NO_2), which then oxidize the silicon (in the electrochemical sense) to form silicon ions:

$$Si + 2NO_2 \rightarrow Si^{2+} + 2NO_2^- \qquad (6\text{-}23)$$

The ions then pick up hydroxyl (OH^-) ions, which are always present in solutions containing water, to form silicon dioxide:

$$Si^{2+} + 2OH^- \rightarrow SiO_2 + H_2 \qquad (6\text{-}24)$$

The oxide in turn dissolves in the hydrofluoric acid according to Eq. (6-21).

Polysilicon etches are mixtures of hydrofluoric and nitric acids, diluted in either water or acetic acid. The proportions of the reagents can vary widely. The etch rates for undoped poly range from 1500 to 7500 Å min^{-1}; phosphorus-doped poly etches even faster. Poly etches will attack single-crystal silicon and oxide. However, by using a dilute mixture (slow silicon etch rate) with a low HF concentration (slow oxide etch rate), good selectivity can be obtained.

The active ingredient in polysilicon etch reaction is NO_2, a trace decomposition product of the nitric acid. The concentration of such trace products, and thus the properties of the etch, are strongly dependent on the history of the etchant. In particular, the etch loses strength after a short time. Practical poly etching may thus involve precise rules for mixing, a complicated "aging" ritual, and rigid time limits, in order to maintain reproducible behavior. Between the variations in the etchant, the etch rate uncertainties of a chemical-vapor-deposited film, and the added complications of grain structure, polysilicon etching is highly empirical. It is usually done on an "etch-to-clear" basis using an operator's visual determination of the end point.

Etchants for Other Microelectronic Materials

We have now looked at a few etchants in enough detail to exemplify some of the complications inherent in wet etching. Of the multitude of additional applications, we have room to mention only an important few.

Single-crystal silicon can be etched isotropically in HF-HNO_3 mixtures, using the same mechanism as for polycrystalline silicon. The etch rates are, however, much slower. Figure 6-8 shows etch rates as a function of HF and HNO_3 composition, for water and acetic acid diluent.[8] Some microelectronic processes require etching silicon preferentially along certain crystal axes. There are a number of formulations for this purpose, a few of which are listed in Table 6-2.[9-11]

Gallium arsenide can be etched in a number of chemicals, but a "standard" etchant has yet to emerge. A good review is given in Ref. 12. Most GaAs etchants are at least somewhat anisotropic. Gallium arsenide etching is also complicated by the different chemical natures of the "arsenic" and "gallium" faces in the $\langle 111 \rangle$ direction. One useful etchant is bromine in methanol. Hydrogen peroxide has also been used in combination

Table 6-2 Some formulations for crystallographically selective etching of silicon

Ingredients	Composition	Temperature (°C)	Relative rate 100	110	111	Absolute rate†	Reference
KOH in water/ isopropanol	19 wt. % KOH	80	—	*400*	1	0.59 μm sec^{-1}	9
N$_2$H$_4$/water	100 g/50 mliters	100	*10*	1	—	0.3	10
Ethylenediamine/	17 mliters	110	*50*	30	3	50	11
Pyrocatecol/	3 g						
water	8 mliters						

† Etch rate for fastest-etching orientation, i.e., the one italicized under relative rate.

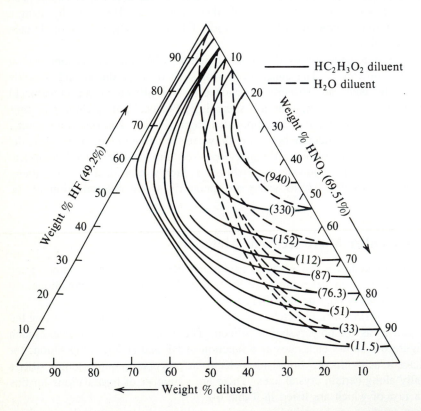

Figure 6-8 Etch rates for single-crystal silicon as a function of HF, HNO$_3$, and diluent concentrations. These etch rates are for dual-sided etching and must be divided by 2 to obtain the etch rate on a single surface. (*Reference 8. Reprinted by permission of the publisher, The Electrochemical Society, Inc.*)

Table 6-3 Common oxidizing agents for etching

Oxidizing agent: (oxidized form/reduced form)	Environment
Cl_2/Cl^-	Acidic
H_2O_2/H_2O	Acid or base
VO_2^+/VO_3^+	Acidic
Cr_2O_7/Cr^{3+}	Acidic
$FeCl_4^-/FeCl_3^-$	Acidic
$CuCl_4^{2-}/Cu_2Cl_2$	Acidic
$Fe(CN)_6^{3-}/Fe(CN)_6^{4-}$	Acid or base
$Co(CN)_6^{3-}/Co(CN)_6^{4-}$	Acidic
I_3^-/I^-	Acid or neutral
H^+/H_2	Acidic
Br_3^-/Br	Neutral
AsO_4^{3-}/AsO_2^-	Basic

Source: After Ref. 13, p. 83. This figure was originally presented at the Spring 1976 Meeting of the Electrochemical Society, Inc., held in Washington, D.C.

with a number of materials, including sulfuric acid in water, phosphoric acid in water, citric acid in water, ammonium hydroxide, and sodium hydroxide.

Metals other than aluminum are etched using an *oxidizing agent* (i.e., an electron acceptor) of sufficient strength to ionize the metal. The oxidizing agent depends on the metal to be etched and on other chemical considerations. The hydrogen ion formed in acid solutions is a good oxidizer and acids are a common choice, but there are a number of others as listed in Table 6-3 (Ref. 13).

Chromium is one metal of particular interest because of its frequent use in fabricating photomasks. Chromium, like aluminum, has a resistant, inert oxide, which must be removed before etching. Zinc oxide (ZnO_2) can be used to break down the oxide. Oxidizing agents such as cerric ion, basic permanganate, or basic ferrocyanide can then be used as etchants.

6-3 PLASMA ETCHING

Wet etching is conceptually simple, using inexpensive reagents and equipment; it requires limited skill and allows processing in large batches. This method has served microelectronics well. Yet it has significant disadvantages. Liquids are given to irreproducible disturbances from bubbles, flow patterns, etc. No really satisfactory wet etchants exist for some materials, such as silicon nitride. And wet etching requires the purchase, storage, handling, and disposal of large amounts of corrosive, toxic materials.

Most important, wet etching limits the minimum dimensions achievable for a microelectronic device. Because of the undercutting inherent in isotropic etching, wet-etched linewidths must be several times the film thickness. Since film thicknesses cannot be reduced indefinitely, submicrometer dimensions cannot be achieved with wet etching and require a new etching method: *plasma etching*.

A plasma is a neutral *ionized* gas; i.e., it contains appreciable numbers of free electrons and charged ions. To remain ionized, a gas must receive constant inputs of energy to offset the recombination of the charged particles in it. It must also be kept at low pressure, to reduce the collision rate and thus the recombination rate of the ions. Plasmas for etching are formed by applying a radio-frequency electric field to a gas held at a low pressure in a vacuum chamber.

Plasma formation begins with some free electrons. The applied field may be sufficient to strip them from gas atoms, or a spark may be applied to help *strike* the plasma. Once available, the electrons are accelerated by the applied field and collide with gas molecules (M) with several effects. In *ionization,* an electron is knocked loose, forming a positively charged molecule or *ion*:

$$e^- + M \rightarrow M^+ + 2e^- \tag{6-25}$$

In *dissociation,* a molecule (M_2) breaks down into smaller fragments, with or without ionization:

$$e^- + M_2 \rightarrow M + M + e^- \tag{6-26a}$$

$$\rightarrow M^+ + M + 2e^- \tag{6-26b}$$

In *excitation,* the molecule holds together, but absorbs energy and enters an excited electronic state (M^*):

$$e^- + M \rightarrow M^* + e^- \tag{6-27}$$

Because the energy of plasma electrons is much higher than a chemical bond energy, molecules in a plasma are essentially randomized, breaking down into all conceivable fragments. For example, a plasma of methane (CH_4) can be expected to include the fragments CH_3, CH_2, CH, H, and C, the equivalent ions (CH_3^+, etc.), plus reaction products among all of them.

Ions, molecular fragments, and excited molecules are all highly reactive and can undergo a variety of reactions. They are removed from the plasma by *recombination:*

$$e^- + M^+ \rightarrow M \quad (\text{or } M^*) \tag{6-28}$$

and by reaction with chamber walls:

$$M, M^+ + \text{wall} \rightarrow \text{products} \tag{6-29}$$

These reactions remove from the plasma energy, electrons, and reactants that must be continually replenished. Energy also escapes as light, giving plasmas a characteristic diffuse glow.

If there are microelectronic substrates in the plasma, they will also react with the energetic species. (Because of the high energy of ions and fragments, nothing is inert

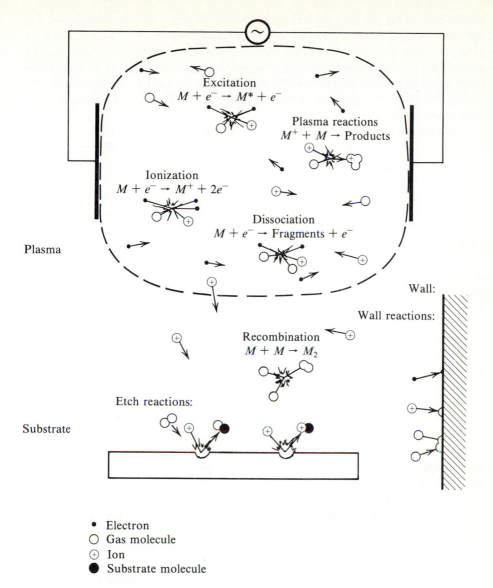

Excitation
$$M + e^- \rightarrow M^* + e^-$$

Plasma reactions
$$M^+ + M \rightarrow \text{Products}$$

Ionization
$$M + e^- \rightarrow M^+ + 2e^-$$

Dissociation
$$M + e^- \rightarrow \text{Fragments} + e^-$$

Plasma

Wall:

Wall reactions:

Recombination
$$M + M \rightarrow M_2$$

Etch reactions:

Substrate

- Electron
- ○ Gas molecule
- ⊕ Ion
- ● Substrate molecule

Figure 6-9 The nature of plasma etching, showing reactions in the plasma, reaction at walls and at the substrate surface, and the desorption of reaction products.

in a plasma.) If the reaction products are volatile under the plasma conditions, they will evaporate, resulting in a removal of substrate material. This is plasma etching. Figure 6-9 shows the nature of the process.

Because plasma etching requires a vacuum chamber and a radio-frequency (rf) generator, it is more complex and expensive than wet etching. Also, the high-energy plasma species are not very discriminating in their reactions, so selectivity is lost. On the other hand, a reactive plasma can be generated from the most safe and inert of gases, so plasma etching avoids the need for volumes of dangerous reactants. The primary advantage of plasma etching, however, is that the downward etch rate can be made much larger than the lateral etch rate. Undercutting can thus be minimized, and dimensions can be reduced independent of film thickness. This type of etching is called *anisotropic* etching.

Anisotropic etching is made possible by the physics, rather than the chemistry, of plasmas. Plasmas are full of mobile charge carriers and thus are conductors. Being a conductor, the interior of the plasma is at a uniform electric potential. However, the plasma cannot exist in contact with a material object, and is separated from the walls and substrates by a *sheath,* as shown in Fig. 6-10. Near the walls, both electrons and ions escape from the plasma to be neutralized on the wall. Electrons, being much smaller and more mobile, tend to escape more easily. This leaves a surplus of positive ions in the plasma, resulting in a positive charge. As the charge increases, the loss rate of electrons falls due to electrical attraction, and the loss of positive ions similarly increases. Eventually the loss rates become balanced, and the charge on the plasma stabilizes at some positive value. This *plasma potential* is usually from 10 to 200 V.

Due to the plasma potential, there is an electric field in the sheath, which lies perpendicular to any physical object immersed in the plasma. Ions leaving the plasma are accelerated across this potential, acquiring velocity parallel to the field. As a result, any physical object in the plasma, such as a microelectronic substrate, is bombarded with a hail of ions perpendicular to its surface. It is this directional bombardment that makes anisotropic etching possible.

Anisotropy is encouraged by ion bombardment in a number of ways, as illustrated in Fig. 6-11. At sufficiently high energies, the bombarding ions simply erode the surface they strike. This process is called *sputtering* and will be revisited in Chapter 7. Sputtering involves the breaking of chemical bonds due to impact and is very unselective. If selective etching is desired, energies must be kept low enough that sputtering is not the predominant process. At lower energies, ion bombardment can promote chemical etching by locally heating the substrate and by loosening chemical bonds. In many plasma systems, an inert residue tends to form on the substrate; in these cases, ion bombardment can promote anisotropy by sputtering the protective residue from exposed areas.[14]

The outcome of a plasma etch process depends on the balance between the chemical reaction of the plasma species with the substrate and the physical process of ion bombardment. Figure 6-12 illustrates some of the complexities involved in this balancing. At the center of the figure is shown the central event in etching: reaction of the film material. To the left is the reaction phase, in which reactants are brought to the film surface. To the right is the removal phase, in which reaction products either volatilize away from the film surface or form residues.

Around the outside of the figure are shown the controllable variables of plasma etching: choice of reactants, gas pressure, rf power, gas flow rates, and design geometry of the etch chamber. Note that most variables appear multiple times in the

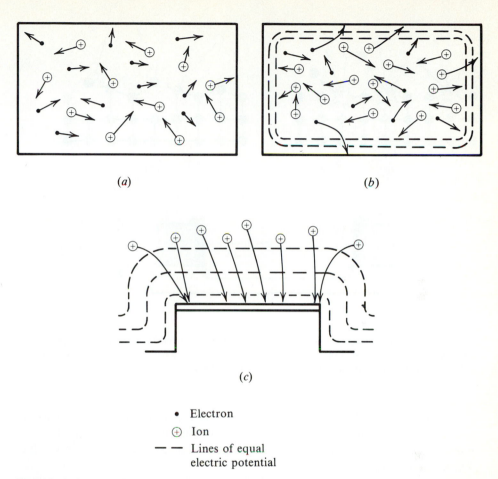

(a)

(b)

(c)

• Electron
⊕ Ion
– – Lines of equal
   electric potential

Figure 6-10 Formation of the plasma sheath and resulting ion bombardment. (*a*) Initial condition: equal number of electrons and ions, no plasma potential, loss of electrons exceeds loss of ions. (*b*) Steady-state condition: excess of ions in plasma, positive plasma potential, equal loss of electrons and ions. (*c*) Detail: perpendicular ion bombardment resulting from potential gradient.

figure, reflecting their multiple effects on various phases of the etch process. It is this interrelation of variables that makes plasma etching a highly empirical undertaking.

In the reactant phase, we balance two legs, the chemical and the physical. Chemical etching, which is selective and isotropic, depends on the nature of the plasma species reaching the film surface. This in turn depends on the gases introduced and their subsequent reactions in the plasma. The evolution of the reactants is governed not only by the electron energy and density in the plasma, but by the residence time of the gases and the number of reaction-inducing collisions. Each of these quantities is driven, often in contradictory directions, by changes in the controllable variables, such as pressure.

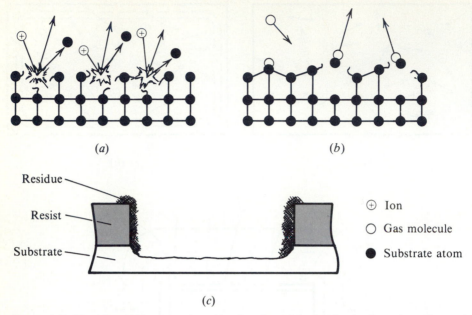

Residue

Resist

Substrate

⊕ Ion

○ Gas molecule

● Substrate atom

(c)

Figure 6-11 Ways in which anisotropic etching is enhanced by ion bombardment in a plasma. (a) Sputtering: physical removal of substrate ions. (b) Heating and bond loosening accelerate chemical reaction with substrate. (c) Sputtering of protective residues in bombarded area enhances reaction rate.

Physical bombardment is governed by the properties of the plasma sheath and the bombarding ions, each of which again varies with the controllable variables. Shown as a dotted line is the interaction between ion properties and the plasma chemistry, which determines the types of ions present.

In the product phase, successful etching requires the formation of volatile products that can be pumped away. A central variable here is the choice of reactants. But volatility of products also depends on the pressure and temperature of the environment. In many plasmas, residues also form, and these must be controlled for successful etching. Here the plasma chemistry interacts with the physical sputtering process to determine the end result. Excessive residue will terminate the etch process, but some residue formation may promote anisotropy in etching. To experience the interplay of effects in plasma etching, let us pose a simple question. In a given reactor carrying out an etch reaction, the overall gas pressure is doubled. Does the etch rate go up, or down?

To begin with, doubling the pressure means twice as much gas is available for the generation of reactive species. This will tend to increase reaction rate. But the higher pressure also raises the electron temperature and density, increases the molecular collision rate, and increases the residence time of species in the reactor. The result will be a change in the mix of reactive species in the plasma. Depending on details of the chemistry, this may promote or reduce etching.

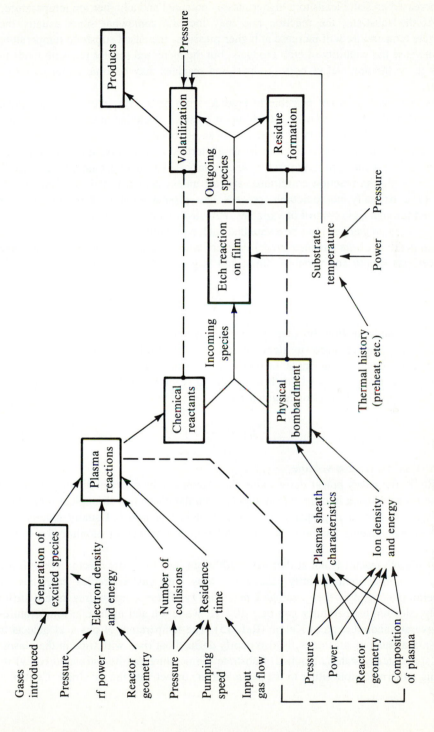

Figure 6-12 Interaction of variables affecting plasma etching.

These chemical effects may also change the mass and concentration of ions available for physical bombardment. The bombardment energy will also change, since the increased pressure leads to a higher plasma potential and a higher ion temperature.

At the substrate, the reaction rate may increase somewhat since usually the substrate temperature will increase at higher pressure. Increased substrate temperature will increase the volatility of etch products, but the increased reactor pressure tends to lower it, so the removal of both products and residues may become either easier or harder.

In summary, it is not possible to predict, from basic principles, even a simple relationship like the dependence of rate upon pressure. Detailed analysis, taking into account the specific geometry of the reactor, can of course produce some answers. But in general plasma processes need to be characterized empirically, not only for each new process but for each new reactor. A corollary is that there are no "insignificant" changes in plasma reaction conditions, at least unless demonstrated experimentally.

For a basically unpredictable process like plasma etching, reproducibility is obtained through rigid control of variables. *Pressure* can be read directly using various kinds of vacuum gauges. (But be aware that such measurements are *not* independent of the composition of the gas measured.) Pressure within a reactor depends on a balance between gas supply and vacuum pump, and is given by

$$P = \frac{F}{S} \tag{6-30}$$

where F is the input flow (in liters per second at standard temperature and pressure), and S is the pumping speed (in liters atm per second). It follows that pressure may be regulated by metering the input (a needle valve and flow meter will suffice) or by throttling the pumping speed (with a throttle valve between the reactor and pump). Residence time in the reactor is given by

$$t_r = \frac{PV}{S} \tag{6-31}$$

where V is the reactor volume.

Radio-frequency *power* delivered to the reactor can be metered directly; rf power can be reflected back into the power supply from the plasma, with unpleasant results, unless the rf system is properly impedance-matched or *tuned*. Usually reactors are supplied with meters for both forward and reflected power to monitor the tuning process.

A plasma reactor is not at thermal equilibrium, and therefore reactor *temperature* is not a well-defined quantity. Electrons, ions, and gas molecules each have a temperature defined by their average kinetic energy. These temperatures usually differ widely: electron temperatures can be $\sim 10^4$ °K, while ion and molecular temperatures are several hundred degrees Kelvin (Ref. 15). The temperature of physical objects in the plasma, such as substrates, is also poorly defined but rises with time in the plasma until some steady state is reached. Substrate temperature can be estimated (only) by placing a thermocouple either in the plasma or in contact with the substrate. Control is usually maintained by first heating the substrates with an inert plasma (e.g., N_2) until the thermocouple reading becomes steady and then plasma etching at a power that maintains the steady temperature.

Gas purity is an important variable to control, since vacuum systems tend to leak and low-concentration contaminants can dramatically affect the reactions in the plasma. For truly scientific composition measurement in plasmas, complex equipment (optical or mass spectrograph) is required. But a good guess can be had from the color of light emitted from the plasma. In particular, nitrogen emits distinctive bright pink light. A pink color in a plasma supposed to contain no nitrogen is a sure sign of a leak from the atmosphere.

One further complication of plasma etching is the *loading effect*. The rates of many reactions are found to depend on the area of substrate being etched: a greater exposed area slows down the reaction. This results from depletion of the reactive species from the plasma by the etch reaction. Consider an etch rate, with rate constant k, dependent on a reactive species with a lifetime in the plasma τ and a volume generation rate (e.g., in molecules per liters^{-1} sec^{-1}) G. Subject to some very plausible assumptions, the etch rate can be written as[16]

$$r = \frac{k\tau G}{(1 + kK\tau A)} \qquad (6\text{-}32)$$

where r is the etch rate, A the area of material being etched, and K is a constant determined by the reaction and the reactor geometry. It follows that R depends on A if the product of reaction rate and lifetime, $k\tau$, becomes sufficiently large. A loading effect can be avoided by choosing a chemistry with a slow reaction rate or a short species lifetime. This, however, results in a decrease in the maximum etch rate, given by $k\tau G$, unless the species generation rate G can be made correspondingly large. The result is an unavoidable trade-off between reaction rate and loading effect.

Loading effect is intrinsically undesirable and becomes especially troublesome when good end-point control is crucial. As the etching becomes almost complete, the area being etched decreases rapidly towards zero. If there is a substantial loading effect, the etch rate rises rapidly, resulting in a runaway etching of the remaining volume. If the reaction possesses any isotropy, the result can be excessive undercutting caused by fast etching of the small exposed area at the sides of the cuts.

Methods and Equipment for Plasma Etching

Methods of plasma etching can be classified by the relative importance of chemical etching (which is selective but isotropic) to physical sputtering (anisotropic and unselective). The term *plasma etching,* which we have used generically, is sometimes restricted only to systems that are almost exclusively chemical.

Etching that is highly dependent on ion bombardment is called *reactive ion etching* or *RIE.* RIE systems frequently apply an additional potential to the substrate electrode, so that ion bombardment energies can exceed the plasma potential. *Ion milling* is a method in which ions are provided by an ion beam, so their energy and direction can be varied independently. We will study ion milling further in Chapter 9. A

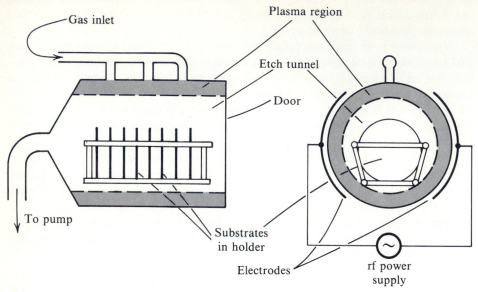

Figure 6-13 Barrel etcher for plasma etching.

number of other variations, with a rich and not always consistent terminology, are available.

Perhaps the simplest plasma etching reactor is the *barrel reactor*. An example is shown in Fig. 6-13. In this system, substrates are loaded vertically, in a boat much like a diffusion boat, into a barrel-shaped horizontal chamber. The door is closed, and the chamber evacuated with a vacuum pump. After all atmospheric gas is removed, the chamber is filled with the reactant gas, metered through a flow meter. Pressure is maintained by manual adjustment of the flow with a needle valve. Typical pressures are around 200 μm. (One micron, or 1 μm, in pressure measurement, is 10^{-3} torr. One torr is 1/760 atm. The use of linear units for pressure derives from the original designation of the torr as a mercury column 1 mm in height.)

Radio-frequency power is applied to the gas by means of electrodes surrounding the chamber. Most of the plasma chemistry takes place away from the substrate surface; reactive species then diffuse between the close-spaced substrates to accomplish the etching. Isolation of the plasma from the substrates is often made complete by use of an *etch tunnel:* a perforated metal shield that confines the plasma to the outer area of the barrel. Barrel etching is analogous to wet etching in that the reactive material is manufactured elsewhere and diffuses to the reaction site.

Barrel etching is essentially chemical and displays little or no anisotropy. Etch rates will vary within the reactor because of the nonuniform environment of the substrates. For example, the outer surface of end wafers is much more exposed to the reacting species than inner surfaces. Similarly, outer edges of substrates are more exposed and etch faster, leading to a characteristic "bull's-eye" pattern. Barrel etching

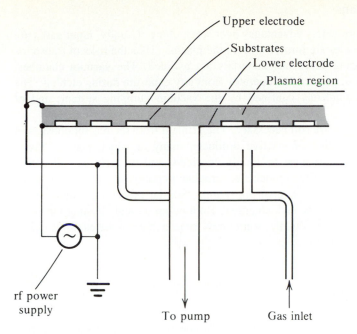

Upper electrode

Substrates
Lower electrode
Plasma region

rf power
supply

To pump

Gas inlet

Figure 6-14 Planar etcher for reactive ion etching.

is thus limited to applications where nonuniformity of etching is acceptable, as when chemical selectivity is high and reactions can be made self-limiting.

For finer control, a *planar reactor* is required. Figure 6-14 shows an example. In this configuration, substrates are laid flat between two closely spaced, planar electrodes. The result is a much lower capacity, but higher uniformity of etch. Our example is fitted with a throttle valve on the vacuum pump, which allows pressure and gas residence time to be varied independently. We have also connected the substrate electrode to the rf power supply, grounding the other electrode. This is the *reactive ion etching* configuration, which allows a high potential between the substrates and the plasma. If we reversed this configuration, we would operate at lower bias, in the *plasma* mode.

Either of our reactors may be fitted with some form of *end-point detector (EPD)*. Plasma etching is rarely predictable enough to be controlled by time, and visual observation of the end point is difficult. Thus it is usually necessary to monitor some characteristic of the reaction and terminate the reaction automatically. Luckily, a plasma is full of excited species that radiate at characteristic wavelengths. End points can be detected by observing the radiation from an etch product: the radiation intensity drops sharply as the last of the film is consumed. Alternatively, if the film overlays another etchable material, we can look for a sudden increase in a species formed by reaction with the underlayer. In either case, the radiation can be monitored with a properly filtered photosensor and the sensor signal used to control the etch operation.

Safety Considerations

Plasma etching has some safety advantages over wet etching. Usually, input gases are quite inert; reactive species are formed only in the plasma. Thus the risks of transporting, storing, and pouring dangerous materials are avoided. The vacuum chamber, being completely sealed, protects the operator from any exposure during etching. The effluent is gaseous, completely confined by the vacuum pumping system, and of negligible mass.

However, precautions are still necessary. The effluent from the reactor consists of an unpredictable combination of reactive products, many of which may be toxic. Therefore, the pump exhaust must be totally confined, and properly treated before discharge to the atmosphere. Etch products accumulate in pump oil, so exposure must be avoided while changing oil or repairing pumps. Similar precautions are in order when cleaning residues from the etch chamber. Both pump oil and cleaning materials should receive proper disposal. Finally, sometimes toxic etch gases are employed: they should be handled accordingly.

6-4 PLASMA ETCHANTS

Plasma etchants are available for each of the common films we discussed in Section 6-2. Etching of nitride, oxide, and silicon is usually done using fluorine-containing compounds, while aluminum is etched in a chlorine plasma.

Fluorine-Containing Plasmas

Both silicon tetrafluoride (SiF_4) and silicon tetrachloride ($SiCl_4$) are gases under plasma conditions; thus either fluorine or chlorine plasmas will etch silicon compounds. The *fluorocarbons*, a family of inert, nontoxic compounds of fluorine and carbon, are popular source gases for forming fluorine plasmas. The simplest member of this family is carbon tetrafluoride, CF_4.

In a plasma, carbon tetrafluoride decomposes into a number of fragments, including fluorine atoms, F. These atoms then react with silicon compounds as follows:

$$Si + 4F \rightarrow SiF_4 \tag{6-33}$$

$$SiO_2 + 4F \rightarrow SiF_4 + O_2 \tag{6-34}$$

$$Si_3N_4 + 12F \rightarrow 3SiF_4 + 2 N_2 \tag{6-35}$$

The etch rates of the compounds varies widely with reactors and conditions. However, under fairly typical conditions in a barrel reactor, the relative etch rates are

$$Si > Si_3N_4 \gtrsim SiO_2 \tag{6-36}$$

In one study,[17] the ratio was 17:3:2.5. Etch rates do vary with film properties (like CVD conditions and phosphorus doping) although not as dramatically as with wet etching.

The ratio of etch rate for the film being etched, to the etch rate of the underlying material, is called *selectivity*. Selectivity is important because process variations usually require at least a 10–20% overetch of the upper film. This must be accomplished without etching too deeply into the underlying material. A selectivity of 10:1 is usually considered adequate; selectivities below 5:1 are often unacceptable. Clearly, the "natural" etch rate ratios of Eq. (6-36) are favorable for etching polysilicon over nitride or oxide, but unsatisfactory for other common processes like etching oxide over silicon.

Fortunately, these ratios can be modified in various ways. Changes in gas pressure and flow or applied power will alter them, as will a different choice of fluorocarbon or a different reactor (see, e.g., Ref. 18). A loading effect is present for silicon, but not its compounds, so that selectivity can also be altered by loading.[19] Etch rate is accelerated by addition of 1 to 10% oxygen (O_2) to the plasma, with the effect strongest for silicon. Empirically, these observations have led to a satisfactory set of recipes for etching silicon compounds in various combinations. More fundamentally, they have led to a better understanding of the etch system.

The carbon in the fluorocarbon turns out to be one key. Carbon-bearing fragments are abundant in the plasma, and they react with fluorine atoms in competition with the substrate film. The magnitude of this effect depends on the fluorine-to-carbon (F/C) ratio. Gases higher in carbon will etch silicon more slowly. Below a F/C ratio of about 8/3, carbon fragments polymerize to form an inert, etch-resistant film that slows etching even further.[20] This residue also encourages anisotropic etching by accumulating on surfaces not exposed to ion bombardment.

Oxygen reacts with carbon-bearing fragments to form carbon dioxide (CO_2), reducing fluorine consumption by carbon and speeding silicon etching. Silicon dioxide supplies its own oxygen during etching and so is less affected by carbon chemistry. Thus the silicon etch rate, relative to oxide, is reduced by a higher carbon concentration in the plasma. (Similar effects can occur with added hydrogen, which also consumes fluorine.)

Other work has shown that SiO_2 etching occurs principally by direct interaction with ions, while Si etching involves more chemistry and is fairly insensitive to electrical effects.[21,22] Thus manipulation of the ion concentration in the plasma can also affect selectivity.

Despite some understanding of the underlying principles, plasma processing remains highly empirical. A new process still requires characterization on the specific reactor and film to be used. Using some starting point obtained from previous work or from the literature, the engineer will characterize etch rates as a function of pressure, power, and any other variables that can be controlled, studying not only the film to be removed, but all other materials present on the substrate during the etch, including photoresist. The engineer will want to measure enough points to determine etch uniformity within each substrate, between substrates, and between successive reactor cycles in order to plot etch rates, selectivities, and uniformities as a function of the controllable variables in order to choose an operating point. The operating point should be chosen not only to obtain satisfactory etching, but for maximum insensitivity to small changes in pressure, power, etc.

In the empirical world of plasma etching, almost everything has been tried. As a result, there are many sources and many applications for fluorine plasma etching multigrade. The most common fluoride etch gas composition is 4% O_2 in CF_4, but there are many fluorocarbons, chlorofluorocarbons, and other fluorine-containing compounds in use. Because many metal fluorides are volatile, fluorine plasmas can successfully etch a variety of metallic compounds. They are particularly valuable in etching metal-silicon compounds (*silicides*), which are otherwise almost unetchable.

Chlorine-Containing Plasmas

Of the four films we studied in depth in Section 6-2, fluorine plasmas will successfully etch three. The fourth, aluminum, forms a nonvolatile fluoride and does not etch in fluorine plasmas. However, aluminum chloride ($AlCl_3$) is reasonably volatile at plasma conditions, and aluminum can thus be etched in chlorine-containing plasmas.

Plasma etching of aluminum is less sensitive to grain structure than wet etching and is without electrochemical effects. But the refractory aluminum oxide film that coats all aluminum surfaces must still be dealt with. This oxide film must be removed before aluminum etching can begin. Although invisibly thin, it etches much slower than aluminum in most plasmas, causing a long and frequently irreproducible *induction period* before the onset of aluminum etching. If oxide thickness or etch rate varies at all, oxide removal is not simultaneous across the substrate, and the subsequent aluminum etching is highly nonuniform. A maximum aluminum oxide etch rate is therefore desirable.

A successful aluminum etcher must also cope with aluminum alloys. Aluminum is frequently alloyed in microelectronics with silicon and/or copper in 1–4% concentrations. Silicon is easily etched in chlorine plasmas, but copper chlorides are not very volatile. Aluminum-copper etching thus requires a reactive ion etching system, in which physical sputtering helps to remove the copper.[23]

Plasmas formed in chlorine (Cl_2) and hydrogen chloride (HCl) will etch aluminum, but give ragged edges. In addition, these gases are toxic and corrosive. A better choice is carbon tetrachloride (CCl_4), a liquid that is still toxic, but much easier to handle than chlorine. Carbon tetrachloride etches aluminum well, but has a significant induction period due to its slow attack of aluminum oxide.[24] Like CF_4, carbon tetrachloride can also form carbon polymers. These can be troublesome, but also promote anisotropy in etching.

Boron trichloride (BCl_3) plasmas etch aluminum with very little induction period, due to their efficient removal of aluminum oxide. However, fragmentation of BCl_3 in plasma produces only a few chlorine atoms, and so the etch rate is relatively slow and exhibits a loading effect. Polymers are not formed, and the etching is more isotropic.[24]

Better results can be obtained by combinations of gases. For example, Saia and Gorowitz[25] discuss the use of $CCl_4/BCl_3/O_2$ mixtures. Sufficient BCl_3 will promptly remove the aluminum oxide, avoiding any irreproducible induction period. CCl_4 contributes additional chlorine and also forms polymers that promote anisotropic etching. Reaction between the oxygen and the BCl_3 enhances chlorine production. Bollinger, Iida, and Matsumoto[26] suggest the use of $BCl_3/CF_4/O_2$. Here the oxygen enhances both Cl and F atom production. The F atoms react with aluminum to passivate those areas not under ion bombardment, producing anisotropy without formation of residues. The enhanced Cl production provides a satisfactory etch rate. Mixtures of this sort provide a reasonable approach to the ideal aluminum etchant: fast-etching, anisotropic, and without an induction period.

Choice of a suitable etch gas is not the only problem in aluminum plasma etching. Control of product residues is also a significant concern. Aluminum chloride is not a very volatile material. It boils at 180°C, and its vapor pressure is only 1 torr at 100°C. Though it may evaporate from warm substrates, it tends to recondense before reaching the vacuum pump. As a result, deposits of aluminum chloride and related compounds accumulate on the etch chamber, and sometimes on portions of the substrate. When exposed to water vapor, these compounds react to form hydrochloric acid. On the substrate, the acid corrodes the metal pattern just etched. In the chamber, deposits can affect both the chemistry and the potential of the plasma.

Etched substrates require treatment to passivate or remove any residues, before they can react with atmospheric moisture. Posttreatments in oxygen or fluorine plasmas can serve this purpose.[27] Sometimes a quick and thorough water rinse after etching will also suffice. Chambers require regular cleaning, and protection from atmospheric moisture. One method is to use a *load lock*. This is a second vacuum chamber that can be isolated from the etch chamber. Substrates are loaded into the load lock, which is then sealed and pumped down. When fully evacuated, the load lock is opened to the etch chamber, and the substrates transferred for etch. After etching, substrates return to the load lock, and the etch chamber is sealed off before the load lock is vented and unloaded. The etch chamber is thus never exposed to atmospheric pressure.

One final complication of chlorine plasmas is their attack on photoresist. Resist erosion is a possibility in any plasma, particularly under reactive-ion-etching conditions, and proceeds readily in chlorine plasmas. As a result, thick resist coatings must be used, and careful pretreatment given the resist to maximize its chemical stability.

As with fluorine, chlorine-containing plasmas can be used to etch additional films, including silicon and titanium. Chlorine-fluorine mixtures can be advantageously applied to silicide etching.[28] Table 6-4 presents data relevant to plasma etching of a number of materials. References 29 and 30 offer further reviews of plasma etching.

6-5 PHOTORESIST REMOVAL

Upon completion of etching, the photoresist pattern has served its purpose. If not now completely removed, it may contaminate diffusion furnaces or deposition furnaces with drastic results. Resist removal may be considered a form of etching and can be carried out by wet chemical or plasma methods. Resist removal is sometimes called *stripping* or *cleaning*.

Wet chemical removal of photoresist can be performed by any compound that decomposes organic molecules. If there is no metal on the substrate, strong oxidizing agents can be used. One possibility is "chromic acid," a solution of chromium oxide in

Table 6-4 Plasma etchants for common microelectronic materials

Material	Common etch gases†,‡	Dominant reactive species	Product vapor pressure (torr at 25°C)	
Aluminum	Chlorine-containing	Cl, Cl$_2$	AlCl$_3$	7×10^{-5}
Copper	(Forms only low-pressure compounds)		CuCl$_2$	5×10^{-2}
Molybdenum	Fluorine-containing	F	MoF$_6$	530
Polymers of carbon	Oxygen	O	H$_2$O	26
			CO, CO$_2$	> 1 atm
Silicon	Fluorine- or chlorine-containing	F, Cl, Cl$_2$	SiF$_4$	> 1 atm
			SiCl$_4$	240
SiO$_2$	CF$_4$, CHF$_3$, C$_2$F$_6$, C$_3$F$_6$	CF$_x$	SiF$_4$	> 1 atm
			CO, CO$_2$	> 1 atm
Tantalum	Fluorine-containing	F	TaF$_3$	3
Titanium	Fluorine- or chlorine-containing	F, Cl, Cl$_2$	TiF$_4$ sublimates at low pressure	
			TiF$_3$	$<10^{-3}$
			TiCl$_4$	16
Tungsten	Fluorine-containing	F	WF$_6$	1000

† Common chlorine-containing gases are BCl$_3$, CCl$_4$, Cl$_2$, and SiCl$_4$.

‡ Common fluorine-containing gases are CF$_4$, SF$_4$, and SF$_6$.

Source: Reference 27. Reprinted with permission of *Solid State Technology,* published by Technical Publishing, a company of Dun and Bradstreet.

concentrated sulfuric acid. Perhaps more popular are mixtures of hydrogen peroxide in sulfuric acid. These "peroxide sulfuric" solutions, formed by adding a few percent of hydrogen peroxide to concentrated sulfuric acid, destroy double bonds in the resist with great enthusiasm. The operating temperatures are usually 100–135°C. Peroxide is lost due to decomposition and evaporation, and must be periodically replenished. Sulfuric-peroxide stripping, though usually trouble-free, can leave a residue on substrates unless vigorous agitation is used during rinsing.

Sulfuric acid attacks metals and thus cannot be used to clean substrates containing metal films. A number of proprietary strippers for metallized substrates are on the market. Most of them contain phenols or sulfonic acids, and operate as strong detergents. These compounds are frequently not totally inert with metals, so thorough rinsing after cleaning is vital to prevent metal corrosion. Many of these compositions also emit toxic fumes, and are subject to environmental restrictions as to disposal.

Resist removal can also be performed in a plasma. An oxygen (O$_2$) plasma will oxidize resist to carbon dioxide and water. This process is often called *plasma ashing*. It is performed in a simple barrel reactor, or *asher*, either with an end-point detector or simply by time. Oxygen plasmas are fairly unreactive with common microelectronic materials and can thus be used on metallized wafers. Ashers are also successful in removing the extremely tough resist layer often generated during ion implantation, which can resist even sulfuric peroxide stripping. Because of its convenience and safety, ashing is fast replacing most other stripping methods. However, cheap and dependable sulfuric-peroxide stripping remains competitive for nonmetallized substrates.

Lift-off Processing

In some cases, resist patterning and stripping can be used to form a physical pattern without recourse to etching, by the means of *lift-off* processing. While not normally used in silicon integrated circuit manufacture, lift-off methods are common in some other applications, as we shall see in Chapter 12.

Figure 6-15 shows how lift-off patterning is accomplished. Figure 6-15*a* shows a resist pattern on a substrate. Particular attention has been given to making the walls of the resist pattern as steep as possible. In Fig. 6-15*b*, the film to be patterned has been deposited *over* the resist. Clearly, this method is not possible with thermal oxides, but can be done by some of the deposition techniques presented in Chapters 7 and 8 if the deposition temperature is low enough. Notice that, due to the steep resist walls, the film does not completely cover the edges at the steps. This failure of *step coverage* makes the lift-off possible.

The resist is now stripped using some wet chemical method. Figure 6-15*c* shows the result. Where the film was deposited directly on the substrate, film remains. Where the film rested on the resist, stripper has entered the gaps in the film and dissolved the resist, lifting away the film. The result is a pattern, the negative of the resist pattern, formed without etching.

Lift-off methods work best with metal films, which can often be deposited at low temperature. In this case, a solvent-type stripper is required to prevent attack on the

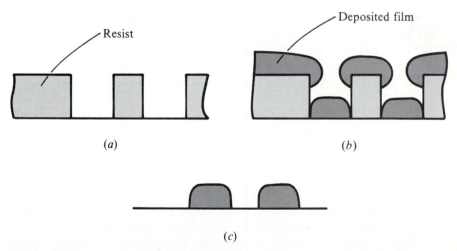

Resist

Deposited film

(a)   *(b)*

(c)

Figure 6-15 Patterning of films by lift-off. (*a*) Resist pattern formed on substrate. (*b*) Deposition of film over resist. Note the gaps caused by step coverage failure over the steep resist walls. (*c*) Resist is stripped, lifting off all film that was deposited on top of resist. The resulting pattern is a negative of the resist pattern.

film. Where a proper combination of film, resist, and stripper can be found, lift-off patterning offers the advantages of steep walls, no undercutting, no need for etching equipment, and applicability to extremely small dimensions.

Safety Considerations

Chemically, photoresist is not unlike human tissue, and resist strippers can do great damage to the careless user. All of the precautions mentioned before for wet etching should be observed. Sulfuric acid liberates an astounding amount of heat when mixed with water or water-containing solutions such as hydrogen peroxide. The general rule is to add acid to water rather than the reverse. When water must be added to acid, as in maintaining a sulfuric-peroxide bath, the addition should be made slowly and carefully, with full precautions against splattering or breakage of the acid container. Sulfuric acid particularly should not be aspirated while hot. When using proprietary strippers, obtain the Chemical Safety Data Sheet on the product and follow it. A qualified safety professional should be consulted for additional information.

6-6 ETCH PROCESS CONTROL

As always, process control in etching is a matter of measuring desired output, detecting deviations, and relating them to variations in input. The output variable in etching is film removal, and etching problems thus consist of *overetching* and *underetching*.

Underetching, or incomplete film removal, is usually detectable by visual inspection. Observation under a microscope will show whether some of the film still remains in the patterned areas. If oxide or nitride is being removed from silicon, the cut will be colored, rather than "silicon-colored" (variously perceived as gray, tan, or white). If metal or polysilicon is incompletely etched, shiny areas or grains will be present. Use of a dark field attachment on a microscope, which emphasizes edges rather than surfaces, is particularly useful for spotting small aluminum or polysilicon grains. Incomplete resist removal appears as a brownish color of the substrate, usually visible to the naked eye. Incomplete ashing has a particularly characteristic appearance: a brown or yellow "bull's-eye" in the center of the wafer, caused by the nonuniform etch rate of a barrel reactor. (White light *must* be used to inspect for residual resist: the brownish hue is invisible under yellow or green light.) If one of several transparent layers is being etched, it may be necessary to measure the film thickness to ensure complete etching. The film thickness analyzers described in Chapter 2 are useful for this purpose.

For some plasma processes with poor selectivity, overetching may be shown by attack on underlying layers. In this case, visual inspection or film thickness measurement will again suffice. But when selectivity is good, downward etching will terminate when the film is completely etched. In this case, inspection of the bottom of the patterned area will reveal nothing. Overetching must then be measured by its effect in the lateral direction.

After downward etching is complete, lateral etching continues (unless a process is

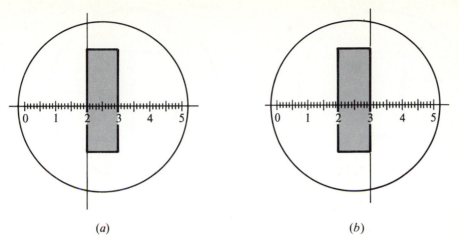

(a) (b)

Figure 6-16 Linewidth measurement by the Filar method. (a) Position of movable needle before measurement. (b) Position after measurement.

perfectly anisotropic). Overetching thus results in line narrowing and is best detected by measurement of the width of the etched line. Overetching is indicated by an unexpected reduction in the patterned linewidth.

Linewidths are measured using one of four methods. The best is a scanning electron microscope (SEM), which provides a detailed look at the line geometry. But a SEM is a complex and expensive instrument, requiring extensive sample preparation that is usually destructive of the device and so is (as yet) unsuitable for routine process use. Several visual methods are thus more common.

Two methods involve manual measurement with a microscope. In each case, the feature to be measured is observed while turning a calibrated knob that manipulates the image to give a measurement. Figure 6-16 shows measurement with a *Filar eyepiece*. A needle moves along a calibrated scale in response to turning the knob. Linewidth is measured from the scale in the eyepiece. Figure 6-17 illustrates *image shearing*. Two colored images of the measured object are produced. Turning the knob moves the two images until the edges to be measured coincide. A calibrated readout translates knob rotation into distance.

The third optical "method" is really a hodgepodge: the automated method. A number of devices are commercially available for automated linewidth measurement using visual images. Typically an operator focuses on the feature to be measured, pushes a button, and receives a readout based on one of several methods. Automatic measurement devices are more reproducible than human observers, but are not nearly as good at pattern recognition. Often such devices cannot measure an arbitrary feature and require specialized structures to be designed into the pattern.

In measuring lines of a few microns width, calibration is a major problem. Only photolithographic techniques can produce such lines, and we have seen that many systematic errors can affect that process. Indeed, if photolithography and etching were

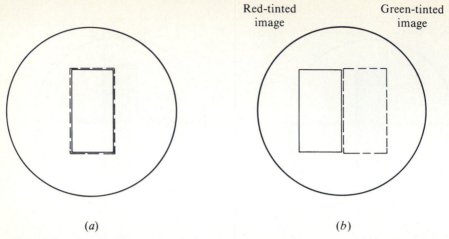

Red-tinted image   Green-tinted image

(a)   (b)

Figure 6-17 Linewidth measurement by the image shearing method. (a) Superimposed images before measurement. (b) Sheared images after measurement.

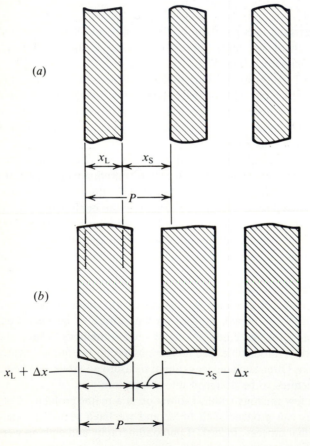

(a)

x_L x_S

P

(b)

$x_L + \Delta x$ $x_S - \Delta x$

P

Figure 6-18 Use of pitch in linewidth calibration. (a) Nominal pattern: $P = x_L + x_S$. (b) Pattern with systematic printing error that increases linewidth: pitch unchanged.

232

Table 6-5 Etch process troubleshooting

Problem	Possible causes	Method of confirming
	Over- or underetch, uniform over substrate	
Wet etching	Operator error	Audit operator
	Improper temperature	Measure temperature
	Wrong etch composition	Audit mixing process; have analyzed; measure etch rate
With timed etches	Film thickness incorrect	Measure pre-etch thickness
With time-dependent etches	Etch exhausted	Check mix date; measure etch rate
	Etch too fresh (not aged)	Check mix time; wait and try again
With wet poly and aluminum	Operator missed "to clear" point	Audit technique
With CVD films and metals	Change in film properties, doping, etc.	Investigate film and deposition process
Plasma etching	Improper end-point detection	Check operation; check sensitivity
	Improper power	Check output, tuning
	Improper pressure or flow	Check pressures, flows throttle valve
	Wrong gas composition	Check all sources and flow rates; observe color; leak check; check for complete pumpdown before backfill
	Change in film properties	Investigate film deposition process
	Localized underetch	
Wet etching	Bubble formation	Circular spots
	Surface tension/poor wetting	Zones, especially small geometries
Plasma etching	Particle on surface blocking etch	Is residue particle-shaped?
	(See below, nonuniformity)	
Metals, CVD films, highly doped oxides	Change in film deposition	Investigate film, deposition process
	Localized overetch	
Wet etching	Poor quench after etch	Check quick dump, rinser, operator technique
Plasma etch	Contaminated prior to etch	Analyze surface
	(See below, nonuniformity)	
Attack on resist-protected area		
All	Resist adhesion failure	Inspect resist post-etch
Aluminum plasma etch	Failure to passivate after etch	Check machine and operator
Changes in uniformity or selectivity—plasma etch	Check everything in Overetch/Underetch section	
	Residue on reactor	Inspect; clean and retry
	Parallel plate spacing off	Measure, adjust
	Pumping speed change	Has pressure-flow relation changed?
		Service pump

dependable enough to make linewidth calibration standards, we would not need to measure the linewidths in the first place! However, reliable calibration can be achieved by use of the concept of *pitch*. Consider an array of equally spaced lines and spaces, as in Fig. 6-18a. The nominal linewidth is x_L and the spacewidth is x_S: the pitch, or width of one line-and-space pair, is $P = x_L + x_S$. In Fig. 6-18b, systematic patterning error has reduced the linewidth by Δx and increased the space by a similar amount. However, the distance between line centers is fixed by the original pattern, and thus P remains unchanged. Linewidth measuring devices can thus be calibrated to the constant distance P, and then used with confidence to measure either x_L or x_S.

Under- or overetch is caused by either variation in film thickness, etch time or etch rate. Film thickness must be controlled as part of the deposition process. Time may be affected by operator error, by improper end-point detection, or by failure to properly start or quench the etch. Etch rate can vary with film properties or with etchant properties. Film properties are a subject for Chapters 7 and 8. Etchant properties in wet chemical etching include chemical composition, temperature, and agitation or flow. Plasma etch rates depend on gas composition, pressure, flow rate, rf power, and the cleanliness and geometry of the reactor walls. Etch process control involves verifying the correctness of each of these parameters. Table 6-5 is a troubleshooting guide, listing a number of more specific causes and some methods of verifying them.

PROBLEMS

6-1 A 0.60-μm film of silicon dioxide is to be etched with a buffered oxide etchant of etch rate 750 Å min^{-1}. Process data shows that the thickness may vary up to 10% and the etch rate may vary up to 15%. (a) Specify a time for the etch process. (b) How much undercut will occur at the top of the film?

6-2 A set of windows are to be etched in a silicon dioxide film of thickness 6000 Å. As patterned in the photoresist, the size of the windows is 6 μm square. (a) Find the dimension of the window, as measured at the top of the oxide, after ideal isotropic etching (i.e., no overetch). (b) Find the dimension of the window as measured at the oxide-substrate interface. (c) Find the average slope of the window edge. (d) Find the upper dimension, (e) the lower dimension, and (f) the average slope after 30% overetch.

6-3 A gradual slope is desired at the edges of a set of windows. Therefore, a 6000-Å layer of silicon dioxide, of etch rate 1000 Å min^{-1} is overlaid with a 1000-Å layer of CVD oxide of etch rate 2000 Å min^{-1}. The window pattern in the photoresist measures 6 μm on a side and is square. (a) Find the etch time required to clear the film. (b) Find the dimension of the window, measured at the top of the film. (c) Find the slope of the window edge. (d) Find the dimension and (e) the slope, for 30% overetch time.

6-4 In a memory device, the cells are accessed using an array of parallel metal lines. The pitch of these lines limits the size of the device. Determine the pitch of a set of parallel lines under the following conditions:

    (a) The lines are made of aluminum, 1 μm thick.

    (b) The minimum space-width resolvable with the photoresist technology in use is 2 μm.

    (c) The lines are wet etched, isotropically, with a 50% overetch necessitated by uncertainties in etch rate and end-point detection.

    (d) The minimum line dimension, after etching, at the top of the line, is required by design considerations to be 4 μm.

6-5 In Problem 6-3, let the wet etch process, condition (c), be replaced by a plasma etch with complete anisotropy, so there is no undercutting. Now find the pitch. If the reduction in linear dimension of the device is proportional to the reduction in pitch, how much can the *area* of the device be reduced?

6-6 In forming contact windows for a bipolar device, a single etch step is used to cut contacts through both the field oxide (6000 Å thick, etch rate 1000 Å min^{-1}) and through the emitter oxide (3000 Å thick, phosophorus-doped, etch rate 1500 Å min^{-1}). (a) Find the etch time required to clear the field oxide windows, with 20% overetch. (b) For this time, find the percentage of overetch and (c) the lateral undercutting per side for the emitter oxide windows.

6-7 An unbuffered oxide etch is prepared by mixing commercially available hydrofluoric acid (HF) and water. The ingredients are mixed in a 1 : 7 ratio by volume, to prepare exactly 8 liters of solution. (a) What is the concentration by weight in the initial mixture of HF? (b) Using the expression for hydrogen ion concentration in a pure weak acid, find the pH of the etchant. (c) How will the pH change if 10% extra water is added to the solution?

6-8 A buffered oxide etch is prepared by mixing commercially available hydrofluoric acid (HF) with ammonium fluoride solution (NH_4F). The chemicals are mixed in a 1 : 7 ratio by volume, to prepare exactly 8 liters of etchant. (a) What is the concentration by weight in the initial mixture of HF? (b) Of NH_4F? (c) Using the equation for buffer solutions, find the pH of the etchant. (d) How will the pH change if 10% extra water is added to the solution?

6-9 A barrel etch system, using CF_4 at certain operating conditions, etches Si, SiO_2, and Si_3N_4 at rates of 200, 40, and 20 Å min^{-1}, respectively. A 10% overetch of the upper film is required by the process variation. Calculate the percentage removal of the lower film for the etching of: (a) 3000 Å of polysilicon over 5000 Å of silicon oxide. (b) 1000 Å of silicon nitride over 10,000 Å of oxide. (c) 5000 Å of oxide over 1000 Å of nitride. (d) If 10% removal of the lower film is acceptable, which of these three processes is workable? Is selectivity the principle determinant?

6-10 In a plasma etching process, reaction gas is metered into a 100-liter reactor at a flow of 10 mliter sec^{-1} (STP). The resulting steady-state pressure is 0.2 torr. (a) Find the pumping speed of the pump in 1 atm sec^{-1}. (There are 760 torr in one standard atmosphere of pressure.) (b) Find the residence time of a gas molecule in the reactor. (c) The flow is increased to 20 mliter sec^{-1}. Find the pressure and residence time. (d) A throttle valve is placed on the pump and the effective pumping speed cut in half. Find the pressure and residence time.

6-11 List appropriate safety precautions for the transport, pouring, and use of liquid acid etchants. State special safety considerations appropriate when using HF or sulfuric acid.

REFERENCES

1. Walter Kern and Cheryl A. Deckert, "Chemical Etching," *Thin Film Processes*, John L. Vossen and Walter Kern (eds.), Academic, New York, 1978.
2. J. S. Judge, *J. Electrochem. Soc.* **118**:1772 (1971).
3. A. S. Tenney and M. Ghezzo, *J. Electrochem. Soc.* **120**:1091 (1973).
4. G. Bell and J. Hoepfner, "Tapered Windows in Silicon Dioxide Pretreated by Low Energy Ions," *Etching*, Henry G. Hughes and Myron J. Rand (eds.), The Electrochemical Society, Princeton, NJ, 1976, p. 47.
5. W. A. Pliskin and R. P. Gnall, *J. Electrochem. Soc.* **113**:263 (1966).
6. C. A. Deckert, *J. Electrochem. Soc.* **144**:869 (1967).
7. W. van Gelder and V. E. Hauser, *J. Electrochem. Soc.* **144**:869 (1967).
8. H. Robbins and B. Schwartz, *J. Electrochem. Soc.* **107**:108 (1960).
9. D. L. Kendall, *Appl. Phys. Lett.* **26**:195 (1975).
10. M. J. Declercq, L. Gerzberg, and J. D. Meindl, *J. Electrochem. Soc.* **122**:545 (1975).
11. R. M. Finne and D. L. Klein, *J. Electrochem. Soc.* **121**:1215 (1974).
12. Sorab K. Ghandhi, *VLSI Fabrication Principles*, Wiley, New York, 1983, p. 482.
13. Don MacAurthur, "Chemical Etching of Metals," in *Etching for Pattern Definition*, Henry G. Hughes and Myron J. Rand (eds.), The Electrochemical Society, Princeton, NJ, 1976, p. 76.
14. D. Bollinger, S. Iida, and O. Matsumoto, *Solid State Technol.* **27**:111 (May 1984).
15. J. R. Hollahan and R. S. Rosler, "Plasma Deposition of Inorganic Thin Films," *Thin Film Processes*, John L. Vossen and Werner Kern (eds.), Academic, New York, 1978, p. 336.

16. C. J. Mogab, *J. Electrochem. Soc.* **124**:1262 (1977).
17. R. J. Poulsen, *J. Vac. Sci. Technol.* **14**:266 (1971).
18. Rakest Kumar, Chris Ladas and Gwen Hudson, *Solid State Technol.* **19**:54 (October 1976).
19. Tatsuya Enonmoto, *Solid State Technol.* **23**:117 (April 1980).
20. J. W. Coburn and Eric Kay, *Solid State Technol.* **22**:117 (April 1979).
21. Ch. Steinbrüchel, H. W. Lehmann, and K. Frick, *J. Electrochem. Soc.* **132**:180 (1985).
22. H. Kawata, K. Murata, and K. Nagami, *J. Electrochem. Soc.* **132**:206 (1985).
23. Daniel HG Choe, Chris Knapp, and Adir Jacob, *Solid State Technol.* **28**:165 (March 1985).
24. Geraldine C. Schwartz and Paul M. Schaible, *Solid State Technol.* **23**:86 (November 1980).
25. R. J. Saia and B. Gorowitz, *Solid State Technol.* **26**:247 (April 1983).
26. D. Bollinger, S. Iida, and O. Matsumoto, *Solid State Technol.* **27**:111 (May 1984).
27. D. Bollinger, S. Iida, and O. Matsumoto, *Solid State Technol.* **27**:167 (June 1984).
28. T. P. Chow and G. M. Fanelli, *J. Electrochem. Soc.* **132**:1969 (1985).
29. S. J. Fonash, *Solid State Technol.* **28**:150 (January 1985).
30. See also the April 1985 issue of *Solid State Technology,* an issue devoted to recent trends in plasma processing.

SEVEN

THIN FILMS: MAINLY METALS

By oxidizing a substrate, etching a pattern of holes into the oxide, and diffusing impurities through the holes, a semiconductor substrate can be transformed into a set of electrical devices. Though functional, these devices must be connected with each other and the outside world before they form a useful circuit. To form these connections requires two types of materials: conductors and insulators. The conductors form the electrical connections; the insulators separate conductors from each other, the substrate, and external contaminants. These materials are deposited on the surface of the substrate as films, micrometer or less in thickness, and then patterned photolithographically. Thin-film deposition is (with diffusion and lithography) the third branch of microelectronic technology.

The next three chapters are devoted to thin-film technology. There are two principal methods of thin-film deposition. In vacuum deposition, the film material is transferred from a solid source, through a vacuum, to the substrate. This method works best for elements or highly stable compounds of moderate melting points, especially when high purity is required; it is most useful with metallic conductors. In the other method, the substrate is exposed to a gas mixture that reacts to form the desired film on the substrate. This is chemical vapor deposition (CVD), a method appropriate for compound films, or where chemical purity is less critical. Most insulators are deposited by this method. In Chapter 7, we examine vacuum methods, along with the features of metallic thin films. In Chapter 8, we will study CVD methods and insulating films. (We should note that the correlation of metals with vacuum deposition and insulators with CVD is only approximate: metals can be deposited by CVD methods and vice versa.) Chapter 9 deals with ion beam applications, which, though not strictly a thin-film method, are usually assigned to the thin-film department.

Vacuum deposition of a film requires two things: a vacuum and a source of film material. In Section 7-1, we examine to the principles and methods of creating a vacuum. In Section 7-2, we look at two methods for transferring material from a source: evaporation and sputtering. In Section 7-3, we focus on specific features of metals used as conductors in microelectronic devices. Metal-semiconductor junctions can also serve as an active electronic device by formation of a Schottky diode. We describe this application in Section 7-4. Finally, in Section 7-5, we summarize methods for measuring and controlling vacuum deposition processes.

7-1 VACUUM SCIENCE AND TECHNOLOGY

The best earthly vacuum chamber is very far from empty. Therefore, vacuum deposition (as well as plasma etching and many CVD applications) is concerned not with a true vacuum, but rather a fairly dilute gas. Vacuum methods are thus based on the behavior of gases at reduced pressure.

At fairly low pressures and high temperatures (far from the boiling point), gas behavior is well predicted by the kinetic theory of gases. In this theory, individual gas molecules are considered to be hard spheres that travel in straight lines at a constant speed until they collide elastically with each other or the wall of the container. Each molecule acts independently of the others except during collisions, which randomly redistribute the kinetic energy of colliding molecules. From this model can be obtained the ideal gas law:

$$PV = nRT \tag{7-1}$$

Here P, V, and T are the pressure, volume, and temperature of the gas, and n is the number of moles of gas present. (Moles and molecular weight were reviewed in Section 6-1.) R is the gas constant, whose value depends on the units used to measure pressure. Table 7-1 gives some common pressure units and the corresponding values of R. The most "official" pressure unit is the pascal (Pa), or newton (N) per square meter. The most traditional unit is the torr (once the "millimeter of mercury")—the pressure exerted by a mercury barometer column 1 mm in height. A very convenient unit is 1 atm, the pressure of the standard atmosphere.

We note that, according to Eq. (7-1), the pressure-volume product of a given amount of gas at fixed temperature is constant (Boyle's law). So 1 liter atm of a gas is (for a given temperature) a fixed quantity proportional to the number of moles of gas present. A gas is said to be at *standard temperature and pressure (STP)* at 1 atm and 0°C (273°K). At STP, 1 mole of any ideal gas occupies 22.4 liters.

The kinetic theory predicts that although molecules move at a variety of speeds that change constantly through collision, the average molecular speed $\bar{v}$ is fixed at

$$\bar{v} = \left(\frac{8RT}{\pi M}\right)^{1/2} \tag{7-2}$$

where M is the molecular weight. The average time between collisions t_{coll} is

Table 7-1 Units of pressure and values of the gas constant in various units

Unit	Value of 1 atm	Basis
Atmosphere (atm)	—	Atmospheric pressure
Pascal (Pa)	133 Pa	1 N m^{-2}
Torr (or mm Hg)	760 torr	Pressure exerted by a mercury column 1 mm high
Micrometer (μm)	760 × 10^3	Pressure exerted by a (hypothetical) mercury column 1 μm high

Value of the gas constant R:
$= 0.08205$ liters atm mole^{-1} °K^{-1}
$= 8.314$ J mole^{-1} °K^{-1}
$= 62.4$ liters torr mole^{-1} °K^{-1}
$= 1.987$ cal molecule^{-1} °K^{-1}
$= 22.40$ liters atm mole^{-1} at 273°K

$$t_{coll} = \frac{1}{\sqrt{2}\pi N d^2 \bar{v}} \qquad (7\text{-}3)$$

where d is the molecular diameter and N is the number density of molecules (e.g., molecules per liter). Between collisions, molecules move an average distance given by dividing $\bar{v}$ by t_{coll}. This distance is called the *mean free path (MFP)*:

$$\text{MFP} = \frac{1}{\sqrt{2}\pi N d^2} \qquad (7\text{-}4)$$

Table 7-2 gives properties of several gases, including molecular diameters. Mean free path is important in vacuum applications, because it determines the distance that molecules move in straight lines, before their motion is randomized by collision. MFP is inversely proportional to N, which is proportional to pressure. Thus, as pressure is reduced, MFP increases.

When a gas flows through a system with dimensions well below the MFP, flow is "fluidlike" or *viscous:* it flows around obstructions, through bends in tubing, etc. However, when the MFP exceeds system dimensions, the flow becomes "lightlike" or *molecular*. Molecules travel in lines of sight and will go around bends only when reflected. Thus, at low pressures, system geometry becomes important, and intuitive ideas about gas behavior become unreliable.

Example The characteristic dimension of a vacuum system is taken to be 5 cm, the diameter of the pump tubing. Use this distance to estimate the pressure at which viscous flow gives way to molecular flow.

We assume that nitrogen gas, the major constituent of air, accounts for most of the gas in the system. From Table 7-2, nitrogen has a molecular weight of 28 g mole^{-1} and a diameter of 0.316 nm. Rearranging Eq. (7-4), we solve for the gas density N, which gives a MFP of 5 cm. We obtain

Table 7-2 Properties of gases of interest in thin-film deposition

Gas (formula)	Molecular weight (g/mole)	Molecular diameter† (nm)	Abundance in air	Boiling point (°K)	Comment
Nitrogen (N_2)	28	0.316	78.08%	77.4	—
Oxygen (O_2)	32	0.296	20.95%	90.2	—
Argon (Ar)	40	0.286	0.93%	87.4	Usual choice for sputtering
Carbon dioxide (CO_2)	48	0.324	0.031%	194.7‡	—
Water (H_2O)	18	0.288	Varies	373.15	Absorbs strongly on most materials
Hydrogen (H_2)	2	0.218	0.5 ppm	20.3	Formed by breakdown of water
Helium (He)	4	0.200	5 ppm	4.2	Used for leak detection

† Derived from gas viscosity (Ref. 14).
‡ Carbon dioxide sublimes rather than boils.

$$N = \frac{1}{(\sqrt{2}\pi \times 5 \text{ cm}) \times (0.316 \times 10^{-7} \text{ cm})^2}$$

$$= 4.51 \times 10^{13} \text{ cm}^{-3}$$

The gas density at one atmosphere is

$$N = \frac{6.02 \times 10^{23} \text{ molecules per mole}}{2.24 \times 10^4 \text{ mole cm}^{-3}}$$

$$= 2.69 \times 10^{19} \text{ cm}^{-3}$$

Thus the pressure separating viscous and molecular flow is

$$P = \frac{4.51 \times 10^{13}}{2.69 \times 10^{19}} = 1.68 \times 10^{-6} \text{ atm}$$

This is 1.27×10^{-3} torr, or 1.27 μm.

For vacuum deposition, we desire low enough pressure that metal atoms will not collide between leaving the source and reaching the substrate, thus ensuring film purity. This is accomplished by sealing source and substrates in a vacuum chamber and then pumping out the gas. Over a wide pressure range, a pump will have a *pumping speed S* measured in liters per second: the pump will remove constant *volumes* of gas. This pumping will result in a *flow Q* to the pump of a certain mass of gas per unit time. As we saw earlier, for an ideal gas, mass is proportional to the pressure-volume product, so that flow can be measured in *liter atmospheres per second*. Flow, pressure and pumping speed are related by

$$Q = PS \tag{7-5}$$

It follows that the lowest pressure achievable by a given pump is reached when

$$P = \frac{Q}{S} \tag{7-6}$$

If flow into the system is significant, ultimate pressure will be determined by the flow. This is the case in sputtering, where a sputter gas is metered into the chamber. It is also true if gas can leak into the chamber. In addition, pressure may be limited by a *virtual leak*.

> **Example** Gas leaks into a vacuum chamber at a rate of 1 cm^3 atm sec^{-1}. The pump has a speed of 1000 liters sec^{-1}. Find the limiting achievable pressure.
>   From Eq. (7-6), the pressure is
>
> $$P = \frac{1 \text{ cm}^3 \text{ atm sec}^{-1}}{(1000 \text{ liters sec}^{-1}) \times (1000 \text{ cm}^3 \text{ liter}^{-1})}$$
>
> $$= 10^{-6} \text{ atm} = 0.76 \ \mu\text{m}$$

which is about a factor of 10^3 higher than the highest acceptable pressure for deposition purposes.

A virtual leak is an effectively unlimited source of gas within the chamber. For example, if chamber pressure falls below the vapor pressure of some solid material in the chamber, the material will evolve large amounts of gas, preventing further drop in pressure. Virtual leaks can result either from poorly chosen construction materials or from contamination inside the chamber. In addition, most metals used in construction strongly absorb water vapor when exposed to the atmosphere: the water will desorb at low pressure, providing large gas volumes. To reduce this effect, chambers are usually vented with dry nitrogen rather than air; sometimes chambers are heated upon pump-down to drive off water.

Equation (7-6) is correct for pressure *at the pump*. However, chamber pressure can be much higher than pump pressure if the flow of gas to the pump is restricted. The effects of tubing, baffles, etc., are measured in terms of *conductance C*, where

$$C = \frac{Q}{P_1 - P_2} \tag{7-7}$$

P_1 and P_2 represent the pressure upstream and downstream from the obstruction, respectively. Vacuum conductances can be combined using the same rules as electrical conductances. For example, two elements in series of conductances C_1, C_2 have a composite conductance C given by

$$\frac{1}{C} = \frac{1}{C_1} + \frac{1}{C_2} \tag{7-8}$$

In the molecular flow regime, particular care must be taken to ensure adequate conductance to the pump.

If not limited by intentional inputs of gas, real or virtual leaks, or obstructions between chamber and pump, Eq. (7-6) implies that ultimate pressure will fall to zero. In practice, however, a pressure will be reached at which the inherent pumping mechanism of the pump is no longer effective. At this point, S is no longer a constant, and pressure reaches a limit. Thus, in a well-designed system, ultimate pressure depends on the nature of the vacuum pump.

Practical Vacuum Systems

Figure 7-1*a* shows a typical system for achieving high vacuum. An airtight chamber is formed by a *bell jar,* which can be raised to allow access or lowered onto a sealing o-*ring* gasket. There are two pumps: a *roughing* pump, which pumps efficiently near atmospheric pressure, and a *high vacuum pump,* which can operate only at reduced pressure. The "high-vac" pump pumps the chamber, and the roughing pump pumps the high-vac pump. A *roughing line* allows the roughing pump to pump the chamber directly down to a pressure at which the high-vac pump can operate. During this period, a *foreline* valve protects the high-vac pump. A *vent line* allows the bell jar to be filled with gas in order to open it. While the chamber is open, a *gate valve* isolates it from the pumps. Figures 7-1*b–d* show the valve arrangement with bell jar open, during rough pumping, and during high-vacuum operation.

Roughing pumps are usually mechanical in nature. A *rotary vane* pump, which uses rotating vanes to compress and eject gas, is a popular and inexpensive choice. It will pump at good volume down to a pressure of about 0.01 torr. A *Roots blower,* which requires backing by a rotary vane or similar pump, will achieve a pressure of 10^{-4} torr. For moderate vacuum applications, such as plasma etching, these pumps will often suffice. Their performance is limited by the vapor pressure of the oil used to seal them and by a tendency of oil vapor to *backstream* back into the chamber or high-vac pump.

High-vacuum pumps usually use nonmechanical means to remove gas from the chamber. The traditional workhorse of high vacuum is the *diffusion* pump, shown in Fig. 7-2. At the bottom of the pump is a reservoir of oil. The oil is heated, vaporizes, and rises through the stack to the nozzle jets. It emerges from the jets in a downward-directed stream of oil molecules. The oil molecules collide with gas molecules, giving them a downward momentum and directing them out of the pump. There are three nozzle stages in the pump shown plus a final ejector stage pointed towards the roughing pump. Water cools the sides of the pump, recondensing the oil vapor whenever it hits the walls. At the top of the pump is a baffle or cold trap, cooled to liquid nitrogen temperature, which prevents oil from escaping from the pump.

If operated at viscous-flow pressures, the diffusion pump would rapidly fill the system with oil vapor, so it must be kept under vacuum, and isolated until the chamber is roughed to about 10^{-3} torr. Diffusion pumps can routinely achieve pressures of 10^{-7} torr or better, and are simple and dependable. Their pumping speed is fairly independent of the gas pumped, although light molecules do pump slower. Their major disadvantage is their tendency to contaminate systems with oil. Small amounts of oil find their way into the system under normal operating conditions, and there is

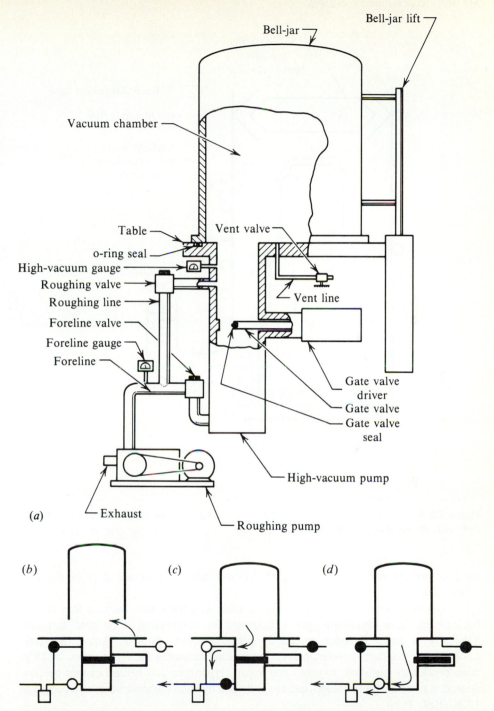

Bell-jar lift

Bell-jar

Vacuum chamber

Table

Vent valve

o-ring seal

High-vacuum gauge

Roughing valve

Roughing line

Foreline valve

Foreline gauge

Foreline

Vent line

Gate valve
driver

Gate valve

Gate valve
seal

High-vacuum pump

(a)

Exhaust

Roughing pump

(b)   (c)   (d)

Figure 7-1 (*a*) A typical apparatus for achieving high vacuum. (*b*) Valve configuration when chamber is open. (*c*) Valve configuration when rough pumping chamber from atmospheric pressure. (*d*) Valve configuration in high-vacuum operation.

243

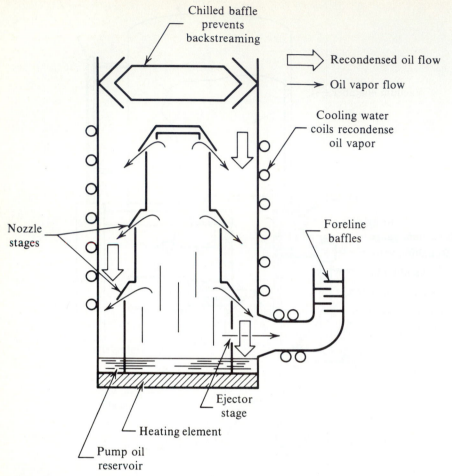

Figure 7-2 A diffusion pump. Oil heated in the reservoir emerges from the jets and collides with gas molecules, driving them into the roughing pump.

an ever present risk of substantial oil backstreaming if pressure rises or the trap warms up.

Cryopumps are increasingly popular in microelectronic use, because they cannot backstream. In a cryopump, gas is collected by condensing it on cold surfaces. Typically, a pump includes several temperature zones. A 77°K (liquid nitrogen–cooled) zone condenses water, preventing it from saturating the other zones. A 15°K zone condenses nitrogen, oxygen, and argon. Hydrogen and helium cannot be condensed at practical temperatures and are collected by charcoal absorbent, also cooled to 15°K (Ref. 1).

Cryopumps collect rather than remove gas and must be periodically *regenerated* by warming them up and pumping away the collected gas. To avoid saturating them, they must be isolated until the chamber is pumped down below 5×10^{-2} torr. Although cryopumping itself is nonmechanical, the associated refrigeration equipment is subject to mechanical failure. And since cryopump pumping speeds depend on gas sublimation temperatures, hydrogen and helium are difficult to pump. Nevertheless, cryopumps are totally oil-free and thus highly appealing for high-purity applications.

Turbomolecular pumps are mechanical pumps in which a turbine rotates fast enough to drive molecules even in molecular flow. High rotation speeds and tight tolerances make them mechanically vulnerable, and they do contain oil and require a cold trap upstream. They can operate at somewhat higher pressure than mechanical pumps and have a speed not too dependent on the gas pumped. They can be useful where continuous pumping of inert gases is required.[2]

Whichever pump is chosen to achieve low pressure, that pressure must be measured. Mechanical pressure measurement can be used down to about 10^{-2} torr using sensitive diaphragms. At lower pressures, either a *thermocouple* (Pirani) gauge or an *ion gauge* can be used.

A thermocouple gauge is useful from about $1-10^{-3}$ torr. A heated element is exposed to the low-pressure gas, which removes heat from it. Pressure is deduced from the rate of heat removal. Heat removal also depends on the heat capacity of the gas measured; thus observed pressure is a function of gas. This fact can be exploited to leak-check a vacuum system: an organic solvent like acetone, with a high heat capacity, is sprayed on the suspected leak site, and the leak is confirmed by a jump in the thermocouple gauge reading.

In an ion gauge, a hot filament emits electrons that ionize gas surrounding it. The ions are collected on an electrode, generating a current proportional to the ionization rate and thus to the gas density, or pressure. Ion gauges work in the $10^{-3}-10^{-8}$ torr range; at higher pressures, the filament would burn out. Thus ion gauges work well with high-vacuum pumps; thermocouple gauges, with roughing pumps. The pressure indicated by an ion gauge will depend on the ionization properties of the gas measured.

The leak is the natural enemy of the vacuum system. An effective tool for finding small leaks is the *helium leak detector,* a portable mass spectrometer designed to detect the presence of helium. The leak detector is connected into the vacuum system, and helium from a bottle sprayed onto possible leak sites. The small, inert, light atom of helium quickly penetrates any leak and quickly diffuses through the system to the detector. Since helium is rare in nature, any positive signal is confirmation of a leak. This method is quite sensitive: if a leak is not found after careful use of a helium leak detector, it is probably from an internal source, such as cooling water piping.

7-2 SOURCES FOR VACUUM DEPOSITION

Once a vacuum has been created within a bell jar, thin films can be deposited on a substrate by creating a supply of film molecules and letting them impinge upon the substrates. There are two principal means of generating this supply: evaporation and sputtering.

Deposition by Evaporation

If a material is melted in a vacuum, it will immediately boil, generating a vapor. Vapor molecules will then be emitted in all directions, traveling in straight lines until they strike a surface. The result will be deposition of the melted material everywhere within sight of the source.

Evaporation sources come in various degrees of sophistication. Sometimes, it is sufficient simply to place the source material on a bar of a refractory material or wrap it around a filament. The bar ("sled") or filament is then heated electrically above the melting point of the material. Emission rate is governed roughly by the electrical power supplied.

For the high-purity application of metallization to a microelectronic substrate, evaporation sources can be quite sophisticated. Figure 7-3 shows a typical source for deposition of aluminum. The aluminum supply, or "melt," is contained in a crucible. Heating the crucible might dissolve crucible material into the aluminum, so direct heating is avoided. Instead, an electron beam impinges onto the aluminum, heating the center of the melt only. The electron beam, in turn, must be generated out of the line of sight of the source to prevent coating the beam apparatus. Also, electrons and resultant

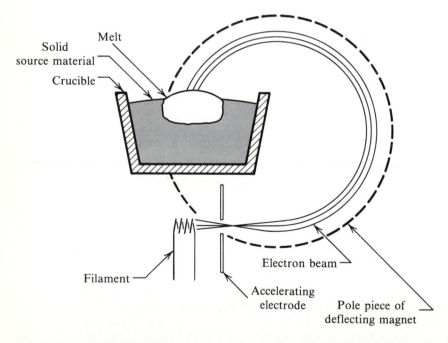

Figure 7-3 An evaporation source using electron-beam heating. The beam is generated out of the line of sight of the source and is focused into it by a magnetic field. A heated filament supplies electrons, and the accelerating electrodes form them into a beam.

radiation may damage the substrates. So the beam is generated under the source and directed between the poles of a magnet that bends the beam 270°.

An *e*-beam evaporation source requires careful control. The acceleration voltage and magnetic field must be balanced to keep the beam from missing the mark and melting holes in the equipment. A high-current beam of charges penetrating a conductive vapor is prone to arcing, so the power supply must deal automatically with arcing conditions. Evaporation rate depends on the energy delivered by the beam, which must be metered. Usually such sources are connected to computerized controllers that automatically monitor deposition rate and adjust beam energy.

An evaporation source is essentially a point source. This has implications for the uniformity of the deposited film. A point source should deposit material onto a surface at a rate r (in thickness per time) with

$$r = \frac{r_{\text{evap}}}{\Omega d^2 \rho} \cos \theta \qquad (7\text{-}9)$$

where r_{evap} is the evaporation rate (in mass per time), Ω is the solid angle over which the source emits, ρ the material density, d the source-to-substrate distance. Here θ is the inclination of the substrate (i.e., the angle between the surface normal and the direction of the source). With flat, perpendicular substrates, the deposition should be quite uniform if d exceeds a few substrate diameters.

However, microelectronic substrates usually possess surface topography, and it is important that film thickness be uniform over steps on the surface. This property is called *step coverage*. For a point source positioned perpendicularly to the substrate, $\cos \theta$ becomes zero at vertical steps, and there will be no deposition. Figure 7-4a shows the resulting situation.

Several methods are used to overcome this problem. One is to place the substrates in rotating *planetaries* that incline them somewhat to the source. As a result, the direction of incidence rotates around the step, as indicated in Fig. 7-4b. This will result in improved coverage, but the step will still block deposition during part of the cycle. Further improvement is observed if the substrates are heated to about 60% of the film melting material. This is believed to promote movement of molecules over the film surface after impact, helping to "fill in" the uncoated area. Figure 7-4c shows this effect. Perhaps the best solution, when possible, is to modify the process to eliminate vertical steps, as shown in Fig. 7-4d.

A disadvantage of evaporation emerges when alloys are to be deposited. When an alloy is melted, the various components will have different vapor pressures and thus different evaporation rates. As a result, the deposited film will be enriched in the lower-melting component, while the melt accumulates the high-melting material. If the melt is continually replenished, eventually a steady state will be reached in which the melt enrichment compensates for the lower vapor pressure and the deposited film has the composition of the feed material. Alloys are sometimes deposited by starting off with an enriched melt. Or continuous feed mechanisms are used so that the melt volume is very small and steady state is quickly reached. But in general, alloy composition is undependable when using evaporation.

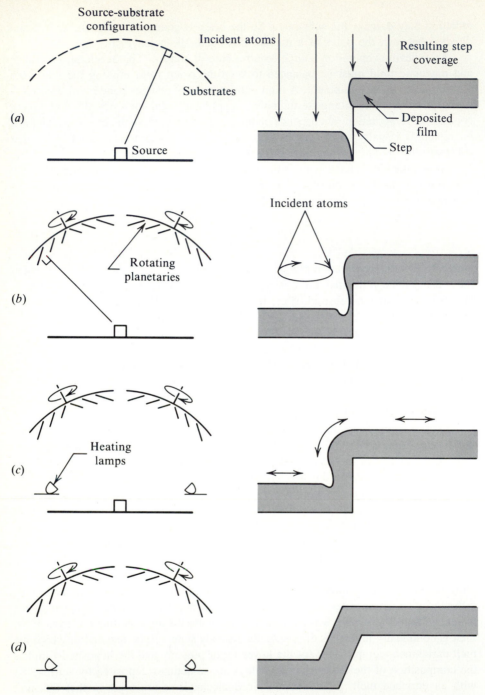

Figure 7-4 Factors governing step coverage in evaporation. (*a*) Perpendicular step on perpendicular substrate. No coverage. (*b*) Rotating planetaries with some substrate inclination. Improved coverage. (*c*) Same configuration with substrate heating. Further improvement. (*d*) Reduced slope of step, plus rotation and heating. No thinning over step.

Deposition by Sputtering

Sputtering is a phenomenon we encountered in Chapter 6 while studying plasmas. Plasmas generate a large number of positive ions that are accelerated by the plasma potential and bombard nearby surfaces. If the bombardment energy exceeds roughly four times the bond energy of a solid, atoms will be knocked loose, providing a source of atomized material.

In sputter deposition, conditions are chosen to minimize plasma chemistry and to maximize the yield of sputtered material. The plasma is generated in an inert gas, usually argon (Ar), which cannot react with the source material. Electrical power is delivered to the plasma in a way designed to maximize the number and energy of ions impacting the source or *target,* to achieve the maximum deposition rate.

A plasma can be created in a dilute gas by passing either dc or ac electrical current through it. Figure 7-5*a* shows what happens when a sufficient dc potential is applied to a gas at pressures around 10^{-1}–10^{-3} torr (Ref. 3). Most of the space between electrodes is filled with a discharge called the *negative glow,* in which roughly equal numbers of ions and electrons maintain a constant voltage. Near the negative electrode, or *cathode,* there is a shortage of electrons, a surplus of ions, and a large electric field. In this region, which does not glow and is called the *cathode dark space,* ions are accelerated to bombard the cathode. Secondary electrons resulting from the impact dominate the space immediately next to the cathode. (As the positive electrode, or *anode,* is moved farther from the cathode, additional glows and dark spaces can appear, but they are not useful in sputtering and usually are designed out of sputtering equipment.) Thus a material can be sputtered by using it to form the cathode of a dc discharge. To serve as an electrode, the material must be conductive, so dc sputtering is not possible with insulating materials.

Both conductive and nonconductive materials can be sputtered using an rf discharge similar to the ones employed in plasma etching. A positive plasma potential, which can be quite large, is generated, driving ions to bombard the electrodes. If rf power is applied to two symmetrical electrodes, as in Fig. 7-5*b,* both will be symmetrically bombarded and sputtered. This is usually not desirable.

We can concentrate the sputtering effect at one electrode by doing two things. The first is to make the electrodes of unequal area. If two electrodes have differing surface areas A_1, A_2, it can be shown that the voltage drops V_1, V_2 across the dark space next to the two electrodes are related by[4]

$$\frac{V_1}{V_2} = \left(\frac{A_2}{A_1}\right)^4 \tag{7-10}$$

Since voltage drop is such a strong function of area, most of the voltage drop and most of the sputtering can be concentrated on one electron by making it relatively small. The second requirement is to add a capacitor between this electrode and the power supply. Since the plasma is a conductor and is thus at a uniform potential, there can be a difference in voltage across the electrodes only if there is a dc bias across the power supply connecting them. The capacitor is required to prevent shorting out the discharge through the power supply.

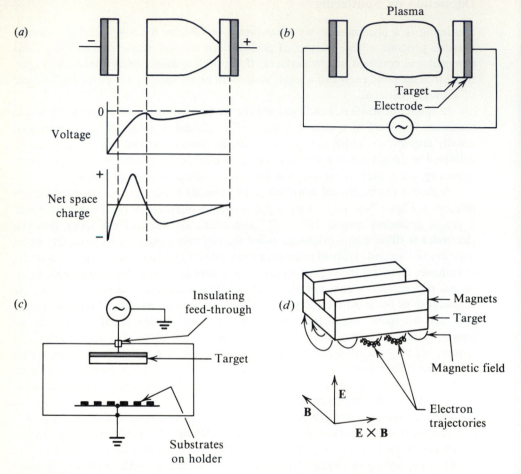

Figure 7-5 Methods of sputter deposition. (*a*) A dc discharge, showing cathode dark space and negative glow along with charge and voltage distributions. (*After Ref. 3, used by permission.*) (*b*) A symmetrical dc discharge. (*c*) A practical sputtering configuration including a target electrode and a substrate holder. (*d*) Magnetron sputtering: A magnetic field lengthens electron trajectories for greater ionization efficiency.

Figure 7-5*c* shows the electrode arrangement in a practical sputtering machine. The large electrode is the entire vacuum chamber. The small electrode (the target) is a plate of the source material, capacitively coupled to the power supply. Also shown is the substrate holder, located parallel to the target and a short distance away. As neutral target atoms fly off in all directions, a large proportion of them will be deposited on the substrates.

In Fig. 7-5*d* one further improvement is illustrated: magnetron sputtering. In conventional dc or rf sputtering, strong fields drive electrons quickly through the sputtering gas, so that they strike and ionize limited numbers of atoms. In a *magnetron*, a magnetic field is applied at right angles to the electric field. This results in a force on electrons perpendicular to their direction of motion, causing them to spiral through the gas and increasing their efficiency in ionization. Magnetron sputtering, done in either dc or rf modes, results in lower power consumption and less target and substrate heating for the same sputtering rate.

Sputtering has several advantages over evaporation. While evaporation sources are pointlike, sputtering occurs over a larger target area. This reduces shadowing near steps and improves step coverage. Material can also be resputtered by incoming atoms, "bouncing around" on the surface and further improving step coverage. Also, since solid bonding energies are at most a few electron volts, sputtering is possible with refractory materials that would be difficult to melt and evaporate controllably. Sputtering also works well with alloys and mixtures. Ion bombardment affects only the outer few layers of atoms near the target surface, smoothly removing them. Therefore, mixtures and alloys cannot fractionate, and the material deposited closely matches the target in composition.

Another advantage is that a sputterer allows the practice of *sputter etching*. Any substrate exposed to air will be coated with a native oxide layer several tens of atoms thick. In sputter etching, the substrate holder is connected to the rf power supply and becomes the target. At fairly low power, a controlled thickness of surface material is sputtered from the substrates at a rate almost independent of the chemistry of the material. The result is an atomically clean substrate that will not reoxidize in the vacuum of the sputtering chamber. By contrast, substrates for evaporation retain their oxide coatings.

7-3 METALLIZATION FOR CONDUCTORS

The principal reason for depositing metal films onto microelectronic substrates is to fabricate conductive interconnections between circuit elements. The substrate is first covered with an insulting film through which windows are etched to allow the metal to contact the substrate. The metal is then deposited and photolithographically patterned into stripes connecting the desired elements. From an electrical point of view, these stripes must be adequate current carriers and make good electrical contact where they touch the substrate. Chemically, the film must be stable, adhere well to the underlying insulator, and allow photolithography and etching. The material most often chosen to fulfill these requirements is aluminum, whose properties we shall study in this section.

Aluminum in bulk is a good conductor, with a resistivity of about 2.7×10^{-6} Ω cm. For thin films, the resistivity can be as little as $5-10\%$ of the bulk value.[5] Thus aluminum conductors can carry quite high current densities—high enough to experience a phenomenon known as *electromigration*. Electromigration turns out to be a major limiter of both the performance of metallic conductors and the lifetime of integrated circuits.

Electromigration, sometimes called the electron wind effect, is the movement of metal "downstream" in response to high current densities. Solid metal will contain some metal ions, which will move under a current flow at a velocity given by[6]

$$v = \mathcal{J}\rho \frac{qZ^*}{kT} D_o \exp \frac{-E_A}{kT} \qquad (7\text{-}11)$$

Here $\mathcal{J}$ is the current density, ρ the resistivity, qZ^* the effective ion charge, k Boltzmann's constant, and T the temperature. The term $D_o \exp (-E_A/kT)$ can be considered a diffusion coefficient containing an activation energy E_A. This activation energy, while about 1.4 eV in bulk aluminum, is around 0.5 eV at a grain boundary and 0.3 eV at the surface. Electromigration is thus encouraged by high current density, high temperature, and the existence of surface or grain boundaries.

When a microelectronic circuit fails during use, electromigration is often the cause. Such failure results from the formation of a void in a conducting stripe, by a mechanism like that in Fig. 7-6. Metal moves under the influence of current, chiefly along grain boundaries. Voids are left as metal is removed from the "upstream" grains, while mounds of extra metal called *hillocks* accumulate "downstream." As the process continues, thinning near the voids will produce resistive heating, accelerating the process. Eventually, a break will form in the conductor and the circuit will fail.

The mean time to failure (MTF) is found to follow the relationship[5]

$$\text{MTF} = K \frac{zx}{\mathcal{J}^2} \exp \frac{-E_A}{kT} \qquad (7\text{-}12)$$

where K is an empirical constant, z is the thickness, and x is the width of the metal stripe.

Excessive electromigration is prevented by controlling grain structure and limiting current density. Grain structure depends on deposition conditions such as deposition rate, substrate temperature, and gas pressure. Larger grains have less boundary area and thus resist electromigration. Consistent grain size is also important. Otherwise,

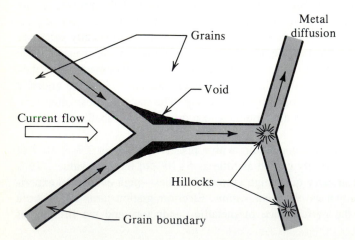

Figure 7-6 The initial step in failure of a metallization stripe due to electromigration. Aluminum flows under the influence of current along grain boundaries, moving faster where the boundaries are parallel to the current flow. Voids are produced upstream of the fast flow region; hillocks, downstream.

different electromigration rates occur in different parts of the conductor and voids form where grain sizes change.

Current density, for a given design, is dependent on the actual cross-sectional area of the conductor stripe. The width of the stripe is determined by the photolithographic process. Either variations in linewidth or notching of lines can lead to excessive current density and eventual failure. Therefore, minimum metal linewidths are strictly specified in circuit manufacture.

The film deposition process contributes to current density by determining the height or thickness of the stripe. Overall film thickness is the most obvious parameter, but it is usually step coverage that turns out to be the concern. Thinning of the film over steps can easily drive the current density above the electromigration limit. This becomes a particular problem for devices with more than one layer of interconnect, because the step heights and shapes grow more demanding with each layer. Step coverage control is a major preoccupation of a microelectronic metallization engineer.

Some alloying agents, such as copper, accumulate in grain boundaries and tend to lock them in place, offering protection against electromigration. Addition of copper also reduces *hillocking* of aluminum due to thermal treatment. Pure aluminum conductors, when heated during processing, form spikes or *hillocks* projecting upward from the surface. If more than one layer of conductor is used, hillocks from the lower layer can penetrate the insulating film and cause shorts to the upper layer. It is thus increasingly common to metallize microelectronic devices with an alloy of a few percent copper in aluminum, especially for the lower layer in multilayer designs.

Devices operating at a few volts, using contacts a few micrometers square, cannot tolerate much resistance where the conductor meets the substrate. Good contact requires both a clean surface and good chemistry between materials. For a silicon substrate, the clean surface is obtained by a short dip in dilute hydrofluoric acid immediately before placing the substrates in the vacuum chamber. The dip etches away the thin native silicon oxide layer in the contact regions (it begins reforming immediately) and also cleans the oxidized areas of the wafer by etching a thin layer from the surface. Obviously, the duration of the dip must be limited to avoid removing too much oxide. After the dip, the surface must be kept clean during pumpdown, which is why oil backstreaming from pumps is a serious concern. When using sputter deposition, it is easier to obtain a clean surface, since the substrate can be sputter etched in vacuum immediately before deposition.

The actual formation of a contact between aluminum and silicon is determined by their mutual chemistry. Figure 7-7 is a *phase diagram* for aluminum and silicon.[7] A phase diagram shows the behavior of aluminum-silicon mixtures at various concentrations and temperatures. Each dark line separates a physically distinguishable form or *phase*. Above the top curving line mixtures are liquid. Below the lower horizontal line they are conglomerations of solid silicon and aluminum. In the intermediate space, liquid mixtures exist in equilibrium with either solid aluminum or silicon. Our interest is in the narrow wedge at the left, labeled (Al), where mixtures take the form of silicon dissolved in aluminum. The wedge is shown in more detail in the inset. The boundaries of this region indicate the solid solubility of silicon in aluminum, which increases with temperature to a maximum of 1.65% at 577°C and

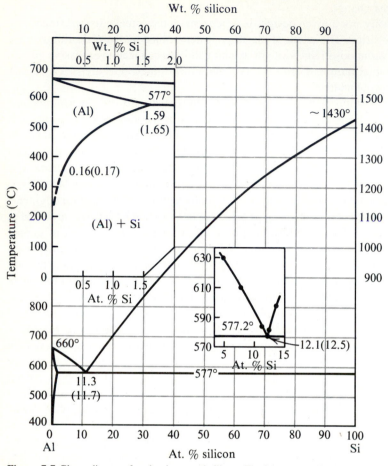

Figure 7-7 Phase diagram for aluminum and silicon. The insets magnify the region showing the limits of silicon solubility in aluminum and the zone around the silicon-aluminum eutectic. Composition values without parentheses are in weight percent silicon by weight; values in parentheses are in atomic percent silicon. (*Reference 7, used by permission. Copyright 1958, McGraw-Hill Book Company.*)

then drops off. Therefore, when hot aluminum is in contact with silicon, the silicon will dissolve into the aluminum.

Aluminum-silicon contacts are therefore *alloyed* by heating to temperatures near 500°C, usually for 30–60 min. As the silicon at the interface dissolves into the aluminum, an intimate aluminum-silicon contact forms. However, the dissolution process also forms pits in the contact area. If these pits grow too large, the aluminum can penetrate right through a shallow junction, shorting it out. Even if alloying is carefully controlled, excessive pitting can be generated during any subsequent operations in fabrication or assembly that heat the substrate. The risk can be minimized by depositing, not pure aluminum, but an alloy containing about 0.5% silicon. The alloy is already saturated at temperatures below about 450°C, and thus does not consume silicon when heated.

Before leaving phase diagrams, we should note an additional feature of Fig. 7-7. At one point in the diagram, at the bottom of the V, the top curving "liquidus" line meets the heavy horizontal "solidus" line. A liquid of this composition has the lowest melting point of any mixture. When cooled, it will condense abruptly into a solid of identical composition, not fractionating like other liquid mixtures. Such compositions are called *eutectic* mixtures. While the silicon-aluminum eutectic is not frequently exploited, a similar silicon-gold eutectic is used to allow attachment of finished devices to packages.

Another feature of aluminum chemistry is its reaction with silicon oxide. Aluminum has a great affinity with oxygen, and will abstract it from silicon dioxide:

$$\tfrac{3}{2}SiO_2 + 2Al \rightarrow Al_2O_3 + \tfrac{3}{2}Si \qquad (7\text{-}13)$$

This has two benefits: First, we recall that any silicon surface (unless sputter etched clean in vacuum) has a thin oxide coating. Aluminum reacts with and consumes this coating so that good electrical contact is made with the silicon. Second, the same reaction occurs at the interface between aluminum films and oxide layers, bonding the aluminum to the insulating coating covering the substrate. Frequently less reactive metals will not adhere directly to oxide, so an adhesion layer of a oxygen-hungry metal such as chromium must be deposited before the conductor metallization.

As circuits become increasingly complex, additional demands are put on the metallization, leading to the use of more complicated metallization schemes. Some of these are reviewed in Ref. 8.

7-4 METALS, SILICIDES, AND THE SCHOTTKY DIODE

As conductive interconnects, metals play an important but passive role in microelectronic devices. However, a metal-semiconductor interface can also be an active electronic element: the *Schottky diode*.

We saw in Chapter 2 what happens when n-type and p-type semiconductors are brought together to form a junction. Because one type of charge carrier is in excess on each side of the junction, there is a concentration-driven diffusion of carriers to the opposite side. This creates a depletion region, causing zones of fixed charge that oppose further carrier diffusion, and in turn creating a built-in potential and rectifying behavior.

A generally similar thing happens at a semiconductor-conductor junction. The conductor acts as a sink for charge carriers, so that they diffuse from the semiconductor near the surface to form a depletion region. A built-in potential and rectifying behavior result. This device is called a *Schottky diode*.

Figure 7-8 shows what happens in more detail. Figure 7-8a shows a metal and an n-type semiconductor in their separated condition. There is a Fermi level in the conductor, with an energy usually defined in terms of a *work function W*. This is the energy difference between the Fermi level and a free electron in vacuum (the vacuum

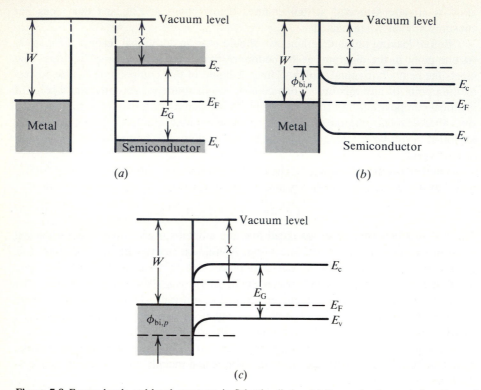

Figure 7-8 Energy levels and band structures in Schottky diodes. (*a*) Energy levels, work function, and electron affinity in isolated metal and semiconductor. (*b*) Band structure of a *n*-type semiconductor-metal junction. (*c*) Band structure of a *p*-type semiconductor-metal junction.

level). The vacuum level is related to energies in the semiconductor by the *electron affinity* χ, which is measured to the bottom of the conduction band.

In Fig. 7-8*b*, the metal and semiconductor have been brought into physical contact. Because the system is in equilibrium, the Fermi level must be the same everywhere. Since the conductor is an infinite source or sink of charge carriers, the semiconductor Fermi level is lowered to the level of the conductor. In the bulk of the semiconductor, the conduction- and valence-band edge locations remain fixed relative to the Fermi level. However, at the interface, the bands bend upward to their original position. The result is an energy barrier of height $\phi_{\text{bi},n}$ given by[9]

$$\phi_{\text{bi},n} = W - \chi \tag{7-14}$$

The height of the barrier can be deduced from an argument based on energy conservation. An electron can be ejected from the semiconductor conduction band to the vacuum at an energy cost χ: this is the definition of the electron affinity. From the vacuum, it can be returned to the metal with a liberation of energy W. It can then pass through the junction and back to its original position after surmounting the energy barrier. Given the impossibility of a perpetual energy machine, energy lost in crossing the barrier must just balance the energy gain in transiting from the semiconductor to the metal. Similar arguments apply in the case of a p-type semiconductor, resulting in the band structure shown in Fig. 7-8c. The resulting barrier height is

$$\phi_{bi,p} = E_G - (W - \chi) \tag{7-15}$$

The band structure of Figs. 7-8b and 7-8c implies a depletion of charge carriers in the semiconductor near the junction, as expected. Unlike the pn junction, there is no significant depletion zone in the metal: a slight rearrangement of the many charge carriers near the metal surface suffices to maintain charge neutrality. The width of the depletion region w can be obtained from the equations of Chapter 2 for pn junctions by taking the limit as the carrier density on the metal side goes to infinity.

The ideal barrier heights given above are modified by several effects. One of these is the *Schottky effect,* which is a lowering of metal work function due to nearby charge, in this case from the depletion layer. This effect is sufficient to give its name to the diode. Surface states can also affect barrier height, just as they affect the capacitance-voltage behavior of MOS structures. In practice, Schottky barrier heights are measured rather than predicted. Table 7-3 gives barrier heights, along with some other properties, for commonly used materials.

The Schottky diode has a number of advantages over the pn junction diode. Because it involves behavior at the metal-semiconductor interface, it requires no recombination of carriers during switching. It is thus very fast. By proper fabrication, the same metal-semiconductor structure used for interconnect can also be used as a diode, so the diode requires no additional design area. Another advantage can be seen by combining Eqs. (7-14) and (7-15) to give

$$\phi_{bi,p} + \phi_{bi,n} = E_G \tag{7-16}$$

This implies that the barrier height, and thus the turn-on voltage, for a Schottky diode is less than the energy gap, and is thus usually less than the turn-on voltage of a pn junction.

This relationship in turn-on voltages is convenient for a major application of Schottky diodes: the Schottky-clamped bipolar transistor. Figure 7-9a shows a bipolar transistor operating as we described it in Chapter 2, with the base-collector junction reverse biased. When driven sufficiently hard, such a transistor will go into *saturation,* as shown in Fig. 7-9b. The large collector current results in a substantial voltage drop through the load resistance, lowering the reverse bias across the base-collector junction.

Eventually, forward bias is attained and the junction turn-on voltage is exceeded, so the collector begins injecting carriers into the base. In this mode, the transistor is slow to switch. Saturation thus reduces circuit speed. In Fig. 7-9c, a Schottky diode has been placed between the base and collector: it is easily combined with the base contact. With its lower turn-on voltage, it acts as a bypass to keep the transistor from saturating, decreasing switching time dramatically.

We have shown that all metal-semiconductor junctions should inherently be rectifying. Yet metal conductors are routinely connected to semiconductor elements

Table 7-3 Some properties of metals and metal silicides

Material (formula)	Resistivity ($\mu\Omega$ cm)	Schottky barrier height (eV) to			Melting point[†] (°C)	Sintering temperature (°C)	Density (g cm^{-2})
		n-Si	p-Si	n-GaAs			
Aluminum (Al):							
Bulk	2.7	—	—	—	660	NA	2.70
Thin film	0.2–0.3	0.7–0.75 (Ref. 16)	0.8	—	—	NA	—
Alloys, change							
per % Si:	+0.7/%Si				—	NA	—
per % Cu:	+0.3/%Cu				—	NA	—
Titanium (Ti)	40.0	0.50 (Ref. 16)	0.61 (Ref. 19)	—	1660	NA	4.54
Tungsten (W)	5.6	0.67 (Ref. 17)	—	—	3410	NA	19.3
TiW (10 wt. % W)	15–50	0.50 (Ref. 18)	—	—	—	NA	—
Gold (Au)	2.44	0.79–0.80	0.25 (Ref. 17)	0.9–0.95	1064	NA	19.32
Silver (Ag)	1.59	—	—	0.88–0.93	962	NA	10.5
Copper (Cu)	1.77	—	—	—	1083	NA	8.96
Platinum (Pt)	10	—	—	0.86–0.94	1772	NA	21.45
Silicides							
PtSi	28–35	0.85–0.87	0.20 (Ref. 17)	—	830	600–800	—
Pd$_2$Si	30–35	0.74	—	—	720	400	—
NiSi$_2$	~50	0.7	0.45 (Ref. 20)	—	966	900	—
TiSi$_2$	13–16	0.6	—	—	1330	900	—
WSi$_2$	~70	0.65	—	—	1440	1000	—
TaSi$_2$	35–45	0.59	—	—	1385	1000	—
ZrSi$_2$	35–40	0.55	0.53–0.55 (Ref. 21)	—	1355	900	—

[†] For silicides, melting point for the lowest-melting eutectic.

Note: References, unless otherwise noted in parentheses in the table:

Resistivities: Aluminum and alloys (Ref. 5); Titanium, tungsten, alloys (Ref. 10); Silicides (Ref. 12); Other metals (Ref. 15).

Schottky barriers: Silicides to *n*-type Si (Ref. 12); To *n*-type GaAs (Ref. 16).

Melting points and sinter temperatures: Silicides (Ref. 12); All others (Ref. 15).

Densities (Ref. 15)

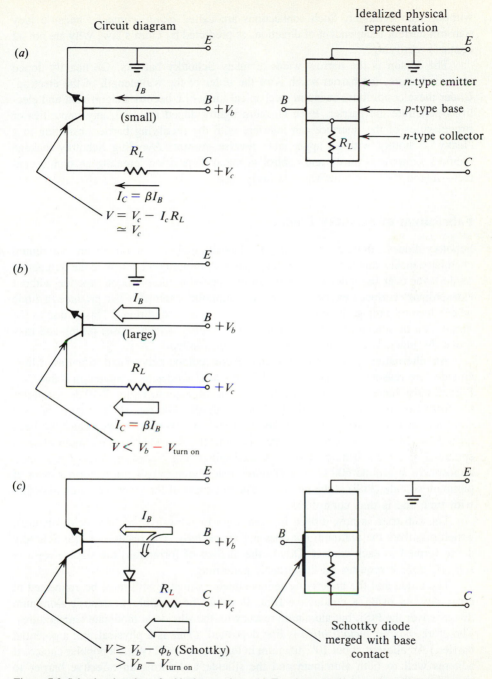

Figure 7-9 Schottky-clamping of a bipolar transistor. (*a*) Transistor, not in saturation. (*b*) Transistor, in saturation. Base-collector turn-on voltage exceeded. (*c*) Schottky diode serves as "bypass" to prevent saturation.

without such behavior. Such connections are called *ohmic contacts,* because they transmit current independent of direction, as predicted by Ohm's law. Why are not all contacts rectifying?

The reason is the narrow width of many Schottky barriers. On heavily doped semiconductors, the barrier width is of the order of the wavelength of the electron. Under these conditions, quantum mechanical tunneling becomes significant and electrons penetrate the barrier. Even on more lightly doped material, any impurities or irregularities of the interface can interfere with the rectifying barrier, resulting in a "leaky" Schottky with an appreciable reverse current. Avoiding Schottky leakage requires scrupulous purity and control in metal disposition: creating an ohmic contact—intentionally or otherwise—is fairly easy.

Fabrication of Schottky Diodes

Schottky diodes can be created simply by depositing aluminum on silicon. Aluminum chemistry makes this process fairly easy, since aluminum will consume the thin native oxide layer over the silicon and form an intimate aluminum-silicon junction without extraordinary surface treatments. However, aluminum chemistry also produces a diode whose turn-on voltage is quite sensitive to processing conditions. This is due to the dissolution of silicon into the aluminum upon heating, which causes pitting and may move the junction down into more lightly doped silicon.

An alternative is the use of metal-silicon compounds called *silicides.* Many silicides are reasonable conductors and form useful diodes as tabulated in Table 7-3. Silicides also have some convenient features for fabrication. Figure 7-10 demonstrates the formation of a platinum-silicide Schottky diode. The top of the figure shows a substrate coated with an insulating layer through which contact openings have been etched. After sputter etching to ensure an atomically clean surface, platinum metal is sputtered onto the substrate. Now the coated substrate is heated or *sintered* in an inert atmosphere at around 600°C. The platinum reacts with the silicon to form a layer of platinum silicide (PtSi) at the interface. The thickness of the layer increases smoothly with time and is thus controllable.

The substrate is now dipped in aqua regia (a nitric-hydrochloric acid mixture), which dissolves the platinum but does not attack PtSi. The result is a PtSi Schottky diode formed in each contact. Due to the method of formation, the silicide layer is *self-aligning:* it requires no lithographic patterning.

The diode and the underlying semiconductor structure still must be connected to other devices using a conductive film. It would be possible to deposit aluminum directly over the PtSi, but aluminum reacts with the silicide at moderate temperatures. Therefore, *barrier metallization* is first deposited. (This is a physical, not a potential barrier.) An alloy of about 10% titanium in tungsten (10% TiW) is a popular choice. It adheres well to both aluminum and the silicide and forms an effective barrier to aluminum diffusion.[10] In our example, a layer of TiW 0.1–0.2 μm in thickness is sputtered onto the substrate. Without removing the substrate from the sputterer, another target is brought into position and 1-μm aluminum is deposited onto the clean TiW surface.

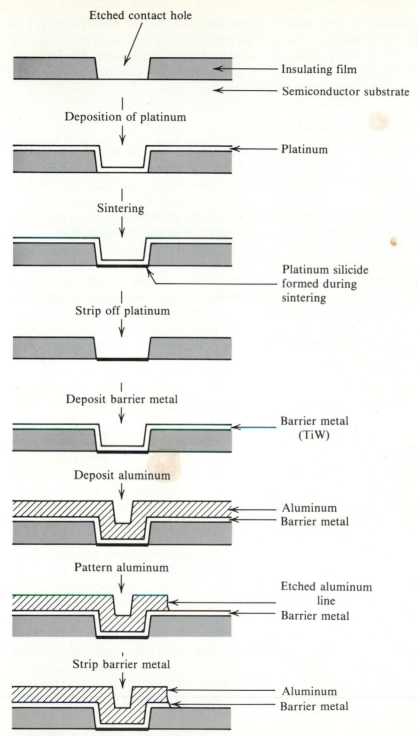

Etched contact hole

Insulating film

Semiconductor substrate

Deposition of platinum

Platinum

Sintering

Platinum silicide
formed during
sintering

Strip off platinum

Deposit barrier metal

Barrier metal
(TiW)

Deposit aluminum

Aluminum
Barrier metal

Pattern aluminum

Etched aluminum
line
Barrier metal

Strip barrier metal

Aluminum
Barrier metal

Figure 7-10 A process for fabricating a platinum-silicide Schottky diode.

The aluminum is now photolithographically patterned to form the desired interconnections. The aluminum etchant does not remove the TiW layer, so the substrate is next dipped in hydrogen peroxide solution to strip the barrier metal. The aluminum serves as a mask for the etching of TiW by peroxide, so the barrier metal is removed everywhere except under the aluminum pattern.

Other Silicide Applications

Silicides can also be used for conductive interconnects. They usually substitute not for aluminum, but rather polysilicon, which is frequently used for conducting elements like MOS device gates and PROM fuses. Since these elements are fairly short, the high (~ 300 mΩ cm) resistivity of polysilicon has been acceptable as a trade-off for various processing advantages. As dimensions grow smaller, however, the circuit time delay resulting from high poly resistivity can no longer be tolerated. Silicides retain many of the process advantages of polysilicon and have only 10–20% of the resistivity. Some applications of silicides and other refractory metals are reviewed in Ref. 11.

A silicide film can be formed on a substrate by a number of methods. It can be formed as in our Schottky example—by depositing the metal over silicon or polysilicon followed by heating to form the silicide. It can be sputtered from a target of the desired silicide composition. Or two sources, one for metal and one for silicon, can be used either in evaporation or sputtering. Finally, silicides can be formed by chemical vapor deposition.[12]

7-5 PROCESS CONTROL FOR VACUUM DEPOSITION

In vacuum deposition, the output variables to be controlled are thickness, step coverage, composition (for alloys), capacitance-voltage drift, and film properties.

There are several methods in use for thickness measurement.[13] One is mechanical: the stylus profilometer. In this instrument, a needle or stylus is moved slowly across a surface. The pressure exerted on the needle varies with the height of the surface. Therefore, a graph of the pressure, measured electronically, gives a profile of the surface that can be calibrated to give height measurements. If the surface contains a strip of the deposited film, the profilometer will indicate the film thickness by the height of the step at its edge. Figure 7-11 illustrates the technique.

Because the profilometer only detects *changes* in height, a pattern must be created in the film to provide a step. Photolithographic patterning can take hours, so prompt measurement requires some type of etching, peeling, or shadowing method to make a step. These techniques in turn may affect the film profile near the step and compromise the measurement. Nevertheless, profilometers are a frequent choice for measuring films 500 nm or more in thickness.

For thinner conducting films, thickness can be determined by measuring the sheet resistance of the film. Thickness is then obtained by dividing resistivity by sheet resistance. The four-point probe described in Chapter 3 can be used to measure resistance. Alternatively, one can use a contactless method that measures resistance by

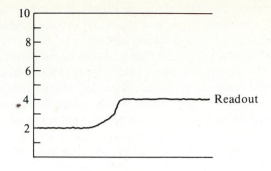

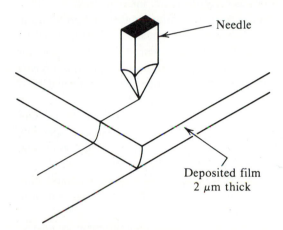

Figure 7-11 Profilometer method of determining thickness.

inducing eddy currents in the film, thus avoiding the complications of probe-film resistance. Resistance methods are quick and simple, but require knowledge of the film resistivity. This is usually *not* the bulk resistivity of the metal and may vary with deposition conditions, so resistivity must usually be determined independently for each film and deposition process.

The composition of deposited alloys can vary with deposition conditions and must be measured. Traditional chemical analysis is exacting and time-consuming, and requires destroying the film. Newer methods use the absorption, scattering, or emission of radiation: optical absorbance, x-ray fluorescence, x-ray diffraction, and beta particle backscattering. Since these methods measure the amount of an element in the radiation path, they can also be used to determine film thickness. Though once considered too complex for everyday process control, they are increasingly being incorporated into convenient, computerized equipment.

Step coverage is affected both by deposition and lithography processes and is observed after patterning using a scanning electron microscope (SEM). Because of the relationship between step coverage, electromigration, and reliability, extensive SEM inspection is often required for military and high-reliability circuits.

The capacitance-voltage (C-V) behavior of a metal film is usually tested periodically. In Chapter 3 we studied the C-V behavior of metal-oxide-silicon (MOS) structures, focusing on the effects of mobile charge in the oxide component. Mobile charge can be generated by contamination not only during oxidation, but during metallization. As a result, the C-V plot technique of Section 3-4, in which an oxidized substrate is metallized to form an MOS device, is a test of the metallization process as well as the oxide. Usually, a batch of oxidized substrates of known quality is kept on hand, to be metallized and plotted periodically as a test of the metallization. The certified metallization machine is then used with newly oxidized substrates to test the oxide quality.

"Film properties" includes such things as grain structure, internal stress, reflectivity, etch rate, adhesion to photoresist, etc., which affect film stability and processability. Frequently, these properties are difficult to measure or quantify, and the variables that affect them are rarely well characterized. To control them, therefore, deposition conditions must be made as reproducible as possible. Inputs to the deposition process include deposition time and rate, system geometry, substrate temperature and bias, substrate cleanliness, and pumpdown and venting conditions.

Deposition rate is a function of the power applied to the source. Sometimes sufficient rate control is obtained simply from power settings. In other cases, especially in e-beam evaporation, an automated rate controller is necessary. This controller constantly measures deposition thickness and adjusts the power to produce the desired rate. When rates are dependable, the desired film thickness will be obtained after depositing for a set time. If automatic controllers are used, they are programmed to shut off when the desired thickness is reached.

The actual rate of deposition onto the substrate depends on equipment geometry. In sputtering, rate falls off near the edge of the target. In evaporation, with its pointlike source, tooling within the chamber can shadow substrates and impede deposition. If rotating planetaries are used, the rotation path and angle of incidence deserve careful attention. Step coverage can be degraded severely if planetaries bend or deform due to use.

Substrate temperature is an important and elusive parameter. Deposition systems are not in thermal equilibrium. In evaporation, heat lamps are frequently used for substrate heating and are controlled with a thermocouple in the chamber. But the temperature of a rotating partially metallized silicon wafer in a planetary is almost certainly not the same as that of a stainless-steel thermocouple sheath. Various subtle effects can cause the relationship between substrate and measured temperatures to change over time. In sputtering, substrate holders are often water-cooled to maintain a stable temperature. But the back of the substrate can be considerably hotter than the holder unless special measures are taken to ensure good thermal contact. The front surface, bombarded by ions, electrons, and deposited material, is certainly still hotter by a hard-to-define amount. The measurement of temperature under deposition conditions continues to encourage both ingenuity and discussion.

Table 7-4 Troubleshooting guide for deposition processes

Problem	Possible causes
I. Film thickness	
A. Deposition controlled by time	Incorrect time used
	Measurement incorrect
	Deposition rate incorrect
	Evaporation
	e-beam out of focus, missing center of melt
	Melt level incorrect
	Sputtering
	Target surface condition
	Gas pressure (or gauge calibration)
	Gas purity (source gas contaminated, leaks, incomplete pumpdown)
	Plasma problems (target grounding, etc.)
B. Deposition controlled by controller	Sensor (crystal) requires changing
	Incorrect controller programming
	Controller malfunction
II. Uneven film thickness	Obstructions between source, substrates
	Problems with substrate movement (planetary rotation, table travel, etc., as applicable)
	Too close to edge of sputter target
III. Incorrect composition (alloys)	Deposition from multiple sources
	(Relative deposition rates incorrect. See above under deposition rates.)
	Deposition from one source
	Target/melt/feedstock composition
	Target surface condition
	Sputtering gas purity
IV. Step coverage	Sputtering
	Too close to edge of target
	Gas pressure (or gauge calibration)
	Substrate bias or temperature
	Evaporation
	Planetary geometry: wrong angle, no rotation, deformed due to use, etc.
	Substrate temperature (including temperature calibration)
V. Change in properties/purity (reflectivity, grain structure, conductivity, adhesion, etc.)	Deposition rate (see earlier)
	Substrate temperature
	Substrate bias
	Gas purity and pressure (leaks, incomplete pumpdown, etc.)
	Premature venting (while hot, etc.)

Substrate cleanliness is important, but difficult to confirm. For confirming oxide removal from silicon contacts, the *cold plate* can be useful. The substrate is placed on a chilled plate and observed under a microscope. Microscopic water droplets soon condense from the atmosphere. Silicon is hydrophobic, and droplets on a clean surface are round and distinct like mercury drops. A surface exposed to air for even 30 min will grow enough hydrophilic oxide that water condenses in a sheet rather than drops.

Care is also required in both generating and venting a vacuum. To ensure purity, the chamber should be evacuated to at least 10^{-6} torr before deposition begins. After deposition, premature exposure of the hot metal surface to air can lead to unwanted reactions. For example, aluminum films can acquire a thicker-than-normal coating of slow-etching aluminum oxide. Substrates should first be allowed to cool; then the chamber be vented with an inert gas.

Table 7-4 lists some causes of variation in vacuum deposition processes.

PROBLEMS

7-1 A very good vacuum is considered attained with a pressure of 10^{-8} torr. Assuming that the residual gas is mainly nitrogen (atomic weight 28 g mole^{-1}), use the ideal gas law to calculate the number of molecules in a liter of this "vacuum." Avogadro's number is 6.02×10^{23} molecules per mole.

7-2 Use the ideal gas law and Avogadro's number L to express mean free path in terms of pressure.

7-3 Find the volume occupied by one mole of the following gases at standard temperature and pressure (STP): (*a*) Hydrogen (*b*) Nitrogen (*c*) Argon.

7-4 The characteristic dimension of a vacuum system for a sputterer is taken to be 4 cm. The system is operated using argon. Estimate the pressure dividing viscous and molecular flow.

7-5 Compute the mean free path of a molecule under the following conditions. (*a*) Air (nitrogen) at 1-atm pressure. (*b*) Argon at 20 μm. (*c*) Nitrogen at 10^{-6} torr. (*d*) Argon at 10^{-6} torr. (*e*) In each of these cases, relative to characteristic system dimensions of 1 cm, is flow viscous or molecular?

7-6 Gas leaks into a vacuum chamber at a rate of 0.8 cm^3 atm sec^{-1}. The pump has a speed of 1200 liters sec^{-1}. Find the limiting achievable pressure.

7-7 A vessel of volume V_0 at initial pressure P_0 is evacuated with a pump of pumping speed S to a final pressure P_f. There is no leakage or outgassing. (*a*) Write an equation for the time required. Hint: Express pumping speed and gas in the chamber in terms of the number of moles n. Then write a differential equation for n. (*b*) If chamber volume is one cubic meter, and $S = 1000$ liters sec^{-1}, find the time to evacuate the chamber to 10^{-8} torr. (*c*) In practice, this evacuation frequently takes an hour or more. Speculate on any deviation from prediction.

7-8 Tooling for a vacuum deposition system is cleaned and reassembled. A drop of water 0.1 cm^3 in volume is left in the bottom of a screw hole. Due to the obstructed path to the pump, a negligible amount of water escapes during the early stages of pumpdown, and the drop becomes a virtual leak at a chamber pressure of 10^{-5} torr. Find the volume of gas formed by the water at this pressure. The density of water is 1 g liter^{-1} and the molecular weight 18 g mole^{-1}.

7-9 An evaporator is controlled to deposit 5 nm sec^{-1} of aluminum on substrates. Substrates are in a rotating planetary, positioned 50 cm from the source and inclined at an angle of $10°$. The source emits over half the sphere, that is, over a solid angle of 2π. (*a*) What is the aluminum consumption rate in grams per second? (*b*) What portion of a melt of 50-cm^3 volume will be consumed in depositing 2 μm of aluminum on the substrates?

7-10 Substrates for sputter deposition are placed 10 cm below the target. Sputtering is performed using argon. (*a*) Find the argon pressure in microns at which the mean free path in the gas becomes equal to the target-substrate distance. (*b*) If sputtering is performed using 20 μm of Ar, is metal-gas interaction likely?

7-11 Tunneling will predominate and contact become ohmic when the Schottky barrier width approximates an electron wavelength. Use Eq. (2-28), with $N_D = \infty$, to estimate the width of an aluminum Schottky barrier to n-type silicon, as a function of N_A. Calculate the doping at which this width equals the wavelength of a 1-V electron. De Broglie's relation for the wavelength of particles is $\lambda = h/mv$, and the electron mass is 9.1 $\times 10^{-28}$ g. One electron volt equals 1.60×10^{-12} erg. For the permittivity of silicon, use 648.2 q V^{-1} μm^{-1}.

7-12 Electromigration is the predominant failure mechanism for a certain circuit. Estimate the effect on mean time to failure (MTF) caused by the following deviations: (*a*) Metal thickness is below design thickness by 25%. (*b*) Metal linewidth is notched in various places, decreasing to 75% of design width. (*c*) Both (*a*) and (*b*). (*d*) Cooling system failure causes operating temperature rises by 25°C above the nominal value of 60°C. (*e*) Improper design results in a current density through a lead 25% above design rule. Let E_A for migration equal 1.4 eV.

REFERENCES

1. John F. Peterson and Hans A. Steinherz, *Solid State Technol.* **25**:104 (January 1982).
2. Donald Krieger, *Solid State Technol.* **22**:62 (December 1979).
3. J. L. Vossen and J. J. Cuomo, "Glow Discharge Sputter Deposition," in *Thin Film Processes,* John L. Vossen and Werner Kern (eds.), Academic, New York, 1978, p. 27.
4. *Ibid.,* p. 28.
5. D. Pramanik and A. N. Saxena, *Solid State Technol.* **26**:127 (January 1983).
6. F. M. D'Heurle and R. Rosenberg, in *Physics of Thin Films,* Vol. 7, G. Hass, M. H. Francombe, and R. W. Hoffman (eds.), Academic, New York, 1973.
7. M. Hansen, *Constitution of Binary Metal Alloys,* New York: McGraw-Hill, New York, 1958.
8. Sal T. Mastroianni, *Solid State Technol.* **27**:155 (May 1984).
9. S. M. Sze, *Physics of Semiconductor Devices,* Wiley-Interscience, New York, 1969, p. 369.
10. Vance Hoffman, *Solid State Technol.* **26**:119 (June 1983).
11. John Y. Chem and Lynette B. Roth, *Solid State Technol.* **27**:145 (August 1984).
12. S. P. Muraka, *J. Vac. Sci. Technol.* **17**:775 (1980).
13. Sheldon C. P. Lim and Doug Ridley, *Solid State Technol.* **26**:99 (February 1983).
14. Walter J. Moore, *Physical Chemistry,* 3d ed., Prentice-Hall, Englewood Cliffs, NJ, 1962, p. 229.
15. *Handbook of Chemistry and Physics,* 56th ed., Robert C. Weast (ed.), CRC Press, Cleveland, 1974.
16. Sorab K. Ghandi, *VLSI Fabrication Principles: Silicon and Gallium Arsenide,* Wiley, New York, 1983, pp. 465–67.
17. S. M. Sze, *Physics of Semiconductor Devices,* Wiley-Interscience, New York, 1969, p. 399.
18. John Shier, personal communication.
19. A. M. Cowley, *Solid State Electronics* **12**:403 (1970).
20. J. M. Andrews and F. B. Koch, *Solid State Electron.* **14**:901 (1971).
21. J. M. Andrews and M. P. Lepselter, *Solid State Electron.* **13**:1011 (1970).

EIGHT

THIN FILMS: MAINLY NONMETALS

In this chapter, we continue our examination of thin-film deposition methods. Most of this chapter deals with the principle alternative to vacuum deposition. The chief of these is *chemical vapor deposition (CVD)*, in which a film is formed by chemical reaction of gases. A CVD process is designed to supply reactive gases to the surface under conditions that encourage surface reaction and discourage reaction elsewhere. CVD processes exist for a vast variety of films, but they are particularly useful for forming refractory, nonmetallic films that are not easily evaporated.

In Section 8-1, we deal with the basic principles of CVD. In Section 8-2, we look at the equipment for CVD, including low-pressure and plasma-enhanced applications. In Section 8-3, we examine CVD processes for three important films: silicon dioxide, silicon nitride, and polysilicon. In Section 8-4, we study a particularly important CVD application: vapor phase epitaxy. In epitaxy, material is added to the substrate itself in such a way that the crystal structure is continued into the new film. In Section 8-5, we present some alternative deposition techniques less widely used than vacuum or CVD methods but still significant. We discuss two alternative epitaxy methods (liquid phase and molecular beam), as well as the spin-on method used for polyimide coatings and (briefly) electroplating. In Section 8-6, we summarize methods for process control of CVD processes.

8-1 CHEMICAL VAPOR DEPOSITION

In chemical vapor deposition, one or more gases react on a surface to form a film. We desire a film of uniform thickness and composition, and of predictable purity and

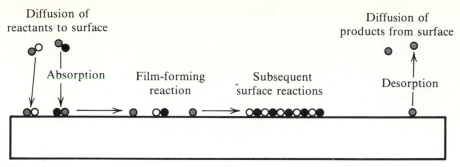

Figure 8-1 The sequence of reaction steps in a CVD reaction.

structure. This requires not only uniform reaction conditions, but the suppression of undesirable side reactions. In particular, the gas must be kept from reacting elsewhere than at the surface to be coated or undesirable particulate contamination will result.

Figure 8-1 summarizes the steps involved in a CVD process. First, the reactant molecules must diffuse through the gas to the reaction surface. There they must be absorbed on the surface and react. After the principal chemical reaction, various surface reactions may follow in order to produce the desired properties in the film. Finally, any by-products formed by the reaction must leave the surface and diffuse away into the gas. Depending on the circumstances, any of these steps may be the slowest, and hence rate-determining, step.

Transport of Reactants to the Surface

At first sight, movement of gaseous reactants to the substrate surface appears to be unnecessary. After all, the surface is directly in contact with the reactant gas stream. But the gas at the surface quickly becomes depleted of reactants, and fresh material must be constantly supplied. This implies that the gas must flow at significant velocities. And this in turn results in the formation of a *boundary layer,* which has important implications for reactant transport.

Figure 8-2 shows what happens when a fluid moves past a solid surface. The solid exerts a drag force on the fluid, which affects the velocity according to Newton's law of viscous flow:

$$F = \mu \frac{dv}{dz} \qquad (8\text{-}1)$$

where F is the force, μ the viscosity of the fluid, v the fluid velocity, and z the distance from the surface. If infinite values of force are to be avoided, the velocity must be zero at the surface and must increase smoothly with distance into the fluid. The result is a slowly moving layer of fluid, called the boundary layer, close to the surface. Since the velocity varies smoothly, the thickness of the boundary layer is a matter of definition; for example, it can be considered to end when the fluid velocity reaches 99% of the bulk value.

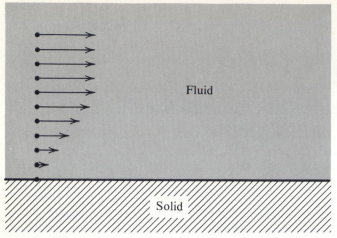

Figure 8-2 Movement of a fluid past a solid surface, illustrating the formation of a boundary layer.

Although fluids span an enormous range of properties, general rules for fluid behavior can be written using certain dimensionless numbers. One of these is the Reynolds number, Re, given by

$$\text{Re} = \frac{dv\rho}{\mu} \tag{8-2}$$

Here ρ is the fluid density, and d is a length characteristic of the flow system, such as the diameter of a flow tube. It turns out that the boundary layer thickness δ is related to the Reynolds number by[1]

$$\delta \sim \frac{l}{\sqrt{\text{Re}}} \tag{8-3}$$

where l is the distance from the front edge of the surface. When a boundary layer is present, fresh reactant must reach the surface by diffusing across the stagnant layer. Therefore, the reactant flux j will be [compare Eq. (3-52)]

$$j = \frac{D}{\delta}(N_G - N_0) \tag{8-4}$$

where D is the diffusivity of the reactant in the gas, and N_G and N_0 are the concentrations at the top of the boundary layer and next to the surface, respectively.

Several features of real systems can complicate this simple picture. One is turbulence. We have tacitly assumed *laminar flow*, in which there is no large-scale mixing of the fluid by turbulence. Laminar flow is the rule for slowly moving fluids; turbulence sets in at velocities such that the Reynolds number exceeds 2000–3000. Most CVD systems are designed to operate with Re $\sim$ 100 and are thus nonturbulent. However, there are strong temperature gradients in most CVD systems, which also

Table 8-1 Some properties of gases relevant to chemical vapor deposition

Gas (formula)	Molecular weight	Viscosity (in micropoises)			
		@0°C	@490°C	@600°C	@825°C
Hydrogen (H_2)	2	83	167	183	214
Nitrogen (N_2)	28	153	337	—	419
Oxygen (O_2)	32	189	400	—	501
Argon (Ar)	40	210	448	—	563

Units of viscosity: 1 poise (P) = 1 dyn sec cm^{-2}
Units of kinematic viscosity: 1 stroke (St) = 1 P cm^3 g^{-1}
Kinematic viscosity is viscosity divided by fluid density.

Scaling of gas properties based on the kinetic theory:
  Diffusivity: Temperature$^{3/2}$/pressure
  Viscosity: Temperature$^{1/2}$
  Density: Pressure/temperature

affect gas flow. Another dimensionless number, the Grashof number Gr, helps to determine the effect of temperature gradients. Spiraling of streamlines, resulting in mixing, is reported[2] to occur when Gr/Re^2 > 0.5.

$$Gr = \frac{\rho^2 g d^3 \beta \, \Delta T}{\mu^2}$$

β is the thermal expansion coefficient, and g is the acceleration of gravity. When effects of this type are reasonably small, they are often incorporated into Eq. (8-4) by considering δ to be an *effective* boundary layer thickness.

Table 8-1 gives some properties of gases used as carriers in CVD reactions and some scaling rules for gas properties derived from the kinetic theory of gases.

Example A CVD reactant is diluted to 1% in hydrogen. The gas flows at 10 cm sec^{-1} down a tube of 15-cm diameter, at 600°C. (a) What is the Reynolds number? (b) What is the boundary layer thickness at the end of a 20-cm sled? (c) From the deposition rate, the flux j is deduced to be 10^{20} molecules per square centimeter per second. Find D, assuming that reaction is quite complete so that $N_0 \sim 0$.

From the ideal gas law, we can find the molar density of hydrogen at 600°C to be

$$\frac{1 \text{ mole}}{22,400 \text{ cm}^{-3}} \frac{273}{273 + 600} = 1.40 \times 10^{-5} \text{ mole cm}^{-3}$$

Multiplying this number by the mole weight gives the mass density ρ:

$$\rho = (2 \text{ g } H_2 \text{ per mole})(1.40 \times 10^{-5} \text{ mole cm}^{-3})$$
$$= 2.79 \times 10^{-5} \text{ g cm}^{-3}$$

We will also need the number density of molecules, which we obtain by multiplying the molar density of Avogadro's number:

$$N_G = (6.02 \times 10^{23} \text{ molecules per mole})(1.40 \times 10^{-5} \text{ mole cm}^{-3})$$

$$= 8.40 \times 10^{18} \text{ molecules per cubic centimeter}$$

From Table 8-1, the viscosity μ of H_2 at 600°C is 1.83×10^{-4} dyne sec cm^{-1}. Thus

$$\text{Re} = \frac{(15)(10)(2.79 \times 10^{-5})}{1.83 \times 10^{-4}} = 22.9$$

The boundary layer thickness at the end of the sled will be

$$\delta = (20 \text{ cm})(22.9)^{-1/2} = 4.18 \text{ cm}$$

We can deduce D by rearranging Eq. (8-4) to give

$$D = \frac{\delta j}{N_G - N_0}$$

With $N_0 \sim 0$, and N_G for reactant to be 1% of the value for the carrier gas, we have

$$D = \frac{(4.18)(10^{20})}{(8.40 \times 10^{18})(0.01)}$$

$$= 4980 \text{ cm}^2 \text{ sec}^{-1}$$

Reaction at the Substrate Surface

Reactant diffusing to the surface is consumed by the film-forming reaction. The flux of reactant at the surface is thus given by [compare Eq. (3-52)]

$$j = k_s N_0 \tag{8-5}$$

where k_s is the surface reaction rate. In turn, k_s can be written as

$$k_s = k' \exp \frac{-E_A}{kT} \tag{8-6}$$

where E_A is the activation energy, k is Boltzmann's constant, and T the temperature. k' is a constant for a given reaction and reactant concentration. The rate of a CVD reaction can be determined by setting the flux equal in Eqs. (8-4) and (8-5) to give

$$j = \frac{DN_G k_s}{D + \delta k_s} \tag{8-7}$$

Normally, we are interested in the growth rate r, in film thickness per unit time, rather than reactant flux. If γ is the number of atoms per unit volume of film, $r = j/\gamma$ and

$$r = \frac{DN_G k_s}{\gamma(D + \delta k_s)} \tag{8-8}$$

The reaction rate is exponential with temperature while the diffusivity varies approximately with $T^{3/2}$, so at high enough temperatures k_s will greatly exceed D/δ. Under these conditions,

$$r \simeq \frac{DN_G}{\gamma\delta} \qquad \delta k_s \gg D \qquad\qquad (8\text{-}9)$$

and the film growth rate will be determined chiefly by the barrier layer thickness. At low temperatures or at slow reaction rates,

$$r \simeq \frac{N_G k_s}{\gamma} \qquad D \gg \delta k_s \qquad\qquad (8\text{-}10)$$

and the reaction rate, which depends strongly on substrate temperature, will determine growth rate. Note that the density of the reactant in the gas, N_G, is significant in both cases.

Equation (8-6) implies that the rate of all reactions increases with temperature. This is correct in fact, but sometimes misleading because it ignores the reversibility of chemical reactions. Figure 8-3a illustrates the energetics of a typical chemical reaction. On the left-hand side of the diagram are shown two reactants, AB and C. On the right-hand side are products, A and BC. This particular reaction is *exothermic:* it liberates energy because the products are of lower energy than the reactants. However, for a reaction to occur, the reactants must usually pass through an unstable state (ABC) that is of higher energy than either products and reactants. The extra energy required to reach this state is the activation energy E_A, which appears in Eq. (8-7).

In principle, all chemical reactions are reversible. Thus AB and C can pass back through the intermediate (or transition) state (ABC) and react in reverse to produce A and BC. Because AB and C are at a lower energy than A and BC, the reaction consumes rather than liberates energy. Such reactions are called *endothermic*. It follows from our diagram that the activation energy for the reverse, endothermic reaction is higher than for the forward, exothermic reaction.

The observed reaction rate will be the difference between the forward and reverse reactions. Figure 8-3b shows the forward and reverse reaction rates and the net rate for an exothermic forward reaction. Note that we have plotted the logarithm of the reaction rate versus $1/T$, giving for the individual reaction rates straight lines with slope E_A. (Such a plot is called an *Arrhenius plot,* frequently used to derive activation energies from rate data.) Therefore, higher temperature is to the left.

For an exothermic reaction, the reaction rate will increase with temperature, but not as fast as the reverse, endothermic reaction with its higher E_A. As a result, at sufficiently high temperature, the reverse reaction may dominate and the observed reaction rate fall off. For an endothermic reaction, shown in Fig. 8-3c, the reaction rate continues to rise with temperature. It follows that, for an exothermic reaction, the reaction rate is often highest in the cooler regions of a reactor, while endothermic reactions will proceed fastest at the hottest spot. This dependence is important in CVD reactor design.

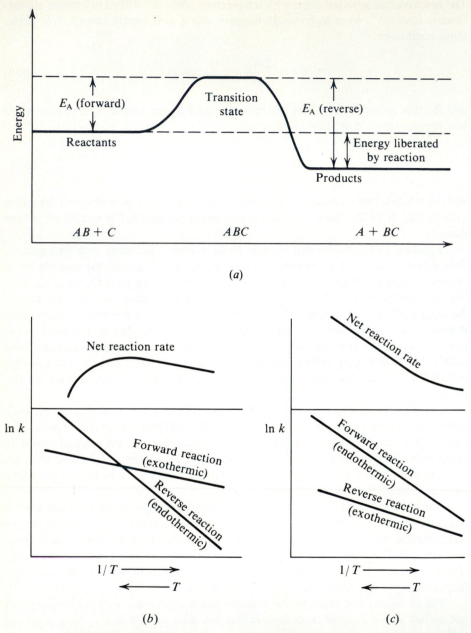

Figure 8-3 Energetics of a chemical reaction. (*a*) Reaction energies and activation energies for a forward and reverse reaction. Note that the activation energy for the exothermic reaction is always less than that for the reverse, endothermic reaction. (*b*) Arrhenius plot of reaction rates for an exothermic reaction, its reverse and the net reaction rate. (*c*) Similar plot for an endothermic reaction and its reverse.

Other Reaction Steps

Once the film-forming reaction has taken place on the surface, various other surface reactions may be required to give the film its final form. The species may have to diffuse on the surface to a desired location, for example. These types of reactions can be dependent on surface properties, including crystal orientation. They may also vitally affect the step coverage of the deposited films.

If products other than the film are formed, they must desorb from the surface, diffuse through the boundary layer, and be swept away in the gas stream. In principle, such steps can be rate-limiting, but usually a reaction limited by product diffusion will not be chosen for a practical CVD process. However, diffusion of products and other species can be important in determining the purity and structure of deposited films.

8-2 REACTORS FOR CHEMICAL VAPOR DEPOSITION

A CVD reactor has two principal purposes. One is to provide a uniform supply of gaseous reactant to the substrate surface. The other is to provide energy (usually thermal) to supply the activation energy for the reaction and allow it to proceed. Reactors can be classified by the way they accomplish these aims. Figure 8-4 gives examples of various reactor types.

Methods of Gas Delivery

When CVD is performed in a batch mode, two principal geometries are used to supply gas to the substrates. Perhaps the simplest is the *horizontal reactor,* shown in Fig. 8-4a. This configuration is quite similar to the diffusion furnace tube introduced in Chapter 3, in that gases flow down a cylindrical tube and over the substrates. In some applications it may be possible to load the substrates as in diffusion, held perpendicular to the gas flow in a quartz boat. More frequently, uniform gas delivery can be obtained only by laying the substrates down on a *susceptor,* so that gas flows over them, as shown in the figure. Gas is depleted in flowing over the substrates, and usually uniform deposition is obtained only if this depletion is compensated. One method is to incline the susceptor as shown; another is to maintain a temperature differential (or *ramp*) across the reaction zone.

Figure 8-4b shows a *vertical reactor* arrangement. Here the basic configuration is that of a bell jar, with gas supplied from a central point and flowing down and out to the substrates. A susceptor supports the substrates and usually rotates to smooth out any nonuniformities in flow. The flow path is usually shorter than in the horizontal arrangement, and depletion is less of a factor, but there are moving parts and thus greater complexity.

If good deposition can be obtained without totally isolating the reaction area, the *continuous* reaction mode becomes attractive. Figure 8-4c shows such a reactor, in which substrates are carried by a conveyor underneath a set of nozzles that provide reactants. Reaction takes place continuously beneath the nozzles, and a set of exhaust

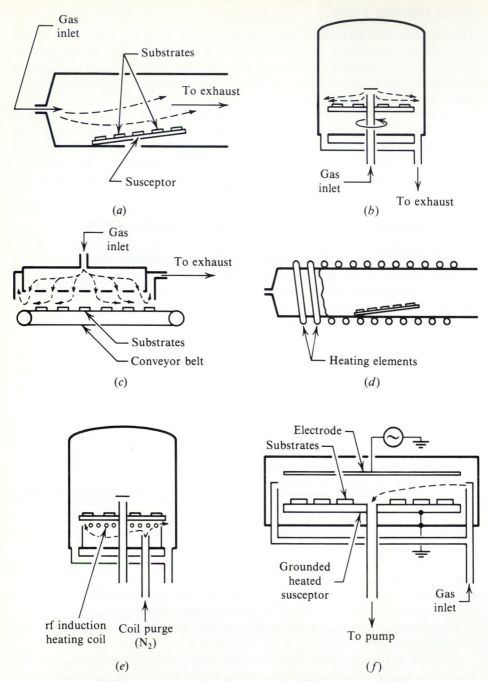

Figure 8-4 Types of CVD reactors. (*a*) Horizontal reactor. (*b*) Vertical reactor. (*c*) Continuous flow reactor. (*d*) Hot-wall reactor (horizontal). (*e*) Cold-wall reactor (vertical). (*f*) Plasma-enhanced CVD reactor.

vents removes products and toxic gases. This configuration is very efficient in production, but provides less control of the reaction environment.

Reduced-Pressure CVD

No matter what reactor configuration is chosen, the boundary layer significantly complicates reactant delivery at atmospheric pressure. It is thus desirable to reduce boundary layer thickness, which can be done by performing the deposition at reduced pressure.

Equations (8-2) and (8-3) show that the gas flux through the boundary layer varies directly with the diffusivity and inversely with the boundary layer thickness, which is proportional to the inverse square root of the Reynolds number. Diffusivity varies inversely with pressure, and the Reynolds number depends on gas velocity, viscosity, and density. Consider the effect of reducing the pressure by a factor of 1000. Diffusivity rises proportionally. Density falls by a factor of 1000. Gas velocities, in practical systems, will rise 10–100 times, while viscosity remains essentially constant. The result is an increase in the layer thickness of a factor of 3–10, but an overall increase in gas flux by a factor of 100–300 (Ref. 3).

Reduced-pressure or low-pressure CVD (RPCVD or LPCVD) has a number of advantages. The most obvious is improved uniformity with closer wafer stacking, leading to large batch sizes and high productivity. Process parameters may also be improved. The reactor configuration can be either horizontal or vertical. The chamber must be leak-free, strong enough to sustain vacuum, and fitted with a vacuum pump and a gas metering system. Because the exhaust gases are reactive, particle-forming, and usually toxic or flammable, particular care is required in vacuum system design and maintenance.

Methods of Heat Delivery

CVD reactions occur at elevated temperatures, requiring heating of the substrates. In the *hot-wall* reactor, the walls of the chamber are heated directly and the heat is transmitted to the substrates. In this configuration, the walls are the hottest part of the reactor, so hot-wall reactors are useful for exothermic reactions. The horizontal configuration is especially convenient for hot-wall operation, because resistive heating elements can be wrapped around the tube. A horizontal, hot-wall reactor, shown in Fig. 8-4d, is thus identical in concept to a diffusion furnace.

For endothermic reactions, the reaction occurs at the hottest spot and thus the substrates should be hotter than the walls. This requires a *cold-wall* configuration, where the substrates are heated more directly. Delivering heat into the corrosive and explosive environment of the reactor requires some ingenuity. Frequently, the susceptor is made of graphite and heated resistively by rf induction. Or heat lamps can be used to heat the substrates themselves. Figure 8-4e shows the induction method. A coil purge sweeps reaction gas away from the heating elements.

Plasma-Enhanced CVD

CVD reactions are run at high temperatures to supply enough heat energy to surmount the activation energy barrier. But high temperatures can damage substrates, especially metallized ones. Therefore, it is useful to supply energy to the reactants more selectively.

We have already seen in Chapter 7, an efficient method for breaking gas molecules into reactive fragments: the rf plasma. When used for etching, a plasma generates gas ions and free radicals, which react with substrates with almost zero activation energy. Similarly, a plasma can break CVD reactants into fragments that can react at substantially reduced temperatures. This process is *plasma-enhanced CVD*. A plasma-enhanced CVD reactor, shown in Fig. 8-4f, is very similar to the planar plasma etcher of Chapter 7. Substrates are placed on a table that is also an electrode, and a plasma is generated in the gas above them. Plasma-enhanced CVD is also reduced-pressure CVD, since the plasma can be sustained only at low pressures.

8-3 CVD APPLICATIONS

Chemical vapor deposition has been used to deposit a vast array of films, including insulators, semiconductors, and metals. Reference 4 gives an extensive list. Generally, the reactants are compounds of the desired film element(s) with hydrogen or chlorine: many such compounds are gaseous. The CVD reaction may be pyrolysis (decomposition by heat), oxidation of the hydride, or reduction of the chloride with hydrogen. In this section, we will look at CVD processes for three silicon-containing films: silicon dioxide, silicon nitride, and polycrystalline silicon.

Silicon Dioxide

On a silicon substrate, silicon dioxide (SiO_2) can be grown by direct oxidation. Where this "thermal" oxide cannot be grown (e.g., on metallized substrates or over other layers), CVD oxide can be used. This film, sometimes called vapox, pyrox, or silox, can be deposited by oxidation of a number of silicon compounds including $SiCl_4$, $SiBr_4$, SiH_2Cl_2, and tetraethoxysilane [TEOS, $Si(OC_2H_5)_4$]. The oxidant may be O_2, NO, N_2O, or CO_2 with hydrogen.[4]

Perhaps the most common process, however, is the oxidation of silane:

$$SiH_4 + O_2 \rightarrow SiO_2 + 2H_2 \tag{8-11}$$

This reaction is typically carried out at 400–500°C and has an unusually low activation energy of around 0.4 eV (Ref. 5). This value of E_A implies that the rate-limiting step is a surface reaction.[6]

CVD silicon oxide is frequently "doped" by adding another gas to the mix. For example, addition of diborane (B_2H_6) produces a silicon-boron-oxygen compound called *borosilicate glass (BSG)* while adding phosphine (PH_3) produces *phosphosilicate glass (PSG)*. One use for these glasses is as dopant sources for diffusion.

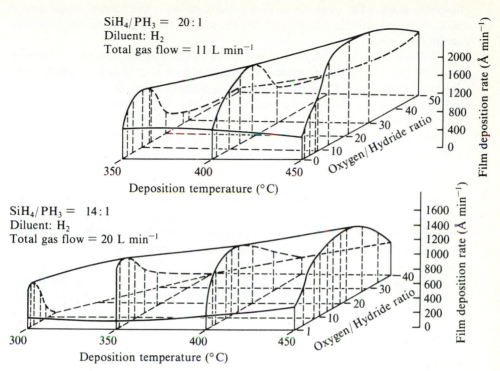

Figure 8-5 SiO$_2$ deposition rate for two CVD reactors as a function of oxygen/hydride ratio and temperature. (*Reference 7, used by permission.*)

However, PSG (or *pyroglass*) is a particularly popular film for use either as an insulator between conductor layers or as a protective *passivation layer* used to coat finished devices. The phosphorus not only reduces stress in the film, but effectively captures or *getters* alkali metal ions, which can cause device instability. When heated to 1000–1100°C, PSG softens and flows, producing a gently sloped topography resulting in good step coverage during metallization. However, excessive phosphorus in the film can react with water to produce corrosive phosphoric acid, so film composition must be well controlled. Phosphorus concentrations above 8% (by weight) encourage corrosion, while concentrations below 6% will not flow when heated.

In choosing process parameters, efficiency calls for a maximum deposition rate. However, process control is best for conditions where the deposition rate changes slowly with small variations in the parameters. Figure 8-5 shows the dependence of PSG deposition rate on reactant gas composition and temperature for two different reactors.[7] It is clear that the general shapes of the curves are similar for both reactors although they are run under different conditions. (However, the details vary significantly, as is typically the case for CVD reactors.) Deposition rate rises with temperature and shows a maximum at oxygen/hydride ratios of 20–30. (The oxygen/hydride ratio is the total oxygen flow in moles per second divided by the sum of the silane and phosphine flows.) The optimum process conditions are in the plateau region around 450°C at oxygen/hydride ratios above 20, where rates are high and the variation with conditions is small.

CVD silicon dioxide is deposited in a variety of reactors, including the continuous type. LPCVD and plasma-enhanced CVD methods have also been successfully used. Like most CVD films, CVD oxide contains significant impurities from the reaction environment (mainly hydrogen for silane deposition), in proportions that increase at lower deposition temperatures. The film can also be nonstochiometric (usually silicon-rich) and is often porous. *Densification* by heating to roughly 1000°C creates a film more similar to thermal SiO_2.

Since CVD silicon dioxide is often used as an insulator between conducting layers, step coverage is an important consideration. Ideally, one desires a *conformal coating*, in which the CVD layer follows the profile of underlying layers. Unfortunately, corners and protrusions are more exposed to the reaction gases in CVD and tend to grow at a higher rate. This results in exaggeration of the underlying topography, complicating the step coverage problem for successive layers. Conformal coatings are obtained only when there is enough surface mobility of film molecules to smooth out the variations in deposition rate. SiO_2 and PSG layers deposited with silane usually have noticeable bulges at steps. Deposition from TEOS is more conformal.

Silicon Nitride

Silicon nitride (Si_3N_4) is an important film because it can serve as a mask for selective oxidation of substrates. It is not easily grown by direct nitridation of silicon, and thus is usually deposited by the reaction of ammonia (NH_3) with a silicon-containing gas. Examples include

$$3SiH_4 + 4NH_3 \rightarrow Si_3N_4 + 12H_2 \qquad (8\text{-}12a)$$
$$3SiCl_4 + 4NH_3 \rightarrow Si_3N_4 + 12HCl \qquad (8\text{-}12b)$$
$$3SiH_2Cl_2 + 4NH_3 \rightarrow Si_3N_4 + 6HCl + 6H_2 \qquad (8\text{-}12c)$$

Activation energies are around 1.8 eV. Typical reaction temperatures are 700–900°C for the silane reaction, 850°C for silicon tetrachloride, and 650–750°C for di-chlorosilane.[8] Silicon nitride CVD is usually performed at reduced pressure in order to get good uniformity with reasonable batch size. Nitride films usually contain substantial hydrogen impurities, up to 8 at. % (Ref. 9). The films can also be silicon-rich and are prone to oxygen impurities. Nitride films usually incorporate very high stress, and thick layers are prone to cracking.

Silicon nitride is a good insulator and an excellent barrier against alkali metal contamination. Also, silicon nitride coatings are usually conformal. For these reasons, it makes a good passivation material for finished devices. However, deposition over metallized substrates requires lower temperatures, which has led to the development of plasma-enhanced processes for nitride.

These processes use silane as the silicon-containing gas, and either ammonia or nitrogen as a nitrogen source. The resulting "silicon nitride" is actually a silicon-nitrogen-hydrogen material of varying properties, usually also containing oxygen and carbon impurities.[10] Hydrogen content can reach 25 at. % (Ref. 11). The silicon/

nitride ratio ranges up to 1.25 versus the nominal value of 0.75 (Ref. 12). Composition, resistivity, etch rate, and film stress all depend dramatically on process conditions, including not only gas ratio, pressure, plasma power and temperature, but also reactor variables such as rf frequency and electrode spacing. When plasma nitride is heated to 900°C, hydrogen is evolved and a film properties become more similar to high-temperature CVD nitride.[12]

Using plasma-enhanced CVD, "silicon nitride" films of widely varying properties can be prepared. Deliberate addition of an oxygen-containing gas gives *silicon oxynitride,* an SiO_xN_y compound whose properties can be varied between those of silicon oxide and nitride. This ability to customize film properties can be quite useful, but the variation of properties with deposition conditions can also complicate process control. Plasma nitride coatings are somewhat less conformal than high-temperature nitride.

Polycrystalline Silicon

Silicon can be deposited by CVD methods in a polycrystalline form in which the film is composed of numerous small crystals. Polycrystalline silicon (*polysilicon* or "poly") can be doped to a low resistivity and can serve as a useful conductor in a number of applications. It is often used for the gate of an MOS transistor, as a resistor, or as a link between metallization and substrate to ensure ohmic contact and prevent pitting of the substrate.

Polysilicon can be deposited by the pyrolysis (decomposition upon heating) of silane:

$$SiH_4 \rightarrow Si + 2H_2 \tag{8-13}$$

For good uniformity, usually low-pressure processes are used. Typical temperatures are 600–650°C, with an activation energy of 1.7 eV (Ref. 5). Hot-wall, horizontal reactors are frequently chosen, employing a temperature ramp to compensate for reactant depletion. Growth rates depend chiefly on temperature, pressure, and silane concentration. Polysilicon coatings are conformal.

Polysilicon is frequently doped by adding another hydride (phosphine, arsine, or diborane) to the reaction mixture. The *p*-type dopant increases the deposition rate, while the *n*-type dopants retard it. Dopings frequently exceed the solid solubility in silicon, with the excess material accumulating at grain boundaries.

The crystal structure of polysilicon depends on deposition conditions. At sufficiently low temperature, the films are amorphous. Crystal formation begins around 600°C, and grain size increases with deposition temperature. Grains are on the order of 0.1 μm in size under typical LPCVD conditions. Preferred crystal orientation changes from {110} to {111} with increasing temperature. When CVD silicon films are heated to around 1000°C, both amorphous and crystalline films are converted to crystallites a micrometer or more in size.[13]

Polysilicon can be processed using many of the methods used on silicon substrates. In addition to in-situ doping during deposition, it can be doped by diffusion or by ion implantation. It can be oxidized by heating with oxygen or steam, forming an insulating oxide. The oxide is less ideal than that grown on single-crystal silicon.

Safety Considerations

CVD processes use reactive gases, which frequently are toxic, flammable, or both. The hydrides phosphine, arsine, and diborane are among the most toxic of gases and are also flammable. Silane is toxic, flammable, and pyrophoric; that is, it burns spontaneously in air. (Because silane is pyrophoric, there is a tendency to neglect its toxicity on the theory that leaks are self-evident. However, at low enough concentrations, silane will not ignite and can be a toxic hazard.) Ammonia and the chlorine-containing compounds are toxic and corrosive. Hydrogen, frequently used as a carrier gas, is highly flammable and forms explosive mixtures. Therefore, precautions against leaks or escape of gases are vital in CVD processing. Exhaust systems, including vacuum systems where used, must be leak-free and the exhaust must be properly treated. Vacuum pumps can accumulate toxic and corrosive materials, and special care must be taken in cleaning them or in changing oil. The CVD engineer should consult a qualified safety professional and thoroughly understand how to operate the process safely.

8-4 VAPOR-PHASE EPITAXY

Epitaxy is the extending of a single-crystal substrate by growing a film in such a way that the added atoms form a continuation of the single-crystal structure. The word is coined from the Greek *epi,* "upon," and *taxis,* "ordered." Epitaxy (or "epi") thus offers a way of adding material without terminating the single-crystal structure of the substrate. *Vapor-phase epitaxy (VPE)* is a special case of CVD in which the film is deposited epitaxially.

Figure 8-6 shows the usefulness of epitaxy. Doping by diffusion requires that the highest dopant concentration be at the surface. Diffusion thus can only produce higher-doped layers over more lightly doped ones, never the reverse. In Fig. 8-6, epitaxy is used to form a *buried-layer* structure, in which the surface is more lightly doped than an underlying layer. First, dopant is diffused into a substrate surface to form the buried layer. Then, additional silicon is epitaxially grown on the surface, in effect extending the substrate above the buried layer. The epitaxial doping can be determined at will and can be either *p-* or *n*-type. Repeated epitaxial layers can be used to produce more complicated doping profiles.

Because each deposited atom must take its proper place in the crystal structure, successful epitaxy requires great care. The surface must be atomically clean before deposition, and the growth rate slow enough that each atom can move into the proper location. High temperatures are usually used to allow good mobility of deposited atoms on the surface.

Several different gases may be chosen as a silicon source in VPE. Table 8-2 lists common reactants, along with reported activation energies[14] and typical temperatures and growth rates.[15] Usually the reactants are diluted in hydrogen. Under these conditions, silane pyrolyzes to yield silicon. The chlorine-containing gases react to give mixtures of $SiCl_2$, HCl, and various silicon-hydrogen-chlorine compounds. Pairs of $SiCl_2$ react at the surface to give silicon and silicon tetrachloride[16]:

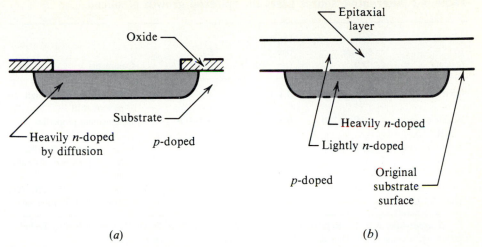

Figure 8-6 Formation of a buried-layer structure by epitaxy. (*a*) Dopant diffusion to form a highly doped layer at the surface. (*b*) Epitaxial growth produces a lightly doped region over the buried layer. This distribution of dopant cannot be obtained by diffusion methods.

$$2SiCl_2 \rightleftharpoons Si + SiCl_4 \qquad (8\text{-}14)$$

and the $SiCl_4$ may react further in the gas mixture. Thus the chlorine-containing ambients, whatever the original reactants, contain a mixture of silicon-hydrogen-chlorine compounds. At these temperatures, HCl will etch silicon, reversing the deposition reaction, so conditions must be properly balanced to obtain a good deposition rate. Epitaxial silicon can be doped by adding appropriate hydrides (phosphine, arsine, diborane) to the reactant mix.

A vertical, cold-wall reactor is usually used. (The "cold" wall is relative: wall temperatures are often well over 500°C.) Figure 8-7 illustrates two configurations, employing different methods for heating the substrates. In Fig. 8-7*a*, substrates lie flat on a circular graphite susceptor, which is heated by rf induction from the coils below. A nitrogen purge protects the coils from the reaction gases. This method produces a temperature gradient across the substrates, which can build stress into the film. In Fig. 8-7*b*, substrates are places on a slanted susceptor and heated directly with heat lamps, thus avoiding temperature gradients. (Note the similarity of flow in this reactor to the horizontal reactor with slanted susceptor.) Direct temperature measurement is impractical, so the temperature of the glowing substrates is usually measured by a *pyrometer* looking through a window in the bell jar. A pyrometer analyzes the emission spectrum of a hot object in order to deduce its temperature.

Table 8-3 shows a typical process sequence for silicon VPE. After substrate loading, the chamber is sealed. Because an explosion will result if atmospheric oxygen comes in contact with heated hydrogen, the first step is a nitrogen purge to remove all air. The nitrogen is then replaced with hydrogen, and the substrates heated. Epitaxy can occur only on an atomically clean surface, so the hot substrates are etched in HCl to expose clean silicon. The deposition cycle follows. Following deposition, the

Table 8-2 Alternative source gases for epitaxial growth of silicon

Reactant	Temp-erature (°C)	Deposition rate (μm min^{-1})	E_A (eV)	Comments
Silane (SiH$_4$)	900–1100	0.1–0.5	1.6–1.7	Wall deposition rate high, no pattern shift
Dichlorosilane (SiH$_2$Cl$_2$)	1050–1150	0.1–0.8	0.3–0.6	Intermediate properties
Trichlorosilane (SiHCl$_3$)	1100–1250	0.2–0.8	0.8–1	Large pattern shift. Insensitive to presence of oxidizers.
Silicon tetrachloride (SiCl$_4$)	1150–1300	0.2–1.0	1.6–1.7	Very large pattern shift. Very low wall deposition rate.

Source: After Ref. 15. Reprinted with permission of *Solid State Technology,* published by Technical Publishing, a company of Dun & Bradstreet. Activation energies taken from Ref. 14.

reactor must be cooled prior to purging with nitrogen, since nitrogen reacts with substrates and reactor materials at reaction temperatures. After the purge removes all hydrogen, the reactor can be opened and unloaded.

Safety Precautions

Hydrogen explodes when heated in contact with oxygen. A process involving a hydrogen ambient at high temperature is thus inherently dangerous. All systems must be leak-free, and the gas handling system must perform flawlessly. Of the silicon-containing gases, silane is pyrophoric, and the chlorides decompose to give toxic, corrosive hydrochloric acid. The dopant gases are extremely toxic, and the silicon gases only somewhat less so. Shortcuts should never be taken with epitaxial equipment. The effluent, containing hot hydrogen and toxic gases, must be treated carefully.

Table 8-3 Representative process sequence for silicon epitaxy

Process step	Ambient	Comments
Load substrates	Atmosphere	
Nitrogen purge	N$_2$	Remove all oxygen
Heat-up	H$_2$	Heat to reaction temperature
Etch	HCl in H$_2$	Expose atomically clean surface
Deposition	Source and dopant, in H$_2$	Deposit epitaxial layer
Cool-down	H$_2$	Cool below nitridation temperature
Purge	N$_2$	Remove all hydrogen
Unload	Atmosphere	

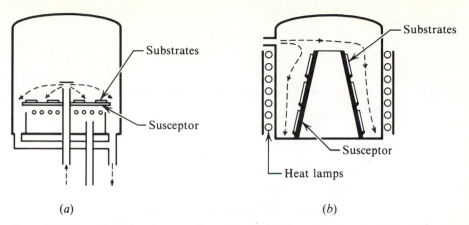

(a)   (b)

Figure 8-7 Two configurations for vapor-phase epitaxy. (*a*) Horizontal susceptor heated by rf induction. (*b*) Inclined susceptor; substrates are heated directly by heat lamps.

Doping and Autodoping

Ideally, an epitaxial layer is of uniform dopant concentration. However, especially when deposited over low-resistivity layers, this ideality is compromised by *autodoping*.

Figure 8-8 illustrates the mechanism. Here an epitaxial layer is being grown over an arsenic-doped buried layer. At the high deposition temperature, a significant amount of arsenic outgasses from the surface and enters the stagnant boundary layer, where it mixes with the reactant gases. This effect is increased by predeposition etching, which directly liberates arsenic from the doped surface. The result is an enhanced arsenic concentration in the epi. Meanwhile, arsenic is also *diffusing* upwards from the buried layer into the epitaxial silicon, further blurring the boundary. These effects can be further aggravated by outgassing from the reactor and susceptor and from the backside of the substrate.

Antimony can be used in place of arsenic for buried layer diffusion to reduce autodoping. When outgassing from the wafer backside is significant, the backside can be sealed with oxide. Lower deposition temperatures also reduce autodoping. Perhaps the most effective measure is the use of reduced-pressure epitaxy, which not only lowers reaction temperatures but minimizes the boundary layer effect. With a thinner boundary layer and higher diffusivity, outgassing dopant enters the gas stream and is swept away.

Pattern Shift

When epi is grown over a buried layer pattern, subsequent patterns must be aligned to the buried layer. This is done by observing the variation in topography, caused by consumption of silicon during the oxidation accompanying diffusion. Figures 8-9*a* and *b* show the formation of this topography. When the pattern after epitaxial growth

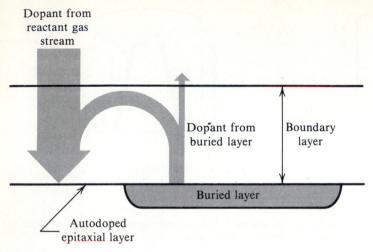

Figure 8-8 Autodoping. Dopant diffuses from the substrate into the boundary layer and affects the epitaxial doping.

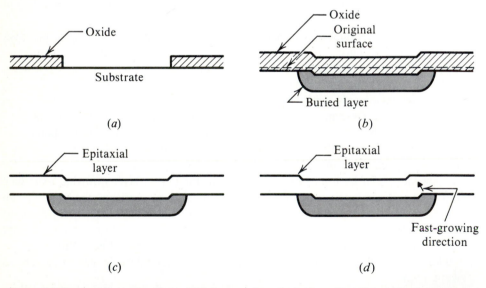

Figure 8-9 Pattern shift and its effects. (a) Substrate prior to buried layer deposition, with window cut in oxide. (b) Formation of topography by oxidation following buried layer deposition. (c) Topography following epitaxial growth with no pattern shift; topographical and doping patterns match. (d) Topography following epitaxial growth with pattern shift; topographical pattern no longer reveals position of buried layer.

accurately reflects the buried layer location, as in Fig. 8-9c, alignment of successive layers is straightforward.

Because epi growth rates are sensitive to the crystal orientation, however, growth rates may vary with the slope of the topography. The result can be *pattern shift*, as shown in Fig. 8-9d. The physical pattern of surface topography no longer matches the electronic pattern of dopant concentration. Variations of this effect, *pattern distortion* and *pattern washout*, can change the shape of the pattern or even make it disappear.

Pattern shift is usually minor in silane epitaxy, but is substantial with chlorine-containing gases. This is due to the constant etching during epi growth caused by the HCl present in the ambient. Pattern shift is quite sensitive to crystal orientation and is minimized on $\langle 111 \rangle$ material by orienting it 3° toward the nearest $\langle 110 \rangle$ direction. (This orientation is now standard for "$\langle 111 \rangle$" material.) A great deal of experimental data has been accumulated on how to minimize pattern shift and distortion. Reduced pressure tends to reduce pattern shift, but can sometimes aggravate pattern distortion.[17]

Crystal Defects

As a single-crystal film, epi is subject to crystal defects. In general, any defects present on the underlying substrate will be propagated into the epi. In addition, *epitaxial stacking faults* may occur, as illustrated in Fig. 8-10a. If growth is disturbed at the surface (e.g., by an impurity atom), an inverted pyramid of material will grow with the different stacking order from the bulk. When etched with a crystallographic etch, these areas will appear as triangles (on $\langle 111 \rangle$ silicon) or squares ($\langle 100 \rangle$ silicon) with a length related to the thickness of the epi. This fact has been used to measure epitaxial layer thickness. Contamination can also result in enhanced growth, producing *growth hillocks* or *spikes*, as illustrated in Fig. 8-10b. Finally, thermal stress in the film, due for example to a temperature gradient across the film during deposition, can result in *slip* of crystal planes and resulting defects. Figure 8-10c shows the distinctive pattern seen upon crystallographic etching of a substrate with slip defects.

Control of epitaxial defects is of particular importance in bipolar processing, in which the base and emitter are usually diffused into an epi layer. Epi defects collect metallic impurities and allow accelerated movement of dopants, leading to "pipes," which can short out the junctions of a bipolar transistor.

Vapor-Phase Epitaxy Involving Other Materials

Several epitaxial processes have been used for gallium arsenide.[18] One involves the direct reaction of trimethyl gallium [$(CH_3)_3Ga$] with arsine (AsH_3) at 650–760°C. Others involve a two-stage process. Metallic gallium reacts with a gas to form an unstable gallium compound in one zone of a reactor. The gallium compound then flows downstream, reacting in another zone of the reactor with an arsenic compound to form an epitaxial layer on the substrate. One scheme uses arsenic trichloride ($AsCl_3$), which not only supplies arsenic, but reacts with gallium at 800–850°C to give unstable GaCl. An alternative process uses arsine as the arsenic source and HCl to react with gallium at 775–800°C.

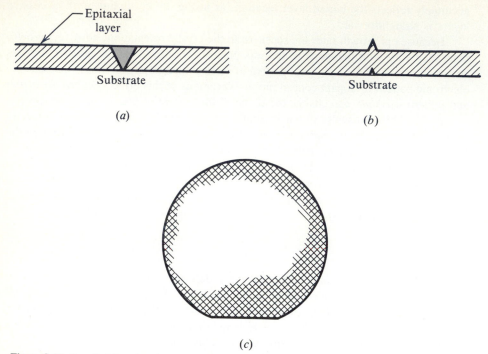

(a)   (b)

(c)

Figure 8-10 Growth defects in epitaxy. (a) Epitaxial stacking fault. (b) Growth hillock. (c) Slip pattern, as it appears macroscopically after crystallographic etch.

Heteroepitaxy is the growth of a single-crystal film of one material over a crystalline substrate of another material. One example is *silicon-on-sapphire (SOS)* in which silicon epitaxy is performed on a sapphire (Al_2O_3) substrate. For many device designs, performance can be improved by raising substrate resistivity. The logical end point of this strategy is to use an insulating substrate, like sapphire. Conventional silicon epitaxy processes can be used. However, the fit between the silicon crystal and the sapphire is not good, and the films are usually highly stressed with significant defects.

Aluminum arsenide and gallium arsenide, on the other hand, have almost identical lattice dimensions. An alloy of the two, aluminum gallium arsenide (AlGaAs), has a similar lattice size and displays interesting optical properties that can be manipulated by changing the aluminum/gallium ratio. Therefore, heteroepitaxy of AlGaAs on GaAs is receiving growing attention.

8-5 OTHER DEPOSITION METHODS

In this section, we will look at several other methods of depositing thin films onto substrate surfaces: molecular-beam epitaxy, liquid-phase epitaxy, spin-coating of polyimides, and electroplating.

Molecular-Beam Epitaxy

Molecular-beam epitaxy (MBE) is epitaxy using evaporation rather than CVD. Figure 8-11 shows an apparatus for MBE.[19] Silicon is evaporated from an *e*-beam source and impinges upon a heated substrate at the top of the bell jar. Antimony dopant effuses from a separate cell. Substrate temperatures of 450°C are sufficient to obtain epitaxial growth. To obtain an atomically clean surface, the substrate can be heated in vacuum (to decompose surface material) or sputtered clean with an inert gas beam.

MBE offers attractive advantages, especially for the deposition of thin layers of unusual materials. The low temperature and the vacuum environment allow quite precise control of doping in very thin layers. Compared to VPE, choice of material and dopant is much less constrained by chemistry. Growth rates are usually slow (~0.01 μm min^{-1}), but higher rates have been reported. The low temperature also causes less diffusion of dopants within the substrate.

Liquid-Phase Epitaxy

Liquid-phase epitaxy (LPE) is the growth of epitaxial films by immersing the substrate in a liquid solution of the film material. It is thus an analogue of the childhood experiment of growing sugar crystals from a water solution. In LPE, a solution of the film material, saturated at some temperature T, is prepared using a suitable solvent. If a substrate is immersed in the solution, and the temperature lowered somewhat below T,

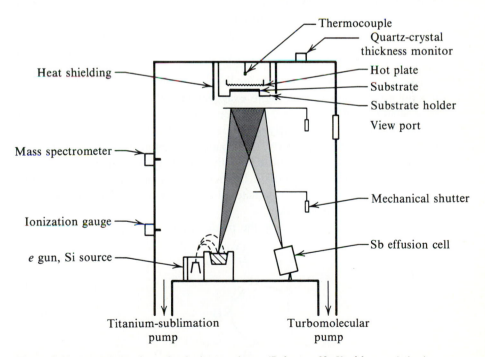

Figure 8-11 An apparatus for molecular-beam epitaxy. (*Reference 19. Used by permission.*)

the solution becomes supersaturated, and film material will be deposited on the substrate. In situ cleaning can be achieved by first *raising* the temperature above T, causing the substrate surface layer to dissolve into the solution.

One problem of LPE is removal of the solvent from the film. If the film material dissolves well in the solvent, it is likely that the solvent will dissolve to some extent in the film. Therefore, solvents are chosen both for low solubility in the film and for electrical inactivity.

LPE is particularly useful in gallium arsenide processing. GaAs substrates grown from a melt (at 1238°C) are arsenic-rich, and GaAs tends to decompose on heating unless kept under an overpressure of arsenic vapor. LPE allows the growth of a stochiometric epitaxial layer of GaAs at 600–900°C, a temperature where decomposition is limited. Devices can then be fabricated in the high-quality epitaxial layer. Liquid gallium (m.p. 30°C) is usually used as the solvent. Being a component of the semiconductor, it has no adverse electrical effects, and it is also an effective getter for deep-lying impurities.[20]

Uniformity of growth in LPE is limited by hydrodynamic and thermal considerations. Maintaining uniform concentrations and temperatures in liquids without creating turbulence or instabilities is a difficult problem. Therefore, thickness uniformity in LPE remains a major limitation.

Spin-Coating of Polyimides

As microelectronic designs grow in complexity, it becomes desirable to have several layers of interconnection between circuit elements. The number of layers is limited, however, by the need for good step coverage over the underlying topography. Each added layer must be formed over an increasingly bumpy surface, complicating the step coverage problem. Figure 8-12a shows a typical cross-section of a device incorporating two layers of metal, which is usually the limit for conventional technologies. Note that topographical features become more exaggerated with each step.

This limitation could be avoided by a deposition technique that reduces, rather than aggravates, the effect of underlying topography. We have already seen a technique that has this *planarizing* property: spin coating of photoresist. A spin-coated film tends to form a level surface over a substrate, filling in valleys and thinning over peaks. Photoresist, of course, does not have the thermal stability or insulating strength of a CVD insulator like PSG. But there is a material that can be spin-coated and possesses good insulator properties: polyimide.

Polyimides are polymers based on the *imide* group:

$$\begin{array}{c} O \\ \parallel \\ C \\ \diagdown \\ C \diagup \quad \diagdown \\ \mid \qquad\quad NH \\ C \diagdown \quad \diagup \\ \diagup \quad C \\ \parallel \\ O \end{array} \qquad\qquad (8\text{-}15)$$

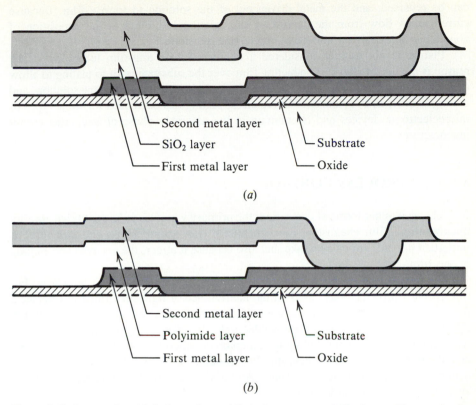

Figure 8-12 Cross-sections of devices using multilayer interconnect and showing resulting topography. (*a*) CVD SiO$_2$ used as interlayer dielectric. (*b*) Polyimide used as interlayer dielectric.

Material is supplied as a liquid solution and applied to wafers by spin coating. The film is then cured by heating to approximately 300°C for several hours. Polyimide layers are both chemically and thermally stable, withstanding temperatures up to 500°C. They can be photolithographically patterned and etched using oxygen or fluorine plasmas.[21] When compared to sputtered SiO$_2$, they were found to have comparable dielectric constants and electric breakdown strengths. Resistivity was $\sim 10^{12}$ Ω cm versus 10^{16} for SiO$_2$. Under typical conditions, polyimide films can smooth out up to 80% of the underlying topography.[22] Several major manufacturers are currently using polyimide as an interlayer dielectric in double-level metal systems. Given the growing need for more complicated interconnection schemes, others are likely to follow.

Electroplating

Electroplating has long been used industrially to apply metallic films to surfaces. Plating is a reversal of metal etching. In etching, metals give up electrons and enter solution as ions. By supplying electric power to a solution of metal ions, the reaction can be reversed, and the metal driven out of the solution to form a film. Because current must flow from the deposition site, plating is limited to the deposition of conductive materials on uniformly conducting substrates.

Plating is not usually considered a microelectronic thin-film technology. One reason is the need to have a conducting film over the substrate prior to plating to allow current to flow. Plated films are also prone to include impurities from the solution, and plating is difficult to automate. Nevertheless, plating has been used for some types of microelectronic devices and is common in the assembly phase of integrated circuit manufacture.

8-6 CVD PROCESS CONTROL

In CVD, as in other forms of deposition, the primary variables to be controlled are film thickness and film chemistry. Film chemistry (i.e., composition and structure) is important not only for its own sake, but also because it affects the masking and etching of the film.

We have already seen a number of methods for measuring film thickness. For transparent films, methods based on the interference of light (described in Chapter 2 for thermal silicon dioxide) can be used. This method can also be extended to epitaxial silicon layers, which are opaque to visible light, by using infrared radiation. Infrared wavelengths will penetrate lightly doped epitaxial silicon, and reflect from the more heavily doped buried layer beneath. For conductive layers, the thickness can be deduced from resistivity measurements. And of course, a profilometer can be used for any film once a step is defined.

All of these methods, however, are harder to use when the layer being measured overlays a number of previous layers, as is often the case for CVD films. Interference patterns and profilometer traces are difficult to interpret if enough underlying layers are present, and resistivity readings are valid only if there are no alternate conduction paths below the film. Under these circumstances, it may be necessary to include blank test substrates with the batch of devices and measure the film thickness using the blanks.

Often the structure and composition of CVD films are not well understood. Powerful methods are now available for the detailed study of structure, but they are frequently expensive, time-consuming, destructive of the film observed, and generally unsuitable for routine use in process control. Therefore, some more easily measured quantity, dependent on composition and structure, is used to regulate the process. For example, refractive index is a good clue to impurity concentration and stoichiometry in CVD nitride. Phosphorus content in PSG my be inferred from the etch rate of the film in an HF solution. Dopant concentration, in both epitaxial and polycrystalline silicon, is controlled by measuring resistivity.

When using secondary variables in this way, one should remember that the property measured may be only empirically correlated with the property to be controlled. It follows that an unexpected change in structure or even composition may occur without a corresponding change in the particular secondary variable being monitored. Even when direct methods are used, results can be misleading. For example, the phosphorus content of PSG is increasingly determined using x-ray fluorescence, which gives a direct measurement of average phosphorus content. Yet two PSG films may have identical *average* compositions and still behave quite differently. If the doping of one film varies with depth while the other is uniform throughout, resist adhesion, etch profiles, and even passivation ability of the two films may differ dramatically. When such subtle variations are suspected, sophisticated film analysis methods must be called upon.

Input variables for CVD can include every feature of the process. Temperature, gas flows and relative concentrations, reactor design and cleanliness, pressure (for reduced-pressure CVD), rf frequency and power (for plasma-enhanced CVD), are all important. Frequently, so are wafer loading patterns, exhaust system backpressure, susceptor history, and a host of other factors. Most CVD reactions are quite sensitive to the presence of water or oxidants, either from leakage or incomplete purging of the reactor after loading.

Therefore, CVD process control is highly empirical and is centered on making all variables as reproducible as possible. While temperature, flows, pressures and powers can be measured directly, more subtle effects are harder to control. Since CVD reactants, by definition, react to generate solids, reactors and exhaust systems must be cleaned regularly to remove unwanted particulate matter. However, a new or recently cleaned tube or susceptor may react differently with the gas stream, affecting reactant gas composition. Therefore, new components may require predoping or conditioning before use. The mixing of gases is frequently sensitive to small variations in nozzles or inlet ports, so care must be taken to keep them clean and reassemble them precisely. Of course, systems must be completely leak-free, both to control the process and for safety reasons.

Each reactor and film will have its own particular sensitivities, which the engineer will learn with experience. Because of this and the wide range of reactions and equipment used for CVD, it is not practical to present a troubleshooting guide for this chapter.

PROBLEMS

8-1 A CVD reactant is diluted to 2% in hydrogen. The gas flows at 8 cm sec^{-1} down a tube of 18-cm diameter, at 600°C. (*a*) What is the Reynolds number? (*b*) What is the boundary layer thickness at the end of a 20-cm sled? (*c*) From the deposition rate, the flux j is deduced to be 5×10^{19} molecules cm^{-2} sec^{-1}. Find D, assuming that reaction is quite complete so that $N_0 \sim 0$.

8-2 Using the value of D found in Problem 8-1, calculate the flux j if all conditions are identical except that the pressure is reduced to 0.005 atm.

8-3 A CVD reaction takes place in a rectangular tube 10 cm high and 15 cm wide. In the tube is a slanted susceptor 20 cm long and is 2 cm high at the back end. The reaction takes place at 825°C. The gas mixture is chiefly hydrogen, and the total gas input is 0.25 standard liters sec^{-1}. (*a*) What is the gas velocity through the tube near the inlet, at 300°C? (*b*) What is the gas velocity at the back edge of the susceptor, with the tube area reduced and the gas temperature at 825°C? (*c*) Using the ideal gas law, what is the density of hydrogen (molecular weight 2) at 825°C? (*d*) Hydrogen at 825°C has a viscosity of 2.13×10^{-4} P (dyne sec cm^{-2}). Calculate the Reynolds number for conditions near the trailing edge of the susceptor. (*e*) Estimate the boundary layer thickness at the trailing edge of the susceptor.

8-4 A CVD reactor for phosphosilicate glass uses 20% silane in nitrogen as a silicon source gas at a reaction temperature of 450°C. Flow rate is 650 standard cubic centimeters per minute (SCCM) and the film growth rate is 70 nm min^{-1}. The reactor is loaded with 12 circular substrates, 10 cm in diameter. The number of molecules per cubic centimeter, γ, for the film is 2×10^{22}. (a) With what efficiency is silane converted into silicon dioxide film on the substrates? (b) The substrates cover only 50% of the susceptor area. If PSG is deposited on the susceptor at the same rate as the substrates and no deposition occurs elsewhere, what is the overall efficiency with which silane is converted to SiO_2? (c) Assume that the reaction rate is fast enough that the reaction is diffusion-limited, and the surface concentration of silane is essentially zero. If the boundary layer is 1 cm thick, what is flux of silane molecules? What is the diffusivity D?

8-5 Using the given activation energies for each process, estimate the relative increase in reaction rate for: (a) Silicon dioxide as temperature rises from 400 to 500°C. (b) Silicon nitride from 700 to 900°C. (c) Polysilicon from 575 to 650°C; silicon epitaxy (using silane) from 1050 to 1150°C.

8-6 The reaction rate of a certain reaction is determined at 600, 650, and 700°C. Temperature effects on rate are found to be negligible. (a) Comment on the relative magnitude of δk_s and D. (b) Predict the effect of doubling the reactant concentration N_G. (c) Of doubling the gas flow velocity? (d) Of reducing pressure by a factor of 10?

8-7 The reaction rate of a certain reaction is determined at 600, 650, and 700°C, and rate is found to increase 50% with each temperature increment. (a) Comment on the relative magnitude of δk_s and D. (b) Estimate E_A for the reaction. (c) Predict the effect of doubling the reactant concentration N_G. (d) Of doubling the gas flow velocity? (e) Of reducing pressure by a factor of 10?

8-8 Phosphorus concentration for PSG may be given either in weight-percent phosphorus (wt. %: grams of phosphorus per gram of glass) or mole-percent P_2O_5 (mole %: number of P_2O_5 units per total P_2O_5 and SiO_2 units). The glass is assumed to consist solely of P_2O_5 and SiO_2 units. Write an expression relating wt. % and mole %. The atomic weights of oxygen, silicon, and phosphrous are 16.00, 28,09, and 30.97 respectively.

8-9 You are assigned to develop a CVD process for atmospheric pressure deposition or phosphosilicate glass on an unfamiliar reactor. You know you will have to establish the exact reaction conditions empirically, but you desire to estimate a reasonable starting point for your first set of trials. (a) Use the ideal gas law to relate gas volume flows, measured on flow meters in cubic centimeters per second to gas delivery rates in moles per second. Gas is supplied to the meters at 15 pounds per square inch (psig) above atmospheric pressure, at 300°K. (b) Pick an operating point from Fig. 8-5 and thereby establish a desired temperature, hydride ratio, and deposition rate. (There is no "right" answer to this part: use your judgment.) (c) Estimate the silane source flow from the desired deposition rate. The supply bottle contains a 5% concentration of silane in nitrogen. The reactor deposition area is 300 cm^2. Assume 10% efficiency in converting silane to SiO_2. (d) Estimate the phosphine source flow needed to deposit a glass of 8 mole % P_2O_5. The supply bottle contains a 5% concentration of phosphine in nitrogen. Assume that the phosphorus concentration in the glass is identical to that in the gas stream. (e) Estimate the oxygen flow from the silane and phosphine flows and your desired hydride ratio. The supply bottle contains pure oxygen.

8-10 An epitaxial layer of $\langle 111 \rangle$ silicon is crystallographically etched and found to have epi stacking faults, observed as equilateral triangles with 5-μm edges. Thse triangles mark the intersection with the surface of [111] planes that meet at the origin of the fault at the interface between the epi and the original surface. Determine the thickness of the epitaxial layer.

REFERENCES

1. A. S. Grove, *Physics and Technology of Semiconductor Devices*, Wiley, New York, 1967, p. 17.
2. Vladimir S. Ban, *J. Electrochem. Soc.* **125**:317 (1978).
3. Werner Kern and Richard S. Rosler, *J. Vac. Sci. Technol.* **14**:1082 (1977).
4. Werner Kern and Vladimir S. Ban, "Chemical Vapor Deposition of Inorganic Thin Films," in *Thin Film Processes*, John L. Vossen and Werner Kern (eds.), Academic, New York, 1978, p. 257.
5. A. C. Adams, "Dielectric and Polysilicon Film Deposition," in *VLSI Technology*, S. M. Sze (ed.), McGraw-Hill, New York, 1984, p. 93.

6. M. Maeda and H. Nakamura, *J. Appl. Phys.* **52**:6651 (1981).
7. Werner Kern and Richard S. Rosler, *J. Vac. Sci. Technol.* **14**:1082 (1977).
8. Adams, *op. cit.,* p. 119.
9. W. A. Lanford and M. J. Rand, *J. Appl. Phys.* **49**:2473 (1978).
10. J. R. Hollohan and R. S. Rosler, "Plasma Deposition of Inorganic Thin Films," in *Thin Film Processes,* John L. Vossen and Werner Kern (eds.), Academic, New York, 1978.
11. W. A. Lanford and M. J. Rand, *J. Appl. Phys.* **149**:2474 (1978).
12. H. J. Stein, V. A. Wells, and R. E. Hampy, *J. Electrochem. Soc.* **126**:1750 (1979).
13. T. I. Kamins, *J. Electrochem Soc.* **127**:686 (1980).
14. C. W. Pearce, "Epitaxy," in *VLSI Technology, op. cit.,* p. 51.
15. H. M. Liaw, J. Rose, and P. L. Fejes, *Solid State Technol.* **27**:135 (May 1984).
16. Sorab K. Ghandi, *VLSI Fabrication Principles,* Wiley, New York, 1983, p. 231.
17. S. P. Weeks, *Solid State Technol.* **24**:111 (November 1981).
18. Ghandi, *op. cit.,* p. 248.
19. U. Konig, H. Kibbel, and E. Kasper, *J. Vac. Sci. Technol.* **16**:985 (1979).
20. Ghandi, *op. cit.,* p. 264 ff.
21. Guy Turbon and Michel Repeaux, *J. Electrochem. Soc.* **130**:2231 (1983).
22. L. B. Rothman, *J. Electrochem. Soc.* **127**:2216 (1980).

NINE

ION BEAMS AND THEIR APPLICATIONS

The last subject in our study of processing techniques will be those applications involving ion beams. When electrons are knocked loose from (or added to) a neutral atom or molecule, the resulting charged particle is called an *ion*. Because of their electrical charge, ions can be manipulated with fields. Thus a population of ions can be formed into a collimated beam of high energy, something that is not feasible with neutral molecules. These high-energy beams are effective tools for modifying the properties of materials.

In Section 9-1, we examine the behavior of charged particles in fields and see how ions are formed into beams. In Sections 9-2 and 9-3, we deal with the leading ion-beam application, *ion implantation* for substrate doping. In ion implantation, the dopant is implanted into the substrate directly, without the need for diffusion, by bombarding the substrate with a beam of dopant ions. We look at implantation equipment in Section 9-2 and at the implantation process in Section 9-3. In Section 9-4, we cover other ion-beam applications. These include ion implantation for purposes other than doping as well as ion milling and ion lithography. Process control of ion implantation is presented in Section 9-5.

9-1 CHARGED PARTICLES AND FIELDS

Charged particles, unlike electrically neutral ones, can be easily manipulated using electric and magnetic fields. This property allows the formation of collimated, high-energy beams of ions or electrons. For our discussion, we will consider a particle of mass m and charge Z. Since the ions we deal with are usually positively charged, we will take Z to be positive. As we shall see, many properties of charged particles depend only on the ratio of charge to mass, so it is useful to designate the charge-to-mass ratio, Z/m by η.

Basic electrostatics teaches us that a particle of charge Z in an electric field $\mathscr{E}$ experiences a vector force $\mathbf{F}$ given by

$$\mathbf{F} = Z\mathscr{E} \tag{9-1}$$

Thus an ion in an electric field is accelerated in the direction of the field. The kinetic energy resulting from this acceleration is

$$\frac{1}{2} mv^2 = \int \mathbf{F} \, ds \tag{9-2}$$

where v is the particle velocity, and the integral is along the path traversed by the particle. Also, since the field is the negative gradient of the electrical potential of the particle, the potential is related to the field by

$$\phi = - \int \mathscr{E} \, ds \tag{9-3}$$

It follows that

$$\frac{1}{2} mv^2 = Z(\phi_0 - \phi) \tag{9-4}$$

where ϕ_0 is a constant of integration. In our applications, we measure not the potential at the location of the particle, but the voltages applied to electrodes. The voltage and the potential are equal wherever a well-defined velocity measurement can be made (i.e., wherever fields are not varying abruptly) and thus it is more meaningful to write Eq. (9-4) in terms of the voltages V and V_0:

$$\frac{1}{2} mv^2 = Z(V_0 - V) \tag{9-5}$$

Now V_0 is seen to be the voltage at which the particle is at rest. If the particles are generated with negligible velocity, then V_0 will be the voltage applied to the electrodes surrounding the generation region.

The energy of a charged particle in a field generated by voltages applied to electrodes is thus determined solely by the charge and the voltages. Knowing the voltage on an electrode surrounding the particle and the voltage in the generation region, we know the energy, independent of the path taken by the particle. The energy of a particle carrying one electron charge, accelerated through 1 V, is an *electron volt* (eV). Note that the energy is independent of the mass. The velocity, of course, is mass-dependent and is given by

$$v = (2\eta(V_0 - V))^{1/2} \tag{9-6}$$

Velocity thus depends only on η and the voltages.

It follows from Eq. (9-1) that a charged particle in an electric field behaves just

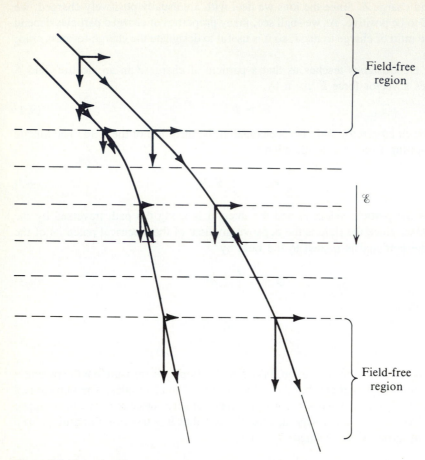

Figure 9-1 The paths of two charged particles in a uniform electric field. The dashed parallel lines are contours of equal electric potential. The vectors show the initial and final velocities and their components relative to the field. Note that both particles are bent toward the perpendicular to the equipotentials, and the slower particle is diverted more.

like a particle in a gravitational field, with the acceleration of gravity g replaced by $\eta\mathscr{E}$. Thus a particle entering a uniform field will describe parabolic paths, maintaining a constant velocity component perpendicular to the field while being accelerated parallel to the field. Figure 9-1 shows the path of two charged particles, of differing initial velocities, in an electric field. The dashed lines are *equipotential lines* connecting equal values of the electric potential. In a uniform field they are parallel and equally spaced. Note that the paths curve to become more nearly perpendicular to the equipotential lines and that the slower particle, spending more time in the field, curves more. This behavior will occur in any field configuration. Therefore, sketching of equipotentials can be used to get a good notion of charged particle paths in fields where mathematical treatment is complicated.

A different kind of force is experienced by a particle moving in a constant magnetic field of strength **B**. In such a field, a charged particle experiences a force given by

$$F = Z\mathbf{v} \times \mathbf{B} \tag{9-7}$$

at right angles to both the original velocity vector **v** and the field. There is no force in the direction of motion, and the magnitude of the velocity is unchanged. However, there will be a constant acceleration transverse to the motion and thus a change in direction. A particle undergoing constant acceleration at right angles to its motion is describing a circle, and the radius R of the circle can be obtained by equating the magnetic and centripetal forces:

$$Z\upsilon B_T = m\upsilon^2/R \tag{9-8}$$

Here B_T is the component of the magnetic field transverse to the particle motion. It follows that

$$R = \frac{\upsilon}{\eta B_T} \tag{9-9}$$

Substituting the velocity from Eq. (9-6) gives

$$R = \frac{\sqrt{2(V_0 - V)/\eta}}{B_T} \tag{9-10}$$

Figure 9-2 shows the forces on particles moving in a uniform magnetic field. Note that a particle moving parallel to a magnetic field experiences no force.

In ion-beam applications, the field is usually made transverse to the particle motion, and ions are singly charged. Acceleration voltage (relative to V_0) is usually measured in kilovolts and ion mass in atomic mass units. Under these conditions, Eq. (9-10) can be written in practical terms as

$$BR = 4.55 \sqrt{mV} \text{ kG cm (kV amu)}^{-1/2} \tag{9-11}$$

The quantity BR is called the *magnetic rigidity* and is constant for a given mass and acceleration.

Example What is the magnetic rigidity of a 100-keV beam of boron (B^{11+}) ions? What field will be required to deflect the beam into an arc with a radius of 10 cm?

The magnetic rigidity is given by Eq. (9-11) as

$$BR = 4.55(11 \text{ amu} \times 100 \text{ keV})^{1/2}(\text{kG cm})/(\text{kV amu})^{-1/2}$$

$$= 151 \text{ kG cm}$$

For $R = 10$ cm:

$$B = 151 \text{ kG cm } (10 \text{ cm}^{-1})$$

$$= 15.1 \text{ kG}$$

which is the field required to produce the desired deflection.

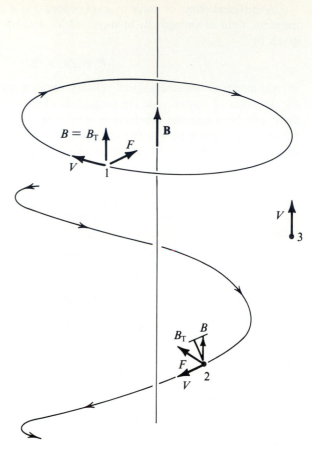

Figure 9-2 Motion of charged particles in a uniform magnetic field. Arrows show the particle velocity, the direction of **B**, the magnitude of B_T, and the resulting force. Note that the particle moving perpendicular to the field describes a circle, the particle parallel to the field experiences no force, and the obliquely moving particle describes a spiral.

Ion Optics

A useful ion beam should be of constant velocity, limited spatial extent, and known composition. Magnetic and electric fields provide the means for forming a group of ions into such a beam. To do this, three functions must be performed: acceleration, focusing, and mass analysis.

Ions are usually formed from molecules with thermal energies of a small fraction of an electron volt. Yet useful ion beams are expected to have energies of some 10–1000 keV. This difference in energy is attained by *accelerating* the beam. Ions are accelerated by allowing them to pass from a region of more positive voltage to a lower one. As Eq. (9-5) indicates, only the relative magnitudes of the voltages matter: ions can be accelerated from ground to a large negative voltage, or from a large positive voltage to ground.

The ions in an useful beam must be limited to a small cross-sectional area. A beam must be concentrated in order to deliver a usable flow of ions through the apparatus to

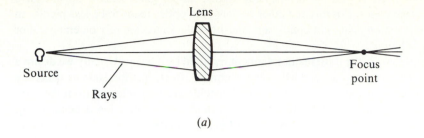

Lens

Source

Rays

Focus
point

(a)

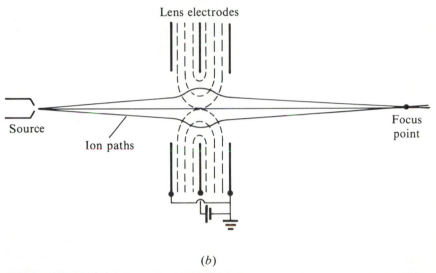

Lens electrodes

Source

Ion paths

Focus
point

(b)

Figure 9-3 Analogy between optical- and ion-beam focusing. (a) Optical beam diverges from a point, is deflected by a glass lens, and focused on an image point. (b) Ion beam diverges from a point, is deflected by an electric field, and focused on an image point.

the target. However, the numerous like-charged particles in a beam are mutually repulsive, so beams naturally diverge. *Focusing* a beam is thus necessary.

As the word *focus* implies, ion-beam and optical-beam focusing are analogous. A beam of photons is made to converge by passing it through media that affect the photon velocity, and focus is obtained if those photons diverging from a point can be converged to another point. Similarly, ion beams are converged by fields that change the particle velocity or direction. In both cases, good focusing of a beam centered on an axis is obtained when the particles are deflected by an angle proportional to the off-axis distance. Figure 9-3 illustrates the analogy between ion and optical focusing.

This analogy is sufficiently useful that the study of beam focusing is called *ion optics,* and the focusing elements are termed *ion lenses.* Charged-particle forms of the paraxial ray equation and other optical principles have been derived. One flaw in the analogy, however, is that electromagnetic fields, unlike glass lenses, do not have sharply defined edges. For this and other reasons, ion optics tends to be less precise in practical application than light optics, and real-life solutions often rely on empirical or numerical solutions.

Mass analysis is the third major treatment required by a beam. Usually we desire a chemically pure beam to accomplish some chemical effect. Yet ion sources generally deliver a mixture of ions, so that the desired species must be selected from the mix. Ions can be "weighed" by deflecting them in a field, since deflection depends on η. Selection of ions of a given mass usually suffices to isolate a pure chemical species.

Various combinations of fields can be used for mass analysis. It should be remembered that "mass analysis" is usually η analysis, so, for example, a doubly charged ion of twice the prescribed mass will pass the analyzer. Furthermore, simple fields select not for mass but for some function of mass: Deflection in a magnetic field, for example, depends on momentum. However, since in beam applications the ions have equal energies, selection for momentum serves to select for mass. Deflection by a magnet is thus an effective means of mass analysis.

Elements for Beam Processing

Figure 9-4 shows some typical elements for electrostatic acceleration and focusing of beams. Figure 9-4a illustrates an *acceleration tube* for beam acceleration. It is composed of a series of parallel plate electrodes, separated by insulating spacers. In the center of each electrode is an aperture through which the beam passes. Each plate in the series is biased at an increasingly negative potential. As the dashed equipotential lines indicate, ions within the tube experience a uniform field accelerating them along the tube axis. The result, in the interior of the tube, is a smooth acceleration.

However, the acceleration tube also has a focusing effect. This is because the ions entering and leaving the tube traverse a region where the field is no longer uniform and the equipotentials bulge out. This effect is most pronounced at the front end, where ions have low velocities. The effect is to bend ion paths towards the perpendicular to the equipotentials, focusing the beam. A similar diverging action at the exit end has less effect due to the higher ion velocity. This example indicates that even in the simple acceleration tube, fields applied to ions can have lens properties that cannot be ignored. The focal length f of an accelerating tube containing an electric field $\mathscr{E}$ can be shown to be

$$f = \frac{4V}{\mathscr{E}} \tag{9-12}$$

where V is the voltage (relative to V_0) of the entrance to the tube.[1]

Figure 9-4b shows a deliberate application of an electric field to focusing: the *Einzel lens*. This lens consists of three tubes, the two outer at ground potential and the inner at a more negative potential. (This is the *accelerating mode* of an Einzel lens. It will also function if the middle potential is more positive.) An electric field is generated in the gaps between the electrodes, with equipotentials as shown. When ions cross the first gap, they are pushed toward the axis, focusing the beam, and then

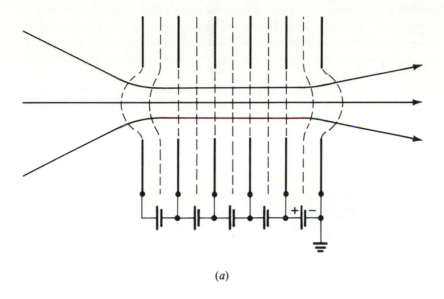

(*a*)

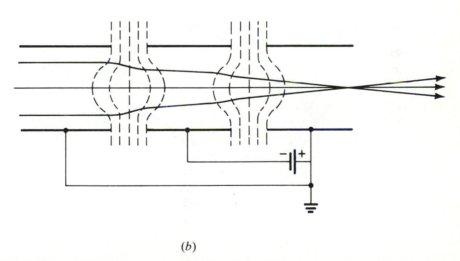

(*b*)

Figure 9-4 Two electric field elements for processing beams. (*a*) Acceleration tube for beam acceleration. (*b*) An Einzel lens for beam focusing. Dashed lines are equipotentials.

pushed away. However, because the ion is accelerated while crossing the gap, it spends less time in the diverging region, and the focusing effect predominates. When crossing the second gap, the same forces are experienced in the reverse order. However, the ion is now being decelerated to its original velocity, so the converging effect again predominates. The result is that the Einzel lens focuses the beam, but leaves its velocity unchanged.

Focusing is combined with mass analysis in a magnetic element, the *magnetic sector analyzer*. As shown in Fig. 9-5, the sector analyzer contains a wedge-shaped,

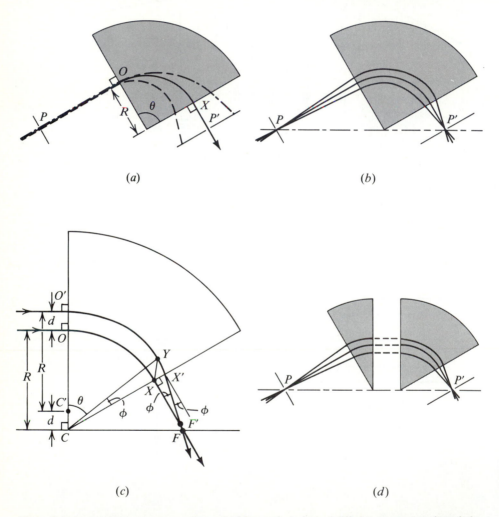

(a)

(b)

(c)

(d)

Figure 9-5 A magnetic sector mass analyzer. (a) Mass analysis: ions of differing masses pass through the aperture at P and are deflected in the field. Only the ion with radius of curvature R passes through the aperture at P'. (b) Focusing: ions of identical mass diverge from P and refocused at P'. (c) Construction for proving that parallel ion paths are focused at F. (d) Combining two sectors in the configuration of Fig. 9-5c into one of the configurations of Fig. 9-5b.

uniform magnetic field, characterized by the sector angle θ, along with two apertures. An electromagnet is used so that the field strength can be adjusted. In Fig. 9-5a, ions pass through an aperture at point P and enter the field perpendicular to the sector edge. The field has been set so that ions of the desired mass describe a radius R equal to the distance from the sector apex to the entry point O. Ions of this mass will thus leave the sector at X, still a distance R from the apex, and pass through the aperture at P'. Ions of other masses will have different radii of curvature in the field and will miss the second aperture. Therefore, only ions of the desired mass will leave the analyzer.

The sector shape of the field and the positioning of the apertures in line with the sector apex are not necessary to achieve mass analysis. Ions of differing mass (actually, momentum) will follow different paths in any magnetic field. However, this geometry has useful focusing properties. Figure 9-5b shows how ions of the same mass that diverge from P travel through the sector and are refocused at P'. This means that the sector analyzer not only will reject unwanted ions, but transmit the desirable ones with high efficiency.

It seems worthwhile to work through the proof that the sector analyzer has this focusing property. While the proof is somewhat lengthy and may never be revisited in practice, it provides an educational example of how physical barriers, applied fields, and ion properties interact in beam equipment. Therefore, we turn to the construction presented in Fig. 9-5c. In this figure, two ions of the same mass approach a sector of θ perpendicular to the edge, on parallel paths separated by a distance d. One ion enters the sector at O, a distance R from the apex C. This ion describes a radius R around the center of curvature C and exits perpendicular to the other edge at X. The second ion enters at O' and describes a radius R around C'. At point Y, it has undergone deflection θ and is still a distance d from the first ion. However, it still has a distance to go before leaving the field and thus is deflected more. It exits the field at X', converging toward the first ion at an angle ϕ. Some simple geometry shows that the other angles marked ϕ in the figure are appropriately labeled.

We now apply the law of sines to triangle CYX to evaluate ϕ. Because the ions are on parallel paths as long as both are in the field, the distance XY equals d. CX equals R, and the angle XYC equals $\theta - \phi$. Thus ϕ is given by

$$\sin \phi = \frac{d}{R} \sin (\theta - \phi) \tag{9-13}$$

The converging ions will meet at a point F', forming a triangle $X'XF'$. We can use the value of ϕ and the length of side $X'X$ to determine the length of side $X'F$. At this point enters the approximation, invariably made in focusing proofs, that D and hence ϕ are small with respect to R and θ. Then Eq. (9-13) reduces to

$$\sin \phi = \frac{d}{R} \sin \theta \tag{9-14}$$

and $X'X$ is very nearly $d \cos \theta$. It follows that

$$XF' = \frac{d \cos \theta}{\sin \phi} = \frac{R}{\tan \theta} \tag{9-15}$$

However, if we look at the triangle *CXF*, formed by a line through *C* and parallel to the initial ion paths, we can see that

$$XF = R \tan (90° - \theta) = \frac{R}{\tan \theta} \tag{9-16}$$

So *XF* and *X'F* are equal, and *F* and *F'* coincide. This result is independent of the value of *D*. Therefore, we have shown that, for small *d*, parallel ion paths entering the sector perpendicularly are converged to the point *F* lying on the line extended from *C*.

However, if the ion direction is reversed, the magnitude of all forces remains the same. Therefore, if ions diverged from *F*, they would exit the sector perpendicular to the edge and on parallel paths. Hence we can combine two sectors as shown in Fig. 9-5*d*. Here ions diverge from point *P*, emerge from the first sector on parallel paths, enter the second sector and are focused at *P'*. Now the size of the gap between the sectors is of no importance, since the ion paths are parallel. In fact, the gap can be of zero length, resulting in one continuous sector that focuses at *P'* the ions leaving *P*. This is in fact the situation found in Fig. 9-5*b*, which we set out to demonstrate.

In ion-beam equipment for production purposes, the simple elements we have seen are elaborated on for greater efficiency. For example, by angling the pole pieces, the magnetic sector can be made focusing in two dimensions rather than one. Apertures and lenses may be designed to produces a beam of rectangular cross-section rather than circular, for better transmission through selecting apertures. Nevertheless, the elements will be built on the principles we have examined.

9-2 THE ION IMPLANTER

We will now combine the elements we have studied into an apparatus for bombardment of substrates with ions—an ion implanter. We will require a high-intensity *source* of ions, a method of *extracting* ions from the source, elements for *analysis* and *acceleration*, and finally a means of *scanning* the beam across the substrates. In addition, to prevent the beam from colliding with neutral gas molecules, we will need a *vacuum system* to keep the implanter at high vacuum. Figure 9-6 shows how these components are combined in an ion implanter. (Our example is for illustration purposes and is specifically *not* intended to reproduce any particular implanter on the market.) It is worth noting that, because an implanter includes a number of electrical and mechanical systems as well as a complex vacuum system, maintaining one in good operating condition is quite complex. Practical implanter operation often requires the daily intervention not only of highly trained technicians, but often of professional engineers as well.

Ion Sources

To form an ion beam, we first need ions. Ions can be extracted from a neutral ionized gas. By definition, this is a *plasma*. We have already seen some ways to generate a plasma, for example by creating an rf discharge in a rarefied gas. A more common

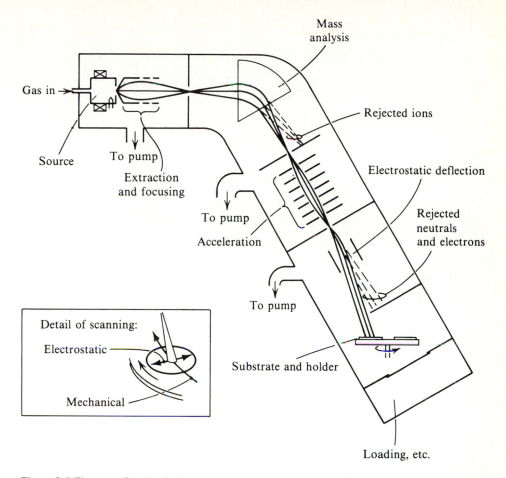

Figure 9-6 Elements of an ion implanter.

method for ion sources is to use a hot filament, which produces electrons to ionize a gas.

Figure 9-7 shows the operation of an ion source. A gas containing the element to be ionized flows into the source chamber. A hot filament, biased negatively to the source chamber, emits and accelerates electrons, which collide with gas molecules and ionize them. The most efficient ionization is usually obtained with electron energies of 50–100 eV. Coils surrounding the source generate a magnetic field, causing the electrons to take a spiraling path through the gas and increasing the ionization efficiency. As a result, the source is filled with a plasma of electrons and positive ions. As we saw in Chapter 6, the plasma will possess a positive plasma potential and be separated from the chamber walls by a plasma sheath.

To collect ions to form a beam, we bring a negatively biased *extractor* electrode close to an aperture in the source chamber. This electrode interacts with the plasma in

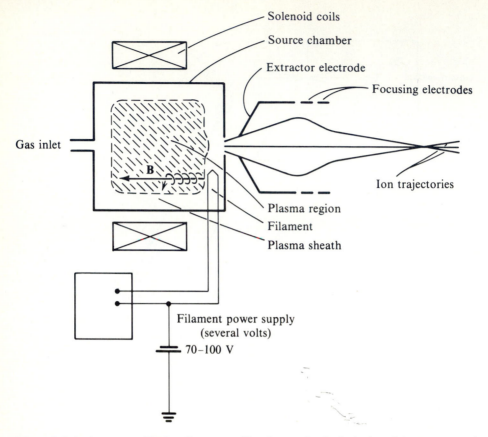

Figure 9-7 An ion source. The hot filament provides electrons for the ionization of the source gas. A magnetic field causes the electrons to spiral, lengthening their paths and increasing ionization efficiency.

an important way. While attracting ions, the electrode repels electrons, so that a plasma (a *neutral* ionized gas) cannot exist near the extractor. As a result, there is a distinct boundary to the plasma, which is repelled by the extractor. Meanwhile, the plasma potential tends to bulge the boundary surface out of the source. Certain combinations of extraction voltage and source conditions can lead to oscillation or instability of the plasma boundary. The boundary is the object surface for all of the ion optics, so beam current will fluctuate unless the boundary is stable in position and energy. Adjustment of conditions to obtain a stable beam can be a time-consuming and frustrating undertaking.

Ions extracted from the source fill a wide cone of diverging directions. Therefore, the extractor is followed by some converging electrodes that focus the beam. In our design in Fig. 9-6, the beam is focused on the mass analyzer.

Mass Analysis

The ions extracted from the source will usually have a variety of masses and charges. For example, if BF_3 gas is used as a source, we would expect to obtain BF_2^+ and B^{2+} as well as B^+. Analysis allows us to select the B^+ ion. In our illustration, a magnetic sector analyzer is used. For implantation, the magnetic field will be set to reject the unwanted masses while focusing the beam into the acceleration region. However, if we want to observe the ion mixture being generated in the source, we can vary the field strength to produce a *mass spectrum*. In this case, we bring each mass into focus in turn, measuring the beam intensity and producing a plot of current (and thus ion abundance) versus selected mass. We could use such a spectrum to decide on source conditions for optimum beam generation.

We bias the analyzer negatively to the source (by perhaps a few thousand volts), accelerating the beam to appreciable velocities. One reason is that the magnetic sector is a momentum analyzer and works as a mass selector only if all ions have essentially equal velocities. Ions are created in the source with a velocity distribution equal to that of the source gas, and we require sufficient acceleration to overwhelm these small variations. Also, a slowly moving beam tends to spread or "blow up" due to the mutual repulsion of the ions.

Mass analysis is complicated by the existence of different isotopes of the same chemical elements. Isotopes have identical chemical properties, which are determined by number of protons and electrons in an atom, but differ in mass because the nuclei contain a different number of neutrons. The masses of isotopes are indicated by superscripts: for example, boron consists naturally of 80.2% B^{11} with a mass of 11 amu, and 19.8% B^{10}. Mass analysis will reject the B^{11} component, reducing achievable beam intensity by 20%. On the other hand, "mass" analysis is really η analysis, so that multiply charged ions may not be rejected. For example, a mass analyzer cannot distinguish between As^+ and As_2^{2+} ions.

Acceleration

Leaving the mass analyzer, the beam is focused into the entrance of an acceleration tube. The bias on the last element of the tube, relative to the source plasma potential, determines the final energy with which ions will impact the target. This voltage is varied according to the demands of the process, and therefore the acceleration stage is designed to produce good focusing and transmission over a range of voltages.

Our implanter is of the *preanalysis* type, meaning that analysis precedes acceleration. This has the advantage that the analysis energy is fixed and fairly low, calling only for small magnets. It also means that the ion optics of the source-extraction-analysis region are not affected by the acceleration energy.

Postanalysis machines, with analysis following acceleration, have also been designed. Their advantage springs from the fact that the target is usually kept at ground potential to facilitate loading and unloading. Therefore, everything upstream from the accelerator must be at a large negative potential and must be carefully insulated from the operator. With postanalysis, fewer components require this protection. However, postanalysis machines require larger analysis fields, and their ion optics are affected when beam energy is changed.

Scanning

Having attained the desired energy, the ion beam issues from the accelerator and is ready to bombard the target. To get uniform coverage over a large substrate by a small beam, the beam must be scanned over the substrate, either by electrostatic or mechanical methods.

In *electrostatic scanning,* voltages are applied to electrodes to deflect the beam, as is commonly done to scan an electron beam. This method not only serves to scan the beam, but also purifies it by removing electrons and neutrals. Electrons are entrained in the beam to partially compensate the mutually repulsive charge of the ions. If they were not present, high-current beams would "blow up" because of the mutual repulsion of ions, with most of the ions striking the tube walls and being lost. Neutrals are formed from ions in the beam when they interact with background gas in the chamber. The ion picks up an electron from the colliding molecule, becoming a high-velocity neutral atom of the beam species.

Electrons and neutrals must be removed to allow accurate measurement of the amount (or *dose*) of material that has been implanted. Implantation dose is controlled by measuring implant current, assuming that each electronic charge in the current represents an implanted ion. Electrons contribute (negative) charge but no implantation. High-energy neutrals are implanted, but contribute no charge. Since only ions are properly deflected to the target, both electrons and neutrals are removed at the deflection step.

In *mechanical scanning,* the beam is kept steady and the substrate is mechanically moved under it. For higher beam energies and currents, it becomes harder to get substantial electrostatic deflection without beam blowup, so mechanical scanning becomes more appealing. In our example, we achieve mechanical scanning in one axis, by placing the substrates on a rotating table that passes them under the beam. In the other axis, the beam is scanned electrostatically. Thus we have a *hybrid* scanning arrangement.

The part of the implanter where the substrates are located is called the *end station*. In a production implanter, the end station incorporates automated equipment for loading and unloading the substrates without compromising the implanter vacuum. The substrate holder is also designed for efficient cooling of the substrates, to remove the heat created by implantation.

Vacuum and Mechanical Design

Ion beams can be formed only under vacuum conditions, so that collisions with gas molecules are rare. Implanters require an elaborate vacuum system because of the varying conditions within the equipment. In the source, gas density is high, to allow for efficient ionization. The extraction aperture allows not only ions, but also gas, to escape into the beam region, compromising the high vacuum required to prevent collisions and scattering of the beam. The end station also leaks gas into the beam region, both from periodic loading and unloading, and from the outgassing of the substrates as they are heated by the beam. Therefore, at least three vacuum pumps are used. Source and end stations are vigorously pumped to remove as much of the gas that they generate as possible. Small apertures separate the source and end station from the beam line, minimizing gas flow, and a third pump maintains the beam line at the highest achievable vacuum.

In implanter design, it must be remembered that all components are subject to bombardment by high-energy ions. Material sputtered from the walls can become ionized and contaminate the beam, as can pump oil and other gaseous species. Also, if any insulating material is deposited onto the walls, charges can build up that will deflect the beam. For microelectronic uses, graphite components are frequently chosen to prevent contamination of the beam with metals. Cryopumps may be used in the beam line to prevent "cooking" of pump oils into refractory insulating deposits.

Safety

Implanters incorporate three potential hazards: toxic gases, high voltages, and ionizing radiation.

The same toxic gases we have seen as dopant sources are frequently chosen for implantation: phosphine, arsine, diborane, etc. All of these gases are extremely dangerous. Gas lines and vacuum pump exhausts must be leak-free and properly exhausted. Cleaning of source assemblies and vacuum pumps must be done with all proper precautions.

A 100-keV implanter carries voltages of 100 kV on its electrodes. Particularly with postanalysis acceleration, these voltages are applied to much of the machine. Modern implanters are carefully interlocked to prevent hazard to personnel, and these interlocks must be respected. Troubleshooting of machines must be done with extreme care.

The acceleration of charged particles produces radiation. X-rays are generated in implanters both from the acceleration of ions and electrons, but the electron effects predominate because of their higher velocity. Radiation is reduced by keeping electrons out of the field regions through the use of additional electrodes, and by selecting construction materials that do not produce x-rays when struck by electrons, such as beryllium.[2] Lead shielding is used to absorb radiation where it cannot be prevented. Personnel working with ion implanters should be given radiation-detecting badges to wear so that exposure can be monitored. As always, the engineer installing an implantation process should consult with a qualified safety professional to ensure that the installation is safe.

9-3 ION IMPLANTATION FOR SUBSTRATE DOPING

In ion implantation, dopants are introduced into a substrate directly, in the form of ions. Implantation is thus an alternative to dopant diffusion, with several advantages. It is a low-temperature method, so dopants already in the substrate do not diffuse, and the crystal is not strained. The low temperature also allows a wide choice of materials for masking: photoresist can be used directly to achieve selective doping without the need to grow and etch an oxide or nitride. And implantation allows more flexibility in the choice of dopant distributions.

A doping process is characterized by the total amount of dopant implanted and by its distribution within the substrate. In implantation, control of the implantation amount or *dose* is particularly easy, because the implanted particles are charged. Therefore, if we ground the substrates, a current equal to the beam current implanted will flow. The dose Q in atoms per square centimeter is given, for singly charged ions, by

$$Q = \frac{It}{qA} \tag{9-17}$$

where I is the measured current, t the duration of implantation, A the area over which current is collected, and q the electron charge.

Errors can arise in dose determination if the charge transfer to the substrate is not the same as the dopant transfer. We have already seen how neutrals in the beam can be implanted without transferring current, leading to an underestimate of the dose. Another type of error is caused by *secondary electron* emission. Ion impact on the substrate may eject secondary electrons, particularly from photoresist used as a mask. This outgoing negative charge is perceived by the dose monitor as additional incoming positive charge, leading to an overestimation of dose. A negatively charged *Faraday cage,* which is an electrode surrounding the substrate, may be used to drive these secondary electrons back onto the substrate. Ions may also sputter the surface of the substrate, ejecting dopant that was already implanted. Finally, current will be underestimated unless good electrical contact is maintained to the substrate. Figure 9-8 shows how dosage is monitored as well as some sources of error.

The distribution of implanted dopant is determined by the mechanisms of *ion stopping.* When a high-energy ion enters the substrate, it collides with the electrons and nuclei of the solid, transferring energy with each collision. The ion will come to a stop when all of its kinetic energy has been transferred. On the average, this will happen at some *projected range* R_P, with a scatter or "straggle" of ΔR. The resulting distribution will be a Gaussian centered at R_P given by

$$N(x) = N(R_P) \exp \frac{-(x - R_P)^2}{2 \, \Delta R^2} \tag{9-18}$$

Integrating this distribution over all x should give the dose. Knowing that

$$\int_0^\infty \exp(-z^2) \, dz = \frac{\sqrt{\pi}}{2} \tag{9-19}$$

we obtain

$$N(x) = \frac{Q}{\sqrt{2\pi} \, \Delta R} \exp \frac{-(x - R_P)^2}{2 \, \Delta R^2} \tag{9-20}$$

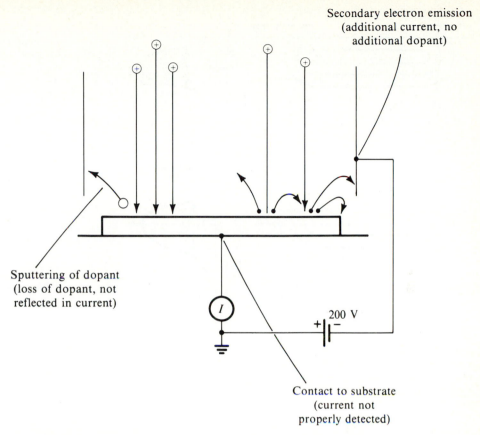

Figure 9-8 Dosage monitoring for ion implantation, with some common causes of error.

Example A silicon substrate is implanted with phosphorus (P^+) ions at 100 keV. The dose Q is 10^{11} cm^{-2}. Find (a) the depth at which dopant concentration reaches a maximum; (b) the dopant density at the maximum; (c) the dopant concentration at a depth of 0.3 μm.

Using Table 9-1, R_P for P^+ ion at 100 keV is 0.135 μm. This is the depth of the maximum dopant concentration.

From Eq. (9-20), the concentration when $x = R_P$ is

$$N(R_P) = \frac{Q}{(2\pi)^{1/2}\,\Delta R}$$

Table 9-1 gives ΔR as 0.0535 μm, so, with Q as given,

$$N(R_P) = \frac{10^{11}}{(2\pi)^{1/2}(5.35 \times 10^{-8}\ \text{cm})}$$

$$= 7.46 \times 10^{17}\ \text{cm}^{-3}$$

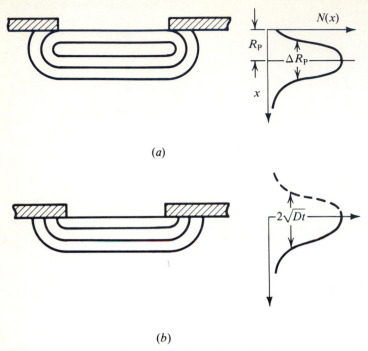

(a)

(b)

Figure 9-9 Distribution of dopant following: (a) implantation through a mask; (b) diffusion through a mask. Note the difference in the depth of the maximum concentration and thus in the point of deepest penetration under the mask.

At a depth of 0.3 μm, the dopant concentration is given by Eq. (9-20) as

$$N(0.3 \ \mu m) = N(R_P) \exp \frac{-(0.300 - 0.135)^2}{(2)(0.0535)^2}$$

$$= (7.46 \times 10^{17} \ cm^{-3})(0.0086)$$

$$= 6.42 \times 10^{15} \ cm^{-3}$$

In general, the collision process will divert the ion somewhat from its original direction, and there will also be some *transverse straggle* $R_\perp$ leading to a spreading of the ion profile. Figure 9-9 shows the distribution of dopant *implanted* through a mask rather than *diffused*. Note that, while in diffusion the concentration is always greatest at the surface, in implantation the concentration has a maximum below the surface. Also note the contours of equal distribution at the edge of the masked area. In the diffusion case, they are quarter-circles, meeting the substrate surface at right angles. For implantation, concentration falls off near the surface. This difference can affect the maximum field and thus the breakdown voltage of junctions.

The value of R_P depends on the details of ion stopping, expressed as the energy

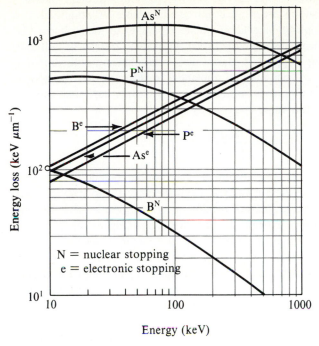

Figure 9-10 Calculated values of the energy loss per unit distance for electronic and nuclear stopping and for three common dopants. (*Reference 4. Copyright 1977, Research Studies Press, Inc., a division of John Wiley & Sons, Ltd.*)

loss per unit distance dE/dx. Ions are slowed by interaction both with electrons and with nuclei. *Electron stopping* is the transfer of ion energy to the "sea" of electrons in the solid with the generation of electron-hole pairs. It is much like viscous drag, with an energy loss proportional to velocity and insensitive to ion mass. *Nuclear stopping* involves ion-nucleus interaction, either displacing nuclei or transferring energy to the lattice. Nuclei are small, and ions will rarely actually strike them; rather the positive charges of the nuclei and ions interact from a distance. Slowly moving ions are more strongly scattered by these events, and so nuclear energy loss decreases with velocity while increasing with mass. This process has been analyzed in detail by Linhard, Scharff, and Schiøtt.[3] Figure 9-10 shows how energy loss varies with energy for three commonly used dopants.[4]

In actual practice, implanted profiles are often *skewed* asymmetrically from the Gaussian distribution of Eq. (9-18). An understanding of stopping helps to explain these skews. For example, boron implants are skewed toward the surface: the light boron ion is primarily stopped by electronic interaction, which is strongest at high velocities. Arsenic implants are skewed away from the surface, since most stopping is nuclear and is strongest at low velocities. Phosphorus implants represent a transitional case where the skew direction varies with implant energy.[5]

Another way implanted profiles can deviate from prediction is through *channeling*. In crystalline substrates, atoms are arrayed into regular planes. If a high-velocity ion enters the array between planes and approximately parallel to them, it may be *channeled* along the path of least resistance between the planes. Instead of colliding directly with atoms and being stopped, it makes numerous glancing collisions that remove little energy. Channeled ions may be implanted to several times the normal depth. Channeling is prevented if the ions bombard the substrate at a sufficient angle to the crystal planes. The minimum angle is given by

$$\psi = \left(\frac{2Z_1 Z_2 q^2}{Ed}\right)^{1/2} \tag{9-21}$$

where Z_1 and Z_2 are the nuclear charges (atomic numbers) of the ion and the substrate atoms, q is the electric charge, E the ion energy, and d the interatomic spacing along the channeling direction.[6] These angles are usually several degrees. Substrates are usually tilted about 10° to the ion beam so that the channeling angle is always exceeded. Our hypothetical implanter, depicted in Fig. 9-6, incorporates this feature.

Doping by implantation involves the same elements as does doping by diffusion. For silicon, these are commonly antimony, arsenic, boron, and phosphorus. Beams of these elements are formed from gases containing the elements, frequently the same gases chosen for dopant sources. One variation worth noting is the use of the gas BF_3 as a boron source. This gas is not only safer than diborane, but has the advantage of forming not only B^+ ions but stable BF_2^+ molecular ions. BF_2^+ has a molecular weight of 49 amu (versus 11 for boron) and hence has a much shorter range at a given energy. However, the additional fluorine atoms have no electrical effect when implanted. As a result, BF_2^+ is a convenient ion when shallow boron implantations are desired. Table 9-1 lists some properties of common dopant ions. Figures 9-11 and 9-12 show calculated projected ranges, and vertical and transverse straggle, for common silicon dopants.[4]

Table 9-1 Ion implantation characteristics of common silicon dopants

| | Isotopes | | Range and straggle (nm) | | | | | | $\Delta R_P/\Delta E$ (approx.) (nm keV^{-1}) |
| | | | At 30 eV | | | At 100 eV | | | |
Ion	Weight (amu)	Abun-dance (%)	R_P	ΔR_P	$\Delta R_\perp$	R_P	ΔR_P	$\Delta R_\perp$	
B^+	10.01	19.78	106.5	39.0	46.5	307	69.0	87.1	3.1†
	11.01	80.22							
P^+	30.97	100	42.0	19.5	16.9	135	53.5	47.1	1.1
As^+	74.92	100	23.3	9.0	6.4	67.8	26.1	18.7	0.6
Sb^+	120.90	57.25	20.8	6.2	4.6	50.7	15.8	10.8	0.45
	122.90	42.75							

† Boron range varies sublinearly with energy above about 100 eV. Range and straggle information from Ref. 5.

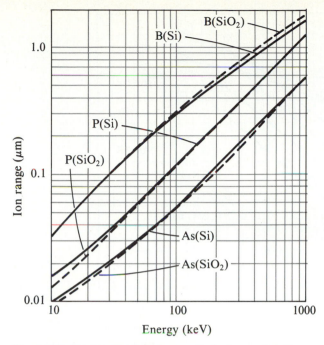

Figure 9-11 Calculated projected ranges R_P for boron, phosphorus, and arsenic as a function of energy, implanted into amorphous silicon (Si) or thermal silicon dioxide (SiO_2). (*Reference 4. Copyright 1977, Research Studies Press, Inc., a division of John Wiley & Sons, Ltd.*)

Implant Damage and Annealing

The impact of highly energetic ions does structural damage to the substrate. Each nuclear stopping event transfers energy to the atoms of the crystal, with an effect that depends on the energy transferred. Below about 15 eV (for silicon or gallium arsenide), the atom retains its place in the crystal, and the energy appears as heat. Between 15 and 30 eV, the atom is displaced, creating an interstitial atom and a vacancy. In this position, the atom can easily move back into the vacancy. Above 30 eV, multiple displacements occur, and stable defects are formed. As transferred energy increases further, very energetic atoms are formed that can, in their turn, cause additional damage.[7] Since each ion displaces numerous atoms, implanted substrates include many more defects than dopant atoms. This results in reduced carrier mobilities and shorter lifetimes. Furthermore, if enough ions penetrate the crystal, essentially all atoms will be displaced, and crystal order will be lost. The result is an *amorphous layer* within the substrate. If the initial stopping is primarily electronic, as with light atoms and high-energy implants, the amorphous layer may be buried below a still-crystalline surface layer.

Amorphous layers do have one advantage: being noncrystalline, they do not allow channeling. (Therefore, one way of avoiding channeling is to do a prior implant to

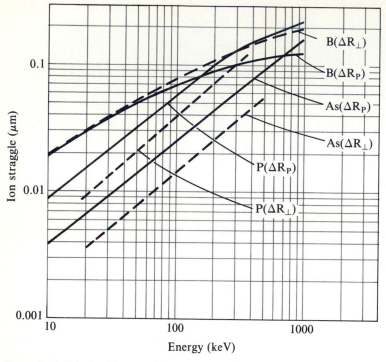

Figure 9-12 Calculated ion straggle in the vertical (ΔR_P) and transverse ($\Delta R_\perp$) directions for As, B, and P ions in silicon. (*Reference 4. Copyright 1977, Research Studies Press, Inc., a division of John Wiley & Sons, Ltd.*)

create an amorphous layer.) But, in general, a highly defective or amorphous layer of silicon is not compatible with good device properties and must be repaired. This is done by heating the substrate to allow displaced atoms to regain their proper position in the crystal, a process referred to as *annealing*. Whether or not an amorphous layer is present, annealing is also required to bring the implanted ions into substitutional sites in the lattice, so that they become electrically active. Implanted ions rarely enter substitutional sites spontaneously, and implantation without annealing has little doping effect.

In conventional annealing, the substrates are heated in a furnace for perhaps 15–30 min. The anneal temperature is usually a compromise: higher temperatures will more completely repair the crystal, but may cause crystal strain or damage to structures already fabricated. In silicon, for example, amorphous layers disappear, and a majority of the implanted dopant is activated, at 600°C. Full activation, with recovery of carrier lifetime and mobility, requires temperatures of 1000–1100°C. However, intermediate results are not always obtained at intermediate temperatures. Boron implants usually show a *decrease* in activation between 500–700°C. This is due to annealing out of damage, causing silicon atoms to displace the boron from substitutional sites in the lattice.

Laser annealing is an increasingly popular substitute for furnace annealing. In this technique, a laser beam sweeps the substrate, causing localized heating in the damaged areas. Damage can be annealed out in a matter of seconds, providing results equivalent to 800–1000°C furnace anneals, without diffusing the dopants or deforming the crystal.

If heating after implant can repair defects, it seems reasonable that heating during implant can *prevent* defects. It is, in fact, true that if a substrate is sufficiently heated, defects will diffuse away at a rate equal to their formation, and an amorphous layer will not form. For boron implants, a 50°C rise in temperature is sufficient to prevent formation of an amorphous layer at any dose.[5] Such a temperature rise can easily result from the heating by the ion beam itself. If substrate cooling is not uniform, the thickness of an amorphous layer can vary across the wafer due to temperature differences. Interference effects of light in this layer can create color bands on the wafer.

Masking for Implantation and Related Effects

Selective doping requires a mask, in implantation just as in diffusion. A pattern of some material must be formed on the substrate, to block incoming ions from reaching the areas where dopant is not wanted. But an important advantage of implantation is the variety of masks that can be used. The mask needs only to physically block the incoming ions; it need not be tightly adherent or able to withstand high temperatures. As a result, photoresist can be used directly as a mask, with no need to etch and pattern an oxide. Also, structures already on the wafer may serve as a mask, so that no additional lithography is needed. In MOS device processing, for example, the polysilicon gate structure is used to mask for the source-drain implant, so the implants are always perfectly aligned to the gate. Figure 9-13 shows this *self-aligning* type of masking.

Several factors must be considered in choosing a masking layer for implantation. Of course, the mask must be sufficiently thick to prevent ions from penetrating it. This is simply a matter of ensuring that the range of the ions in the mask is well short of its thickness. Given the range R_P and the straggle ΔR of ions in the mask, the effectiveness of a mask is easily obtained, because the ions follow a Gaussian distribution. For example, a thickness of $R_P + 4 \Delta R$ will give a "four-sigma" reduction in transmitted ions: about 0.03%. Table 9-2 gives some convenient values of the Gaussian distribution for such calculations.

Example R_P for a certain photoresist is 400 nm, and ΔR_P is 200 nm. The resist thickness is 1 μm. Find the effectiveness of the mask in preventing the transmission of ions.

The thickness of resist below R_P is equal to $3\Delta R_P$. From Table 9-2, this attenuates the ion flux to 1.11% of its original value.

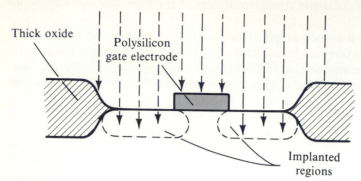

Figure 9-13 Use of a device structure as a self-aligning mask for an implant layer. The gate of an MOS structure is used to define the location of the source and drain implants.

When silicon oxide is used as a mask, ion bombardment can cause considerable damage to the oxide. Also, significant numbers of oxygen and silicon atoms will be displaced by the impact and implanted into the substrate, a process known as *knock-on*. When photoresist is used, the resist is chemically altered by the bombardment. Implanted resist is heavily cross-linked and becomes difficult to strip. Extended plasma ashing is usually required to remove such layers. Resists also outgas substantially during implantation, and emit many secondary electrons. Metals can be used as masks, but implantation of the metal due to knock-on may be hazardous to the circuit.

Implantation of ions into insulators (either masking layers or surface coatings) leads to *charging:* the buildup of charge on the surface. This charge may eventually arc to the substrate, causing defects in the insulating layer. The resulting field may also

Table 9-2 Useful values of the Gaussian distribution

$x/\Delta R$	$N(x)$ as a proportion of peak value
1	0.61
1.18	0.50
2	0.14
2.14	1×10^{-1}
3	1.11×10^{-2}
3.04	1×10^{-2}
3.72	1×10^{-3}
4	3.35×10^{-4}
4.29	1×10^{-4}
4.80	1×10^{-5}
5.25	1×10^{-6}
5.67	1×10^{-7}

deflect the incoming ion beam, leading to nonuniform implantation. As circuits become less tolerant of doping variations and of oxide defects, the control of charging during implant becomes more vital. One method is to coat the wafer with a thin conducting layer, such as aluminum or polysilicon, before implantation.[8] Of course, this adds a step to the process, and the coating may have to be removed later. Another alternative is to flood the wafer surface with low-energy electrons from a filament.[9]

9-4 OTHER ION-BEAM APPLICATIONS

Ion beams are unique in their ability to direct onto a surface a chemically pure collection of atoms of high kinetic energy with precise placement. Therefore, they are a powerful tool for altering the properties of materials and are being used for an increasing number of purposes other than doping. Ion beams can be used to modify substrate properties, add or substract material from the substrate, or for lithography.

Modification of Substrate Properties

We have already seen how ion implantation disturbs the structure of a crystalline substrate, even forming a buried amorphous layer. Instead of a side effect of implantation, such structural modification can be made a primary aim. Figure 9-14 shows three such applications.

The first, shown in Fig. 9-14a, is simply to damage and disturb the surface of the target material. For example, ion implantation of silicon dioxide damages the oxide, causing the surface layer to become faster-etching in fluoride-based etches. An ion implant of the proper dose and energy will alter the oxide layer so that its etch rate decreases with depth. Windows etched in such an oxide will have smoothly sloped edges, which produce good step coverage at subsequent deposition steps. Argon is frequently chosen for the implanted ion, due to its mass (40 amu) and chemical inertness.[10]

Another use of substrate damage is found in the production of magnetic bubble memories. In these devices, data bits are encoded by means of small bubble-shaped magnetic domains in a substrate of magnetic rare-earth garnets. Implantation of the garnet compresses the implanted area. Sufficient compression can change the easy direction of magnetization from perpendicular to the surface to parallel, with various beneficial effects. Implantation of the surface can make bubbles less stable (and thus easier to collapse when desired), by decreasing their height or providing a "keeper" layer, which provides closure for the magnetic flux within the bubble. An implanted pattern creates "rails" along which bubbles will move. A lower-dose implantation of the entire garnet reduces the stability of "hard bubbles," which are unwanted defects. Garnets also show a dramatic etch rate change on implantation, up to a factor of 1000 (Ref. 11).

Implantation damage is a powerful method for stirring or churning the atoms of solids. Figures 9-14b and 9-14c show the use of this stirring, or *ion mixing*, effect to create new materials. In Fig. 9-14b, a layer of metal is deposited over a silicon

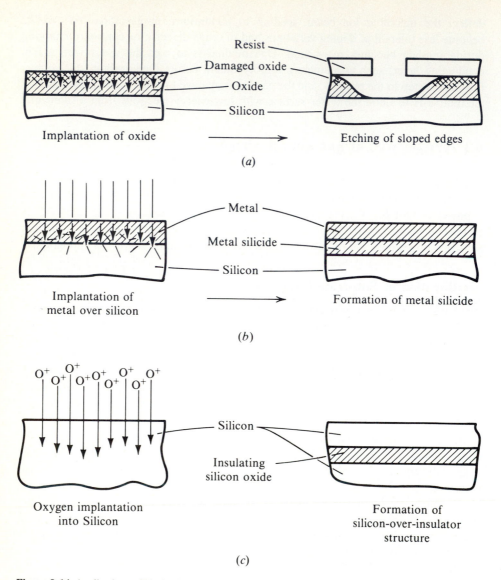

Implantation of oxide   ⟶   Etching of sloped edges

(a)

Implantation of metal over silicon   ⟶   Formation of metal silicide

(b)

Oxygen implantation into Silicon   Formation of silicon-over-insulator structure

(c)

Figure 9-14 Applications of ion implantation damage to fabrication. (a) Formation of a fast-etching oxide surface layer to allow etching of tapered contacts. (b) Formation of a metal silicide by implantation of a metal film over silicon. (c) Formation of an insulating layer beneath a semiconducting surface layer by implanting oxygen ions into the bulk semiconductor.

substrate, and ion implantation is used to mix the materials at the interface. By this means, a metal silicide can be formed. This method has been used to form molybdenum silicide (MoSi) from Mo films on silicon using phosphorus implantation.[12]

In studying the formation of amorphous layers in silicon, we saw that high-energy or light-atom implants can stir atoms within a material while leaving the surface layer reasonably intact. In Fig. 9-14c, this fact is exploited to create a *silicon-on-insulator (SOI)* structure from a semiconducting crystal. In many technologies, better device properties are obtained if a high-resistivity bulk substrate is used below a lower-resistivity epitaxial layer in which the device is fabricated. SOI, in which the substrate is insulating, is the logical end point of this strategy. So far, such structures have principally been fabricated by growing a semiconducting epitaxial layer on an insulator, a difficult process that has not proved popular commercially. In Fig. 9-14c, implantation of oxide ions into silicon is used to create a silicon oxide insulating layer perhaps half a micron below the surface. Devices can then be fabricated in the thin, electrically isolated silicon surface layer.[13]

Ion Plating and Ion Milling

Ion beams can also be used to add to or substract from the amount of material on the substrate surface. Addition of material involves a process known as *ion plating*, in which low-energy high-volume ion currents are used to deposit material onto the substrate. The ions are provided by a gas discharge, usually with no focusing or separation. This technique, also known as physical vapor deposition, has found little application in microelectronics.[14]

By contrast, the removal of substrate material by *ion milling* is an increasingly popular ion-beam application. Ion milling is sputter etching using an ion beam. The beam supplies ions of high and well-defined energy which are highly directional. When they impact the surface, they knock off surface atoms, resulting in extremely anisotropic etching. Such a beam also etches photoresist, but if a thick enough resist film is deposited, it can survive the etching process, and patterned etching can be achieved.

Figure 9-15 shows an ion milling apparatus.[15] The ion source is similar in design to that of an implanter, but there is no focusing or mass analysis. Instead, ions are extracted and accelerated by a pair of grids covering an entire "side" of the source. The first grid is near the plasma potential, and separates the plasma from the acceleration field. The second grid is at the acceleration voltage. The holes in the two grids are aligned so that a "beamlet" of ions is formed by the focusing action of each perforation. The source diameter, and hence the beam diameter, is of similar size to the target plate, so there is no need for scanning.

Ion milling rates are determined by the *sputtering yield,* which is the ratio of atoms sputtered to the number of incident ions. Yields depend on ion energy, impact angle, the material etched, and the mass and chemistry of the ions. Ion milling energies are on the order of several hundred to several thousand electron volts and in this range sputtering yields are on the order of 0.1–10 atoms per ion. Yields increase with increasing energy up to several keV. Above that energy, implantation becomes significant and sputtering yields fall.

Sputtering yields are a minimum for perpendicular impact by the ions. As the angle or incidence θ becomes smaller, the ion strikes substrate atoms more obliquely and has a better chance of ejecting them from the surface, so the yield per ion increases. However, the number of ions striking a given surface area varies as $\cos \theta$ and thus decreases with θ. As a result, most materials have a maximum sputter etch

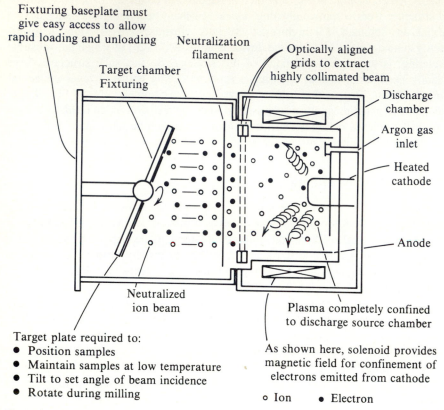

Fixturing baseplate must
give easy access to allow
rapid loading and unloading

Neutralization
filament

Target chamber
Fixturing

Optically aligned
grids to extract
highly collimated beam

Discharge
chamber

Argon gas
inlet

Heated
cathode

Anode

Neutralized
ion beam

Plasma completely confined
to discharge source chamber

Target plate required to:
● Position samples
● Maintain samples at low temperature
● Tilt to set angle of beam incidence
● Rotate during milling

As shown here, solenoid provides
magnetic field for confinement of
electrons emitted from cathode

○ Ion   ● Electron

Figure 9-15 An apparatus for ion milling of substrates. The beam is extracted from the plasma and impacts the substrates without mass analysis or focusing. A filament provides neutralizing electrons (*After Ref. 15. Reprinted with permission of* Solid State Technology, *published by Technical Publishing, a company of Dun & Bradstreet.*)

rate (in terms of material removed per unit time) at some value of θ from 40° to 60°. The target table in Fig. 9-15 can be tilted so that the incident angle of the ion beam can be varied.

Sputter yield does vary with material etched, but the variations are fairly small. Most common microelectronic films, including resists, etch at rates within 25% of each other. Therefore, there is little selectivity in ion milling, and resist masking is achieved only by making the resist thick enough to survive the milling process.

Since there is no mass analysis in an ion milling apparatus, the ion beam will contain a sample of the ion distribution in the source. Ion milling is commonly performed with an inert monatomic gas like argon (atomic mass = 40), in which case the beam is almost entirely composed of Ar^+ ions. For inert gases, sputtering yield increases with ion mass. However, in *reactive ion milling,* the chemistry of the gas ion plays a part in the etching. For example, adding oxygen ions to an argon beam can reduce the etch rate of certain metals, allowing ion milling to be made more selective.

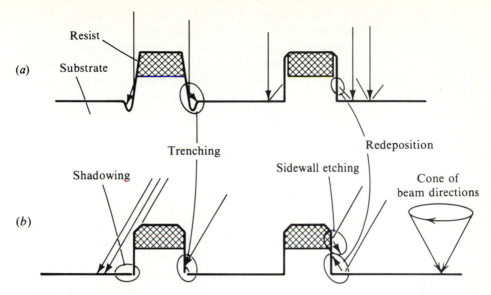

Figure 9-16 Geometric effects on step formation in ion milling. (*a*) Perpendicular beam direction: formation of trenches due to reflection off sidewalls, and redeposition on sidewalls to form "ears." (*b*) Oblique, rotating beam direction: shadowing can balance trenching; sidewall etching can balance redeposition. Also shown is faceting of the resist.

Ion milling is a physical process dominated by geometrical effects. Figure 9-16 shows some of these effects. Figure 9-16*a* shows a substrate bearing a resist pattern, which has been partially etched by a perpendicularly incident beam. Impact of the beam removes material from the substrate, which flies off much like a billiard ball struck by the cue ball. If the resist step intercepts this material, it will be redeposited on the edge, forming "ears." Similarly, ions ricocheting from the step can cause enhanced etching of "trenching" of the substrate near the step. These effects can be controlled by choosing the proper incident angle, by inclining the target table. To prevent asymmetries, the target table is rotated, so that the ion incident angle, relative to the substrate, describes a cone of directions, as shown in Fig. 9-16*b*. The wall of the resist step now has a finite etch rate, and redeposited material is removed. Trenching is controlled since, over the cycle of rotation, the trenching effect on "upsteam" steps is balanced by the shadowing effect when the edge is "downstream."

Another geometric effect is called *faceting*. Because resist, like most materials, has a maximum in etch rate at a particular angle of incidence, it tends to form *facets*—sloping corners—at that angle to the beam. Figure 9-16*b* also shows faceting of the resist edge. Because faceting occurs at the angle of maximum etch rate, it results in faster etching and shortens the survival time of the resist coating.

Ion milling is advantageous when etching involves small patterns, very steep walls, or inert materials. The milling process imposes no resolution limits, so that minimum dimension is limited only by the printing process. Use of a collimated beam at a controlled angle of incidence makes ion milling highly anisotropic, so very steep profiles can be obtained. Also, the physical nature of sputtering and the high energies available means that materials can be etched independent of their chemical reactivity. Ion milling has been particularly popular in magnetic applications such as bubble memories and thin-film recording heads. These devices use Permalloy, a very unreactive nickel-iron alloy, which is difficult to etch by other means. Also, the extremely vertical walls obtained with ion milling lead to sharp definition of magnetic domains and thus increased performance.[15]

Lithographic Applications

Photolithography engineers, always in quest of radiation of shorter wavelengths, have not overlooked ions. Ions are efficient exposers of resists: PMMA, a leading resist for use with electrons or x-rays, is about 10^4 times more sensitive to ions than electrons.[16]

Several lithography schemes have been proposed in which ions are used to replace light sources. In *ion projection lithography,* an ion beam passes first through a mask or reticle, and then through an ion optics system that produces an image on the substrate. This method is thus an analogue of conventional projection lithography.[17] In *channeled ion lithography,* the proximity method is used, with a collimated ion beam passing through a mask and then directly into the substrate. This technique is proposed as an alternative to x-ray lithography. It has the advantage that ions can be focused and collimated into a parallel beam, while x-rays diverge from a point source.[16]

Ion beams have obvious similarities to electron beams, suggesting their use for direct writing in the photoresist as in *e*-beam lithography. However, the speed of writing is limited by the speed of the ion. The beam cannot be steered to a new location any faster than the ion transit time between the deflecting plates. Since ions travel much slower than electrons of the same energy, ion writing is much slower than electron writing. The higher sensitivity of the resist to ions might offset this effect, but only if the doses per picture element were reduced to a few tens of ions. At this level, statistical variations become too large for reliable performance.[18]

However, an extension of direct ion-beam writing continues to arouse interest. A principal purpose of lithography is to control the placement of dopant. However, if the dopant could be directly implanted into selected locations by a controlled ion beam, the entire masking sequence would be unnecessary and the dopant pattern itself could be written into the wafer. This is the *writing beam implanter.* Companies have been formed to develop this concept, but the current consensus is that cost and slow throughput will make it uneconomical for general production.[19] Of course, the current consensus is not always right. Direct writing might prove particularly useful for doping selected high-value parts of a device. For example, read-only memories (ROMs) on microprocessor chips contain permanent codes that are generated as part of the fabrication process. Each customer specifies his or her own code, requiring a new mask, with the accompanying delay and expense of a mask-making cycle. Here the flexibility of the beam writing method and the ability to give customers quick turnaround on new codes might offset its slowness and cost.

This completes our survey of ion-beam applications. In summary, the uses of ion beams can be classified in terms of ascending energy and thus of penetration depth. At the lowest energies, ions simply coat the target, as in ion plating. At higher energies, sputtering predominates, as in ion milling. Higher energies yet lead to implantation, first near the surface and then more deeply. Implantation also causes damage to the substrate structure, due to displacement and mixing of the substrate atoms. This damage may be exploited to modify material properties, or to form new compounds through ion mixing. Finally, ions can be regarded as a type of radiation, and used in photolithographic applications.

9-5 CONTROL OF IMPLANTATION PROCESSES

Ion-beam applications, like other processes, are controlled by maintaining their input and output variables within desired ranges. We will concentrate on the control of implantation processes, since they are presently the major application of ion beams. The output for an implantation process is a dopant distribution in the substrate, characterized by the amount of dopant and its distribution both laterally and with depth. The input variables include the beam current, energy, chemical and electrical purity, and the way the beam is scanned over the substrate. Of course, these inputs depend on a multitude of equipment parameters. Some of these parameters are explicitly determined, like acceleration voltage, mass analyzer settings, beam impact angle, and scanning frequencies. Others are implicit, like background pressure and accuracy of mechanical assembly.

Since the output of implantation, like diffusion, is a dopant distribution, similar methods are used to evaluate both processes. The major difference is that implantation allows much lower doping, so that the sensitivity of the methods used must be greater.

The overall amount and lateral distribution of dopant can be determined by measuring the resistivity of the implanted layer. The four-point probe described in Chapter 3 is an effective tool with doping concentrations of 10^{13} cm^{-3} or more. However, this method assumes that all of the measurement current flows through the lower-resistivity surface layer. This approximation is good when doped-layer resistivity is low, but begins to fail at high resistivities as an increasing amount of current travels through the bulk substrate. This source of error can be controlled using the *six-point probe,* which is useful into the high 10^{11} range. As shown in Fig. 9-17, this method adds two additional probes on the back of the substrate. These probes measure the leakage current through the bulk, given by

$$I_l = \frac{\pi w V'}{\rho \ln (3s/w) + 0.5772} \tag{9-22}$$

where w is the thickness of the substrate, V' the voltage drop across the backside probes, ρ the substrate resistivity, and s the topside probe spacing. This leakage current can then be subtracted from the measurement current to obtain the current flowing in the implanted layer.[20]

Measurement of very high-resistivity implants remains an area of active research, and a number of electrical and physical methods are currently available. Capacitance-voltage (C-V) measurements, which we explored in Chapter 3, can be used to measure

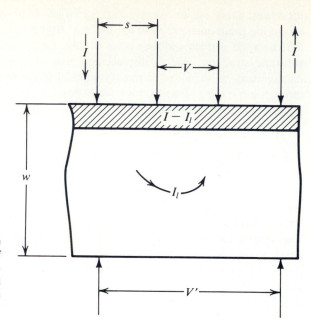

Figure 9-17 The six-point probe technique for measuring resistivity of implanted layers. To the conventional four-point probe are added two additional probes that determine the leakage current I_l.

both the amount and profile of dopant. In this method, the doping as a function of depth is deduced from the variation in the depletion layer thickness with changing voltage. However, this method is limited in depth resolution to the order of the *Debye length* L_D given by

$$L_D = \sqrt{\epsilon_s kT/q^2 N} \qquad (9\text{-}23)$$

where ϵ_S is the permittivity of the substrate and N the doping density. L_D exceeds a micron at doping densities below 10^{13} cm^{-2}. Unfortunately, it is shallow implants in this doping range we would often most like to investigate. The *C-V* technique also requires great care in performance and interpretation. In practice, it does little to extend the range achievable with the six-point probe.

Other electrical methods involve the fabrication of devices, such as resistor arrays or MOS transistors, followed by a determination of doping from the device parameters. These methods are useful in research and for periodic (say weekly) process monitoring, but the delay time resulting from the required fabrication is not acceptable in routine process control. Physical methods include such sophisticated surface science techniques as secondary-ion mass spectrometry (SIMS) and Rutherford backscattering. These methods are sensitive, but the testing is tedious and sometimes destructive, the equipment is elaborate, and the results can be difficult to interpret.

In the absence of channeling or other such effects, the depth profile of implanted dopant is a predictable function of ion energy. Ion energy is determined by acceleration voltage, an easily controlled variable. Therefore, dopant depth does not usually require day-to-day monitoring in production environments. For research, or when the characteristics of a new process are to determined, depth distribution can be measured using the *C-V* method, device parameters, or destructive techniques such as spreading resistance probing.

Much more than diffusion, implantation is subject to systematic variations in dosage across a substrate. Thus the lateral distribution of dopant is a major concern in implantation process control. Uniformity can be checked by making resistivity measurements at a number of locations on a substrate. If these results are presented in a contour map, the resulting patterns can reveal various types of error.

The input variables of ion implantation are easily understood physical quantities: beam energy, purity, current, and scanning. However, they are determined by equipment of great complexity, and thus the challenge in implantation process control is to maintain mastery of both the obvious and the hidden equipment parameters. This is done by ensuring a good machine setup, monitoring the available machine performance indicators, and using doping uniformity as a diagnostic tool.

Of the input variables, beam energy is usually not an issue given proper machine calibration. Chemical purity and "electrical purity"—the absence of neutrals, multiply charged ions, etc.—depend on maintaining a good vacuum and avoiding collisions of the beam with wall and apertures. Excursions in vacuum will show up in gauge readings. Instruments are usually provided to measure current flows between machine components. A misaligned beam will show up in an unexpected current.

Beam current can be measured directly as well as by inference from dose. In setting up an implanter, a *Faraday cup* is usually positioned at the target location to collect and measure the beam current. Various electrical and mechanical adjustments are made until a stable beam of sufficient current is obtained. To ensure that the beam is of the proper ion, a mass spectrum may be taken by varying the mass analyzer setting and measuring the current at the Faraday cup. This allows determination of the output of the ion source and ensures that the analyzer is passing the proper species.

Figure 9-18 shows a Faraday cup, and the methodology and appearance of a mass spectrum. Once a stable beam is obtained, substrates replace the cup at the target location and dose control is turned over to the dose processor. (We have already mentioned the role that neutrals, secondary ions, and poor contact can play in deceiving the processor.)

To determine scanning accuracy, a contour map of resistivity is made. Various discrepancies in implanter behavior will show up as characteristic patterns in the map.[21,22] Figure 9-19 shows some idealized contour maps displaying typical patterns. Figure 9-19a is a desirable distribution: there is very little variation in resistivity across the wafer. In Fig. 9-19b, the effect of neutrals in the beam is shown. The neutrals, which are not subject to electrostatic deflection, produce a "lump" in the dopant distribution. Such distributions point to a vacuum failure in the implanter.

Various problems with scanning can cause nonuniform doping. In Fig. 9-19c, the scan is simply off center, so that an edge is missed by the beam. In Fig. 9-19d, the scan is nonlinear, resulting in an overdose along the left and right edges. Such a pattern results in the beam spending more time "turning around" at the ends of the scan than it does traversing the middle. Such a pattern would result, for example, if a sinusoidal voltage was applied to the deflection electrodes.

In Figure 9-19e, the contours show a striped pattern angling across the wafer. This

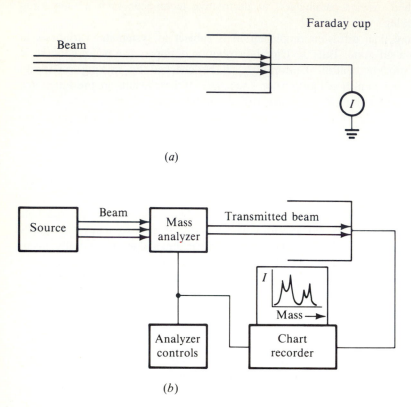

Figure 9-18 The Faraday cup and its use in preparing a mass spectrum. (*a*) Faraday cup for collection of beam current. (*b*) A mass spectrum results from a plot of transmitted current versus mass analyzer setting.

is characteristic of "lockup," in which the *x*-scan and *y*-scan frequencies are in resonance so that the beam repeatedly traces the same pattern. Such patterns are usually much more complex than simple stripes, but are characterized by regularly repeating peaks and valleys. Nonlinear scan and lockup are best prevented by attention to scanning waveforms and cycle frequencies both in implanter design and setup. Modern machines use crystal-controlled scanning frequencies with irrational quotients, which makes lockup unlikely but not impossible.

Figure 9-19*f* displays a characteristic charging pattern. Because of the accumulation of charge on the substrate surface, ions are deflected, and the center is underdosed. Although electron flooding can control this phenomenon, excessive flooding can lead to an overdosed center.

Several factors complicate the diagnosis of dose nonuniformities in real life. Of course, several causes may be present simultaneously, obscuring the pattern. Also, heat treatment is always a variable: resistivity methods measure *electrically active* dopant, whose ratio to implanted dose depends on the thermal history of the substrate.

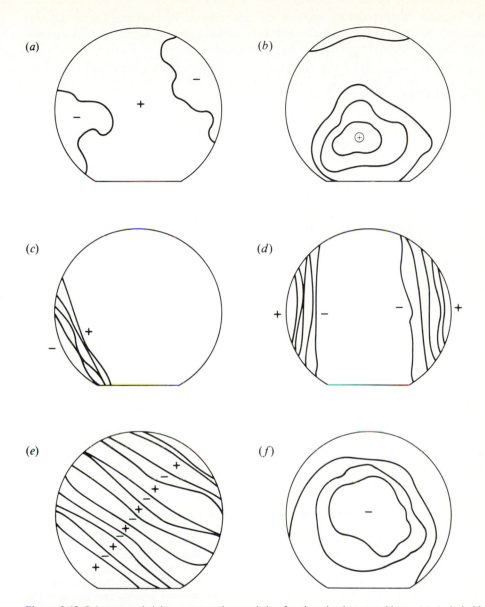

Figure 9-19 Substrate resistivity contours characteristic of various implanter problems. (*a*) A desirable profile; little variation in resistivity across the substrate. (*b*) Overdoping in a portion of the substrate, caused by implantation of neutrals. (*c*) An edge is underdosed because of an off center scan pattern. (*d*) Two edges overdosed because of nonlinear scan in the *x* direction. The orientation and polarity may both change. (*e*) A striped pattern due to "lockup" of *x*- and *y*-scan frequencies. Such patterns may be much more complex. (*f*) Underdosing of the center area due to charge-up of the substrate. *Note:* In these figures, a plus sign indicates a maximum in dopant concentration (minimum in resistivity), and a minus sign, a minimum in doping. This is done for clarity: normal practice is to plot contours of resistivity rather than dopant concentration, so that the plus sign would represent a minimum in doping.

Differential heating during implant can add thermal nonuniformities to those of dosage. Nevertheless, dopant concentration contours are a powerful tool for identifying and diagnosing implanter process excursions.

PROBLEMS

9-1 Give the energy and velocity of the following charged particles accelerated through a potential of 1000 V. Each carries one electron charge. (*a*) An electron. (*b*) An arsenic ion. (*c*) A boron ion.

9-2 A group of ions are traveling in a uniform electric field. The ions are brought to a focus at point *P*, and then diverge from *P* with directions, relative to the field, of $45° \pm \alpha$, where α is small. Show that these ions are refocused at a point *P'*, and describe its location. (That is, show that all ions pass within a distance δ of *P'*, where δ is second order or higher in α.)

9-3 Sketch the equipotentials of an Einzel lens operated in the *decelerating mode* (i.e., with the middle electrode more positive). Illustrate why the lens is still a converging lens despite reversing the center electrode potential.

9-4 Show that the magnetic sector analyzer is actually a momentum, rather than a mass analyzer (i.e., that ions of equal *momentum* are equally deflected). Assuming that all ions have been accelerated through an equal potential, express the deflection angle in the magnetic field as a function of mass.

9-5 A magnetic mass analyzer is constructed to pass ions deflected with a radius of curvature *R* of 10 cm. Find the appropriate field strength for selecting: (*a*) 100-keV B^{11+} ions; (*b*) 30-keV B^{11+} ions; (*c*) 100-keV B^{10+} ions; (*d*) 100-keV P^{31+} ions.

9-6 A 0.5-mA beam of 100-keV P^{31+} ions is scanned across a silicon substrate 100 mm in diameter. (*a*) How long a time is required to implant a dose of 10^{16} atoms per square centimeter? (*b*) What is the peak concentration of dopant? (*c*) At what depth will the peak concentration occur? (*d*) The background doping is *p*-type at a concentration of 10^{14} atoms per square centimeter. Assuming all implanted dopant is activated and ignoring diffusion effects, at what depth will the *pn* junction be formed?

9-7 Do Problem 9-6 for the B^{11+} ion with *n*-type background doping.

9-8 The number of atoms per cubic centimeter in crystalline silicon is 5×10^{22}. It takes about 30 eV to permanently displace each atom. Estimate the dose required to form an amorphous layer as a function of beam energy, range, and straggle.

9-9 In a certain resist, 70-keV phosphorus ions have a projected range of 1.1 μm and a straggle of 0.15 μm. (*a*) What proportion of ions will penetrate a resist masking layer 2.0-μm thick? (*b*) What is the minimum resist thickness to reduce ion transmission by 10^4?

9-10 A mass spectrum is taken of the output of an ion source, by varying the voltage applied to the solenoid coils of a magnetic mass analyzer and measuring the current of transmitted ions. (*a*) Assume that the current in the coils and thus the field are proportional to voltage. Derive an expression for the mass of the ions transmitted as a function of coil voltage. (*b*) The spectrum is graphed on a chart recorder using current collected as a *y* input and solenoid coil voltage as an *x* input. Will the *x* coordinate be linear in mass? Lay out the coordinates and label the *x* coordinate in atomic mass units in the range 1–100. (*c*) This source emits B^+, BH^+, and BH_2^+ ions in the ratio 70:50:100. Sketch the part of the mass spectrum containing these ions. (Remember that boron has two naturally occurring isotopes.)

9-11 An ion beam is electrostatically scanned in a sinusoidal manner over a substrate. Show that the doping is nonuniform, and write an expression describing the dopant concentration as a function of displacement of the beam.

9-12 Consider each of the following deviations in an implantation process, and describe the effect of each upon dopant concentration, dopant depth distribution, lateral doping uniformity, and chemical purity of the beam. (*a*) A pressure rise in the beam line. (*b*) A pressure rise between the last deflection plates and the substrate. (*c*) Miscalibration of the acceleration voltage. (*d*) Mechanical misadjustment causing the beam to strike an element within the implanter. (*e*) Lockup between *x*- and *y*-scanning frequencies. (*f*) Failure of the elements that provide electrons to neutralize the beam during acceleration. (*g*) Failure of the electron flood filament used to prevent charging of the substrate.

REFERENCES

1. J. R. Pierce, *Theory and Design of Electron Beams,* D. Van Nostrand, New York, 1954, p. 96.
2. Heiner Ryseel and Karl Haberger, "Ion Implantation: Safety and Radiation Considerations," in *Ion Implantation Science and Technology,* J. F. Ziegler (ed.), Academic, New York, 1984.
3. J. Linhard, M. Scharff, and Schiøtt, *Mat. Fys. Medd. Dan. Vidensk. Selsk.* **33**:1 (1963).
4. B. Smith, *Ion Implantation Range Data for Silicon and Germanium Device Technologies,* Research Studies, Forest Grove, Oregon, 1977.
5. T. E. Seidel, "Ion Implantation," in *VLSI Technology,* S. M. Sze (ed.), McGraw-Hill, New York, 1984.
6. D. V. Morgan (ed.), *Channeling: Theory, Observation and Applications,* Wiley, New York, 1973.
7. Sorab K. Ghandi, *VLSI Fabrication Principles,* Wiley, New York, 1983, p. 322.
8. M. Nakatsuka, K. Tanaka, and T. Kikkawa, *J. Electrochem. Soc.* **125**:1829 (1978).
9. C. P. Wu, F. Kolondra, and R. Hesser, *RCA Rev.* **44**:61 (1983).
10. J. C. North, T. E. McGahan, D. W. Rice, and A. C. Adams, *IEEE Trans. Electron Dev.* **ED-25**:809 (1978).
11. J. C. North, R. Wolfe, and T. J. Nelson, *J. Vac. Sci. Technol.* **15**:1675 (1978).
12. S. W. Chiang, T. P. Chow, R. F. Reihl, and K. L. Wang, *J. Appl. Phys.* **49**:5826 (1978).
13. S. Thomas Picraux and Paul S. Peercy, *Sci. Am.* **252**:102 (March 1985).
14. Lienhard Wegmann, "The Historical Development of Ion Implanters," in *Ion Implantation Science and Technology,* J. F. Ziegler (ed.), Academic, New York, 1984.
15. D. Bollinger and R. Fink, *Solid State Technol.* **23**:79 (November 1980); and **23**:97 (December 1980).
16. G. Stengl, R. Kaitna, H. Loschner, R. Rieder, P. Wolf, and R. Sacher, *Solid State Technol.* **25**:104 (August 1982).
17. D. B. Rensch, R. L. Seliger, G. Csanky, R. D. Olney, and H. L. Stover, *J. Vac. Sci. Technol.* **16**:1897 (1979).
18. W. L. Brown, T. Venkatesan, and A. Wager, *Solid State Technol.* **24**:60 (August 1981).
19. Wegman, *op cit.,* p. 45.
20. P. L. F. Hemment, "Measurement of Electrically Active Dopants," in *Ion Implantation Science and Technology,* J. F. Ziegler (ed.), Academic, New York, 1984.
21. Jeffrey R. Golin and James A. Glaze, *Solid State Technol.* **27**:289 (September 1984).
22. Matthew Markert and Michael I. Current, *Solid State Technol.* **26**:101 (November 1983).

TEN

THE FABRICATION OF BIPOLAR DEVICES

In the first nine chapters of this book, we have studied the tools used in microelectronic processing. We have discussed diffusion for driving dopant into substrates, lithography and etching for the replication of patterns, and thin-film methods for the application of coatings. In this final section, we will see these tools applied to construct a variety of practical electronic devices. In Chapter 10, we describe the fabrication of devices using bipolar transistors. In Chapter 11, we deal with the fabrication of devices based on the MOS transistor. In Chapter 12, we show how the same methods can be used to manufacture a variety of other devices.

Today the most widespread microelectronic application is the manufacture of silicon integrated circuits. Integrated circuits are composed of transistors, and there are two competing methods for transistor fabrication: bipolar and MOS. This chapter explores the older technology: the bipolar method. In Section 10-1, we review some key properties of the bipolar transistor, with emphasis on how they influence fabrication methods. In Section 10-2, we describe in detail a process sequence for construction of a bipolar device. In Section 10-3, we deal with some popular variations on the basic process, and in Section 10-4 we summarize important points about bipolar processes.

10-1 THE BIPOLAR TRANSISTOR

A bipolar integrated circuit is one composed mainly of bipolar transistors. An integrated circuit may contain thousands of transistors, which can interact, using various schemes of logic, to produce the complex types of behavior expected of modern electronics. However, the process engineer is usually concerned only with the properties of single transistors. His mission is to produce transistors with specified characteristics. If that mission is carried out, the interactions will be as predicted and the device will function as designed. Therefore, it is important (and usually sufficient) that the process engineer understand the relationship of process variables to the characteristics of isolated transistors.

Equations for *pn* Junctions

The *bipolar transistor* derives its name from the fact that both doping polarities— positive and negative—are used in its construction. We saw in Section 2-3 that the bipolar transistor consists of three regions, doped to create two oppositely facing *pn* junctions. The properties of *pn* junctions, which we first studied in Section 2-2, thus underly the properties of the bipolar transistor.

We saw in Section 2-2 that, when *n*-doped and *p*-doped silicon regions are brought together, the following occurs:

1. Charge carriers diffuse across the junction, driven by the difference in concentration, until a space-charge region is created that opposes further diffusion.
2. This space-charge region results in a depletion region and a built-in potential across the junction.
3. When the junction is forward-biased, charge carriers are injected across the junction, and current flow results.
4. When the junction is reverse biased, the depletion zone widens and charge carriers are swept away from the junction, preventing current flow.
5. At a sufficiently large reverse bias, the junction will break down, leading to large reverse current flows.

We obtained some results for an *abrupt* junction, in which two uniformly doped regions are in contact. For this case, the built-in potential ϕ_{bi} was given by Eqs. (2-29) and (2-16) as

$$\phi_{bi} = qE_{Fn} - qE_{Fp} = kT\left(\ln\frac{N_A}{n_i} + \ln\frac{N_D}{n_i}\right) \tag{10-1}$$

The width of the depletion region is a function of doping, of ϕ_{bi}, and of the dopant distribution. Equation (2-28) gave the result for two regions of uniform dopant concentration N_A and N_D as

$$w = \left(\frac{2\epsilon_S}{q}\phi_{bi}\frac{N_A + N_D}{N_A N_D}\right)^{1/2} \tag{10-2}$$

where q is the electron charge and ϵ_S is the permittivity of the semiconductor. An important practical approximation to this equation is the *single-sided abrupt junction* approximation. In this case, we assume that one side of the junction is much more highly doped than the other. In this case, the depletion region exists mainly on the *more lightly doped* side of the junction, and

$$w = \left(\frac{2\epsilon\phi_{bi}}{qN_l}\right)^{1/2} \qquad \text{single-sided abrupt} \qquad (10\text{-}3)$$

where N_l is the lesser of the two dopant densities. When an external voltage is applied to the junction, the depletion region width will change. In this case, ϕ_{bi} should be replaced in the equations by $\phi_{bi} + V_r$, where V_r is the applied voltage in the *reverse-biased* direction. Many actual *pn* junctions are far from abrupt. In this case, it is useful to use the *linearly graded junction* approximation. The dopant density variation, in the region of the junction, is approximated by its slope c, where $c = \partial N/\partial x$ in the vicinity of the junction. In this case

$$w = \left(\frac{12\epsilon_s\phi_{bi}}{qc}\right)^{1/3} \qquad \text{linearly graded} \qquad (10\text{-}4)$$

The built-in potential of a linearly graded junction also depends on c, via a transcendental equation. The relationship is shown in Fig. 10-1.[1]

Example A silicon substrate has a uniform background doping of 10^{16} cm^{-3}. In this substrate, a junction is formed using a predeposition that establishes a Q of 10^{16} cm^{-2}, followed by a drive-in diffusion with $\sqrt{Dt}$ equal to 2.0 μm. Use the linearly graded junction approximation and find w and ϕ_{bi}.

By differentiation of the Gaussian expression [Eq. (3-39)], we can show that

$$\frac{dN}{dx} = \frac{x}{2Dt} N(x)$$

Since at the junction depth x_j, $N(x)$ equals the background doping N_b, at x_j we can write

$$\frac{dN}{dx} = \frac{x_j}{2Dt} N_b = c$$

We can find x_j by solving Eq. (3-39) for x, with $N(x)$ equal to N_b:

$$x = \left(-4Dt \ln \frac{\sqrt{\pi Dt N_b}}{Q}\right)^{1/2}$$

$$= \left(-4 \times 4 \times \ln \frac{\sqrt{\pi} \times 0.0002 \times 10^{16}}{10^{16}}\right)^{1/2}$$

$$= 11.3 \times 10^{-4} \text{ cm}$$

It follows that

$$c = \frac{(11.3 \times 10^{-4} \text{ cm})(10^{16} \text{ cm}^{-3})}{8 \times 10^{-8} \text{ cm}^{-2}}$$

$$= 1.41 \times 10^{20} \text{ cm}^{-4}$$

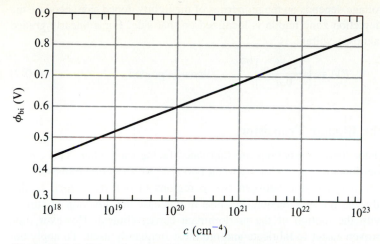

Figure 10-1 Built-in voltage as a function of the impurity concentration gradient c for linearly graded junctions in silicon at room temperature (*Reference 1, used by permission. Copyright 1967, John Wiley & Sons.*)

From Fig. 10-1, ϕ_{bi} is 0.6 V. The depletion layer width w follows from Eq. (10-4):

$$w = \left(\frac{12 \times 11.7 \times 55.4 q V^{-1}\ \mu m^{-1} \times 10^4\ \mu m\ cm^{-1} \times 0.6\ V}{q \times 1.41 \times 10^{20}\ cm^{-4}} \right)^{1/3}$$

$$= 6.92 \times 10^{-5}\ cm$$

$$= 0.629\ \mu m$$

Another important characteristic of a junction is the current flow as a function of forward bias. We would expect the current to be a sensitive function of bias, since the depletion layer width begins to approach zero at quite low voltages. The relationship turns out to be exponential and is given by

$$I \propto A \exp \frac{q|V_f|}{ikT} \tag{10-5}$$

where A is the area of the junction and V_f is the forward bias. The value of i depends on the mechanism of current flow, being 1 for pure recombination current and 2 for pure diffusion current.

Junctions break down, losing their rectifying property and passing large reverse currents, when the maximum electric field in the junction, $\mathscr{E}_{max}$, exceeds some critical electric field $\mathscr{E}_{crit}$ for the material. For a single-sided abrupt junction the breakdown voltage BV is given by

$$BV = \frac{\epsilon_s \mathscr{E}_{crit}^2}{2qN_l} \qquad \text{single-sided abrupt} \tag{10-6}$$

The breakdown voltage depends on the properties of the lightly doped side, because the field-containing depletion zone region is localized on that side. For a linearly graded junction, the breakdown voltage is

$$BV = \left(\frac{32\epsilon_s \mathscr{E}_{crit}^3}{9qc}\right)^{1/2} \qquad \text{linearly graded} \qquad (10\text{-}7)$$

Properties of the Bipolar Transistor

In a bipolar transistor, two *pn* junctions are fabricated facing each other. This can be done either by inserting a *p*-type region between two *n* regions (an *npn* transistor), or by sandwiching an *n*-type region between two *p* regions (a *pnp* transistor). In this section, we will discuss the *pnp* version, because in this case the direction of current flow coincides with the motion of the predominant carrier (holes). However, *npn* transistors have proven easier to fabricate and are most frequently used. To apply our discussion to an *npn* transistor, reverse the biases, the roles of holes and electrons, and the direction of current. (You will find that the direction of charge carrier flow remains the same.) In either case, the intermediate layer is called the *base*, and the two outer regions the *emitter* and the *collector*. Figure 10-2 shows idealized versions of both types of transistors, along with their electrical symbols.

Transistors can perform an amplifying function, producing a large variation in collector current from a small current variation at the base. In this mode of operation, called the *normal* mode, the emitter-base *pn* junction is forward biased, and the base-collector junction reverse biased. This results in the situation illustrated in Fig. 10-3. A current carried by holes flows from the emitter terminal, across the forward-biased junction and into the base. Some of this current flows to the base terminal and is lost. But if the base is narrow enough, most of the holes reach the base-collector depletion zone and are swept across it into the collector and to the collector terminal. The currents at the emitter, base, and collector terminals are designated I_E, I_B, and I_C respectively. As Fig. 10-3a indicates, they are related by

$$I_E = I_B + I_C \qquad (10\text{-}8)$$

Figure 10-3b shows components of the current in more detail. Current flows from the emitter terminal and across the emitter as a hole current. The base current flows in the base in the form of electrons, moving toward the emitter. The electrons contributed by the base current recombine with holes flowing from the emitter terminal, either in the base, the emitter, or in the emitter-base space-charge region. The excess holes wind up in the collector as the collector current. If I_B can be made sufficiently small, then I_C can be made very nearly equal to I_E, and large swings in collector current can be mediated by small variations in base current. Thus the amplifying capability of the transistor depends on minimizing the base current.

The amplifying power of a transistor can be expressed in terms of ratios of the currents. The *common-base current gain* denoted h_{FB}, or α, is defined as

$$\alpha = \frac{I_C}{I_E} \qquad (10\text{-}9)$$

pnp

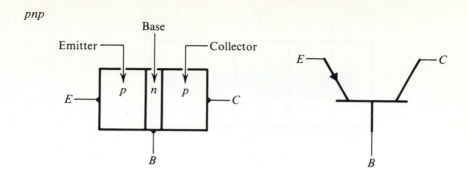

npn

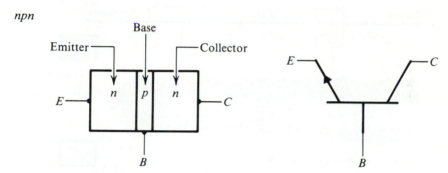

Figure 10-2 Idealized *pnp* and *npn* transistors, with their symbols.

According to Eq. (10-8), this ratio can never exceed one; for a good amplifier, it will approach one closely. The *common-emitter current gain*, denoted h_{FE} or β is

$$\beta = \frac{I_C}{I_B} = \frac{\alpha}{1 - \alpha} \tag{10-10}$$

For good amplifiers, β will be many times unity (typical values are 30–200). The value of β can be related to the physical properties of the transistor by examining how the base current behaves in each of the three regions: base, emitter, and space-charge region. The result can be expressed, for a transistor with large β, as[2]

$$\frac{1}{\beta} = Xw_B^2 + Y\frac{N_B w_B}{N_E w_E} + Z\frac{N_B w_B w_{EB}}{\tau \exp{(qV_{EB}/2kT)}} \tag{10-11}$$

Here N_E and N_B are the doping densities; w_E and w_B, the widths; and D_E and D_B, the carrier diffusion lengths in the emitter and base, respectively. w_{EB} is the width, and τ the recombination lifetime of the space-charge region. X, Y, and Z are properties of the material. (X is the diffusion length of minority carriers (here, holes) in the base; Y is the ratio of the diffusivities of minority carriers in the base to the emitter, and Z is the reciprocal of twice the intrinsic carrier concentration.)

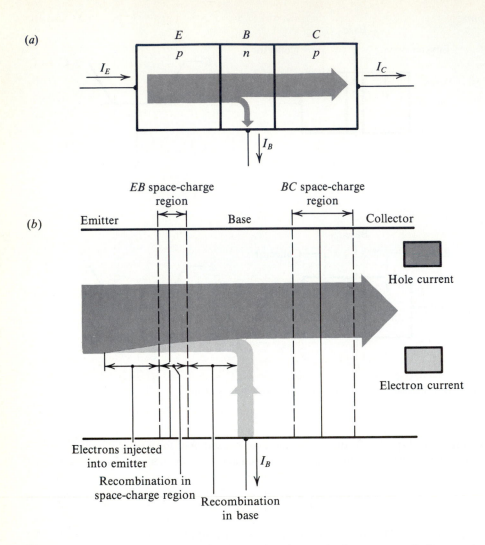

Figure 10-3 Current flow in a *pnp* transistor. (*a*) Emitter, base, and collector current. (*b*) Components of the base current.

The first term of Eq. (10-11) accounts for the amount of emitter current diverted to base current while crossing the base and depends strongly on w_B. The second term measures current resulting from injection by the base into the emitter and depends on their relative dopings. The third term, dependent on τ, accounts for recombination in the space-charge region. We can see that small w_B, small N_B/N_E, and large τ are all conditions encouraging large betas. It is frequently desirable that the gain β be independent of emitter-base voltage V_F, a condition also encouraged by long lifetimes

τ. Therefore, transistor fabrication usually strives for heavily doped emitters, lightly doped, narrow bases, and maximum recombination lifetimes.

Success or failure in achieving these aims can be determined from the electrical characteristics of the transistor. Figure 10-4 shows how the electrical characteristics of a transistor would appear on a curve tracer. The ordinate shows collector current I_C and the abscissa shows emitter-to-collector bias V_{CE}. Shown are several different traces, each for a given value of base current I_B. Figure 10-4a shows how the quantities are defined. Note that the emitter is connected in common to the input and output circuits, hence the term common-emitter gain. Figure 10-4b shows idealized traces. We see that, once a turn-on voltage V_{offset} is exceeded, the collector current rises to a level value, determined by I_B. We can determine β by measuring $\Delta I_C / \Delta I_B$.

Example What is β for the transistor whose trace appears in Fig. 10-4?

By inspection, a 10-μA increment in base current induces a 1-mA increment in collector current. Therefore,

$$\beta = \frac{1 \text{ mA}}{10 \ \mu\text{A}} = 100$$

Figure 10-4c shows a more realistic trace. We observe several deviations from ideality. One is the upward slant of each trace with increasing E-C voltage. This is the *Early effect* discovered by J. M. Early.[3] Increasing voltage widens the collector-base depletion region, and shortens the base width, increasing β. We also observe a bunching of traces at high and low base current, so that β is a function of current. The falloff in gain at low current occurs since most of the current is being consumed by fixed current losses due to surface recombination, etc. The gain falls off at high current because these currents result in significant voltage drops within the transistor. In real transistors, as we shall see, current flows parallel to the base-emitter junction, and the voltage drop in this region "debiases" portions of the emitter, reducing the effective junction area. Finally, the curves behave strangely at the left and right sides of plot, at low and high voltage. Each of these regions deserves some further exploration.

Breakdown

Being composed of *pn* junctions, a bipolar transistor will break down under sufficient reverse bias. This behavior appears at the right-hand side of Fig. 10-4c, where the voltage on the reverse-biased collector-base junction becomes large. The point where the zero-base-voltage trace first leaves the abscissa is called BV_{CEO}. Subscripts in breakdown voltages follow this convention: the first two letters indicate the terminals across which voltage is applied (collector and emitter) and the third whether the unlisted terminal (the base) is open (*O*) or shorted (*S*). Here the terminal is effectively open, since no base current is allowed to flow. Other breakdowns often measured for

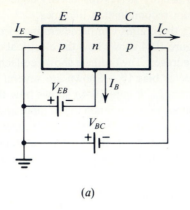

(a)

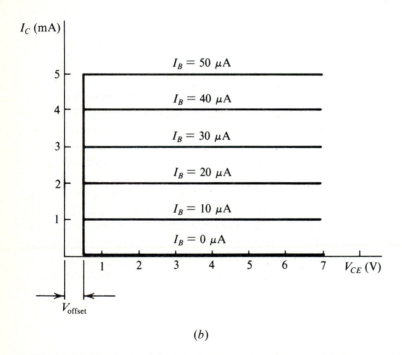

(b)

Figure 10-4 A plot of collector current versus emitter-to-collector voltage for a bipolar transistor for various values of base current. (a) Definition of quantities. (b) An idealized plot, showing the determination of β. (c) A more realistic plot, showing the Early effect, high- and low-current deviations, saturation at low voltage, and breakdown at high voltages.

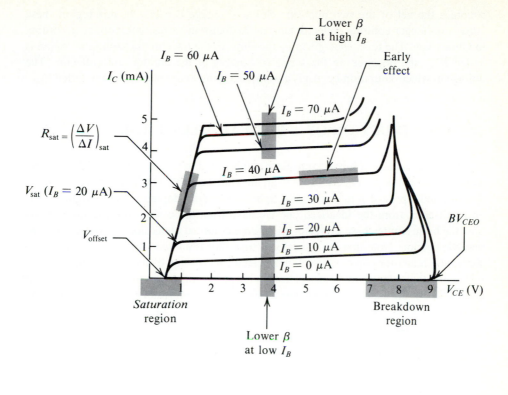

(c)

Figure 10-4 (continued)

transistors are BV_{CBO}, the base-collector breakdown, and BV_{CES}, in which the base and emitter are shorted together.

The BV_{CEO} trace behaves strangely, sweeping back towards the origin with a "negative resistance" characteristic. This results from the fact that, as current flows from the emitter to collector, it biases the base region, which is free to float. The result is a forward bias across the emitter-base junction, resulting in transistor action and amplification. Thus, once breakdown begins, the breakdown current is amplified and the voltage drops further. Breakdown voltages can be used to determine various process and electrical parameters.

Saturation

The left-hand region of Fig. 10-4c is the *saturation* region of operation. As emitter-collector voltage approaches zero, it becomes comparable to the emitter-base voltage required to maintain I_B. When this happens, the base-collector junction becomes *forward* biased. It now injects charge carriers into the base, and the collector current

becomes the net of the two opposing flows of charge carriers. In this region, base current no longer controls collector current, and current varies with voltage according to Ohm's law. The resistance given by the slope of the trace in the saturation region is called R_{sat}, and depends on the series resistances of the emitter and collector. The voltage at which saturation begins is a weak function of base current and is called V_{sat}.

10-2 CONSTRUCTION OF A BIPOLAR TRANSISTOR

We will now see how a bipolar transistor is fabricated in practice. Figure 10-5*a* shows a cross-section of a typical transistor in an integrated circuit. The letters *p* and *n* refer to doping, and a plus or minus indicates lighter or heavier doping. The configuration is quite different from the idealized one we have been discussing, but the inset shows how the basic emitter-base-collector arrangement, turned on its side, is contained in the realistic version. Note that the typical real transistor is an *npn* transistor. Figure 10-5*b* shows the transistor construction in more detail, with some of the factors determining minimum transistor dimensions. These include not only the minimum dimensions attainable by the mask and etch processes, but also allowances for diffusions and tolerances for alignment. Figure 10-5*c* shows a top view of the transistor.

The real transistor contains some components we did not see in the ideal version. On either side of the transistor is an *isolation* diffusion, which serves to isolate this transistor electrically from its neighbors in the device. The collector is two-layered, made of a heavily doped n^+ *buried layer* and a lightly doped n^- overlayer. The buried layer serves as a high-conductivity path connecting the "bottom" of the high-resistivity

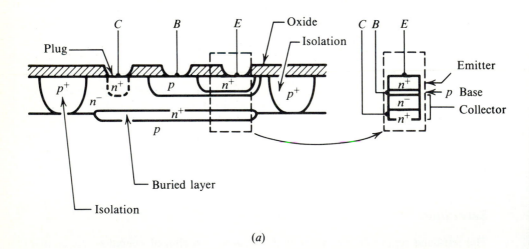

(*a*)

Figure 10-5 A realistic bipolar transistor fabricated in a semiconducting substrate. (*a*) Cross-section of a typical transistor; the inset shows the relationship to the idealized transistor. (*b*) More detailed view showing how the minimum dimensions achievable by the masking process (*D*), alignment tolerance (*A*), layer thicknesses (*t*), and junction depths (x_j) define the dimensions of the transistor.

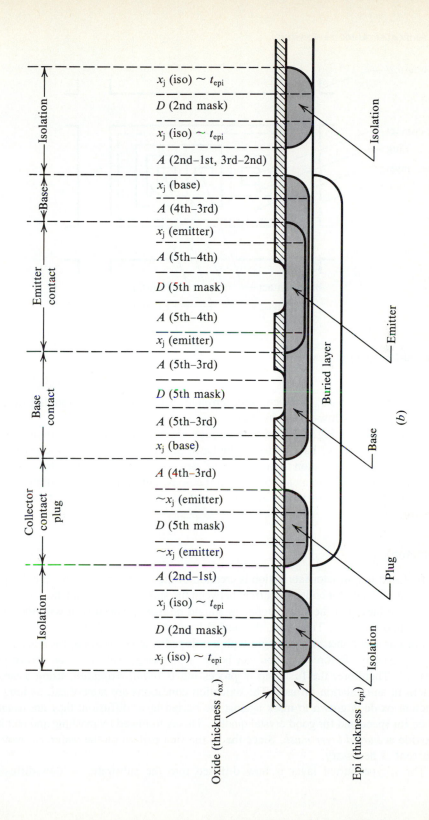

(b)

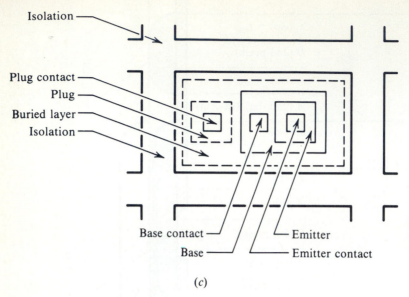

Isolation

Plug contact
Plug
Buried layer
Isolation

Base contact — Emitter
Base — Emitter contact

(c)

Figure 10-5(c) Overhead view.

collector under the base, to the collector terminal. The *plug* diffused from the surface helps complete this low-resistivity path.

Table 10-1 and Fig. 10-6 take us through the construction of the transistor in detail. In Fig. 10-6, we have added to the basic transistor in Fig. 10-5 the necessary metallization and passivation to make a useful device. As we go through the construction process, we will comment not only on methods, but on process and dimensional limitations, on control methods, and on the effects of each step on the final device behavior.

Collector

The first step in transistor fabrication is creation of the collector structure. We desire a low-resistivity path for collector current between the surface contact and the active area beneath the base, but the active collector should be lightly doped so it will not inject carriers into the base. Therefore, the buried-layer strategy is adopted.

For our *npn* transistor, we begin with a *p*-doped silicon substrate, for reasons we shall see shortly. Before diffusion can begin, a masking oxide must be grown and patterned. Therefore, the first step in processing is *initial oxidation,* using a steam ambient in an oxidation furnace. The oxidation conditions are noncritical, as long as sufficient oxide is grown to mask against the buried-layer diffusion: they are usually chosen for speed and for good crystal quality. This is followed by masking and etching the oxide at *buried layer mask.* Since this is the first pattern on the wafer, no pattern alignment is necessary.

The *n*-type buried layer is now diffused into the substrate. A slow-diffusing material such as arsenic or antimony is used, so that the buried layer will "stay put" during subsequent diffusions. The junction depth is typically a few microns, with sheet resistivities of perhaps 20 Ω per square. Because of the large x_j and the slowly diffusing dopants, the diffusion cycle will take many hours. During this process, there is oxidation where the silicon surface is exposed, consuming silicon and resulting in a lowering of the silicon surface under the buried layer openings. No significant oxidation takes place where windows have not been opened.

In order to bury the buried layer, an epitaxial layer of lightly doped n^- silicon is now grown. This rest of the diffusions will take place into this "epi" layer. Prior to epitaxy, the initial oxide is stripped off to give a clean silicon surface. Some physical evidence of the location of the buried layer is needed so that subsequent structures can be aligned; this is provided by the difference in silicon height caused by oxidation during buried layer diffusion. Epitaxial growth of silicon, doped with a slow-moving dopant (usually arsenic), follows. During epitaxial growth, there is some up-diffusion of the buried layer. Obtaining an epi layer of the proper thickness and doping with high crystal quality is perhaps the most formidable challenge in bipolar processing. Epitaxy reactor performance typically gates the output of bipolar fabrication areas.

The buried-layer structure carries collector current to the terminal and thus chiefly determines the resistance of the transistor in saturation. Buried layer properties are thus directly reflected in R_{sat}. Epi thickness can determine breakdown properties, and epi resistivity can effect a number of parameters including β.

After epi growth, a new masking oxide is needed for subsequent diffusions. Therefore, an *epi reoxidation* takes place. A steam furnace will probably be used.

Isolation

The integrated circuit we are making will contain many transistors. The collectors of all of them will be electrically connected, unless we isolate them from one another. This can be done by creating a *pn* junction completely surrounding the transistor. Figure 10-7 shows how this is done using a p^+ diffusion that completely penetrates the epi layer. It thus joins with the *p*-type substrate below the epi to form a *p*-type moat completely surrounding the *n*-type epi island. To form the isolation, the epi oxide is patterned and etched at the second masking step, *isolation mask*. Alignment now becomes an issue. The isolation structure, after diffusion, must not touch the buried layer, or a low-resistivity path will be formed. Therefore, the isolation structure must be separated from the buried layer by the combined junction depths, plus an allowance for alignment tolerance. These dimensions are shown in Fig. 10-5*b*.

The isolation diffusion takes place in a furnace using a boron source. The resistivity is noncritical, unless the isolation also serves some other purpose in the circuit. However, the junction depth must at least equal the epi thickness in order to obtain complete isolation. In practice, a small gap is sometimes left, so that further diffusion during subsequent high-temperature processes will complete, rather than overdrive, the isolation. Also, some circuits employ an iso-pinch resistor, leaving a small conducting gap between the isolation and the bottom of the epi, which serves as a resistor in the circuit. Following diffusion, an LTO step is used to remove the

Table 10-1 A process sequence for fabrication of a bipolar integrated circuit

Structure/step	Process	Purpose/considerations	In-process	Completed device
				Related parameters
Substrate	p-type		Resistivity	
Buried layer				
Initial oxidation	Thermal oxidation	Mask against buried layer diffusion: must have sufficient thickness to mask against long diffusion	Thickness	
Buried layer mask	Lithography/etch (oxide)	Determines location of buried layer		
Buried layer diffusion	n^+ diffusion	Forms buried layer. Low resistivity to reduce R_{sat}, slow-diffusing to control epitaxial autodoping	Resistivity	R_{sat}
Epitaxy				
Strip oxide	Dip in dilute HF	Form clean bare surface for epitaxy		
Epitaxy	n-doped epitaxial growth	Bury buried layer; provides silicon in which later structures are made	Resistivity, thickness	Various
Epi reoxidation	Thermal oxidation	Mask against isolation diffusion: must be thick enough to mask	Thickness	
Isolation				
Isolation mask	Lithography/etch (oxide)	Isolation must not touch buried layer	Pattern, alignment	
Isolation diffusion	n^+ diffusion (may include predep, LTO, drive-in)	x_j must reach through epi to substrate to provide complete isolation	x_j, resistivity	Isolation leakage
Isolation reoxidation	Thermal oxidation (may be part of isolation diffusion)	Mask against base diffusion	Thickness	

				Current drawn
Base				
Base mask	Lithography/etch (oxide)	Base must not touch isolation, resistor value must meet specifications	Pattern, alignment, dimension	
Base diffusion	p diffusion, includes predep, LTO, drive-in	Uniform and predictable resistivity and x_j vital to produce good betas, resistors	x_j, resistivity, BV_{CBO}	Beta, current, BV_{CBO}
Base reoxidation	Thermal oxidation (usually part of base diffusion)	Must mask against emitter diffusion	Thickness	
Emitter				
Emitter mask	Lithography/etch (oxide)	Emitter must not touch base	Pattern, alignment	
Emitter diffusion	n^+ diffusion (may include predep, drive-in)	x_j must closely approach base x_j to produce high betas. Resistivity often not as critical	x_j, resistivity, β	Beta
Emitter reox	Thermal oxidation (often part of emitter diffusion)	Provides oxide to insulate metal leads from bare silicon of emitter	Thickness	
Contact				
Contact mask	Lithography/etch (oxide)	Emitter contact must not cross edge of emitter	Pattern, alignment	
Metallization				
Metallization	Evaporate or sputter aluminum	Metal must be continuous, cover steps well, and have sufficient thickness to carry required current	Thickness	Contact resistance, Schottkies
Metal mask	Lithography/etch (aluminum)	Leads must be aligned to contacts, be wide enough to carry required current	Pattern, alignment, dimension, step coverage	Shorts/opens
Alloy	Furnace heat treatment	Forms Al-Si alloy for good contact		Contact resistance, Schottkies
Electrical test	Parametric test	Identify process deviations		
Passivation				
Passivation deposition	CVD (e.g., doped silicon dioxide)	Good integrity against moisture, alkalis, mechanical damage	Thickness, composition	
Pad mask	Lithography/etch (CVD layer)	Cut holes in passivation for attachment of package leads	Observe etch completeness	No contact

boron-rich layer, and another oxidation step forms a new oxide layer over the isolation diffusion.

If the oxide is stripped after isolation, the isolation process can be confirmed electrically by probing two adjacent transistor structures. If isolation is complete, no current will flow unless the voltage exceeds BV for the isolation junction, at which point a sharp breakdown occurs. If a small gap exists at the bottom of the isolation, a leakage current will be seen.

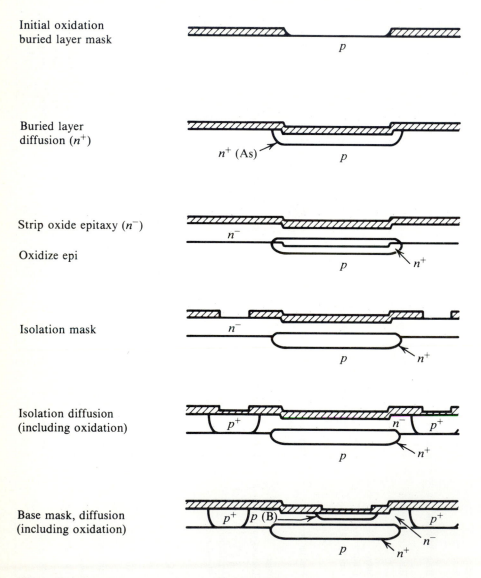

Figure 10-6 Steps in the construction of a bipolar transistor.

Emitter mask, diffusion
(including oxidation)

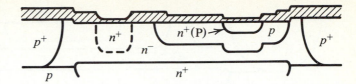

Contact mask

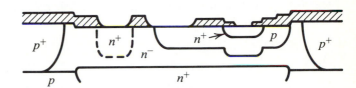

Metallization
Metal mask
Alloy
Electrical test

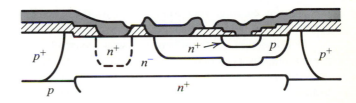

Passivation
Pad mask

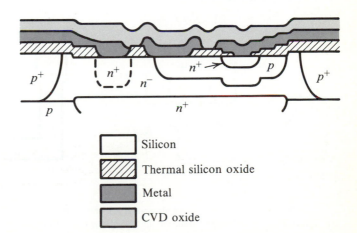

☐ Silicon

▨ Thermal silicon oxide

■ Metal

▨ CVD oxide

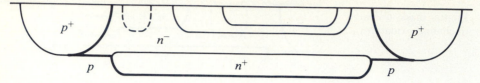

Figure 10-7 Isolation. The isolation diffusion joins with the substrate to produce a *pn* junction surrounding the transistor.

Poor isolation results in catastrophic device failure as all transistors are shorted together. Test programs usually begin with a "shorts" test, which will detect poor isolation.

Base

Fabrication of the base is a critical step in the construction of a bipolar transistor. Base fabrication begins with the third masking step, *base mask*. The base must be aligned so that, after diffusion, it does not touch either the isolation or the plug. Again, the design must allow tolerance for misalignment. Frequently, the base diffusion is also used to fabricate diffused resistors for the circuit, as shown in Fig. 10-8. The value of these resistors depends not only on the diffusion conditions, but on the width of the opening made during etching. Therefore, resistor width is one of the *critical dimensions (CDs)* for the device, and is measured as part of the masking process.

Because base width influences β so much, the base junction depth and resistivity must be tightly controlled. The base sheet resistivity should be fairly high (perhaps 200–500 Ω per square) so that the base does not inject carriers into the emitter. In our example, we will diffuse the base in a furnace using a boron source. Higher-resistivity bases may require ion implantation. We will use a base predeposition, followed by a low-temperature oxidation to remove the boron-rich layer, and a base drive-in. The drive-in is done in an oxidizing ambient, so that an oxide is grown over the base region for subsequent masking.

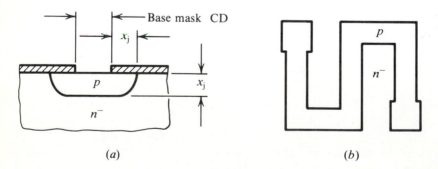

(a)

(b)

Figure 10-8 Fabrication of diffused resistors as part of base diffusion. (*a*) Cross-section, showing influence of lithography and diffusion parameters on resistor area. (*b*) Top view, showing a possible geometry.

Following base diffusion, we can make an electrical measurement of the collector-base breakdown voltage, BV_{CBO}. Since the base resistivity is still much lower than the epi resistivity, this breakdown voltage is heavily influenced by the properties of the epi. The measurement can be taken by stripping oxide from the surface or by waiting until after emitter mask and probing through the emitter and plug windows.

The most direct electrical effects of base parameters are on the diffused resistors. In many designs, all the current drawn by the circuit flows through these resistors, and so a variation in resistor width, junction depth, or resistivity will show up in the total circuit current. Base processing also affects β, but these effects are confounded with those of the emitter.

Emitter

Emitter diffusion is the last step in making the transistor itself. The *emitter mask* is the fourth mask. The emitter opening must lie wholly within the base, so there must be one alignment tolerance between the base and emitter edges. Emitter masking not only opens windows for the emitter, but for the plug, which provides a low-resistivity path for the collector current. The plug, after diffusion, must remain clear of the base and the isolation.

The emitter diffusion is a heavy n-type diffusion, producing a low-resistivity layer that will efficiently inject charge into the base. We will choose a phosphorus source so the diffusion times can be short and the previous layers will not diffuse much further. The emitter is diffused into the base, so that the emitter junction depth very closely approaches the base junction depth. The active base is then the region between these two junctions, which can be made very narrow by adjusting the emitter diffusion time.

Clearly, control of the emitter junction depth, in relation to that of the base, is a key step in determining bipolar transistor behavior. Unless the emitter junction closely approaches the base, the β of the transistor will be too low. However, if the emitter penetrates the base, we no longer have a transistor. Therefore, the base junction depth must be in a predictable value, and the emitter diffusion time be carefully chosen. Minor process variations will show up as significant changes in β.

Control of β is complicated by the *emitter push effect*. The heavy doping in the emitter creates strain in the crystal, which enhances diffusion of the base structure below the emitter. The base under the emitter therefore diffuses further than the base in other regions. The emitter push tends to keep the base width wider than it would otherwise be and is shown in Fig. 10-9.

Various diffusion and drive-in cycles can be used to fabricate the emitter. The resistivity of the emitter is usually not too critical, except as the doping concentration affects junction depth. The junction depth location is very critical. The same diffusion cycle is used to form the n^+ plug; the plug, being of n^+ type in the n^- epi, does not form a junction. Some part of the cycle will be done in an oxidizing ambient, to grow an oxide over the emitter region. Oxides grown from heavily phosphorus doped silicon grow at a faster rate than normal, so the oxidation time can also be short.

Because of the sensitivity of transistor β to process variations, it is wise to measure β as soon as possible after emitter processing. This allows quick correction of any

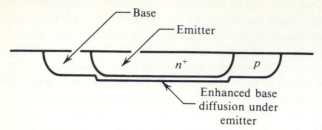

Figure 10-9 Emitter push. Strain from the emitter doping enhances diffusion of the base.

process drift. Therefore, β is sometimes measured directly after emitter diffusion by stripping oxide from an area of the substrate and probing the bare silicon. This measurement is somewhat compromised by various surface effects, but does provide immediate process feedback.

Contact Mask

After emitter fabrication, a working transistor has been formed. However, it exists in the substrate under an assortment of oxide layers and is not accessible to the outside world. The fifth mask, the *contact mask,* cuts holes in the oxides to allow electrical connection of the device. Contact formation is a mask-and-etch step only; henceforth, the masks associated with deposition steps will follow, rather than precede, the deposition.

The hole created for emitter contact must lie, after etching, wholly within the emitter, so the emitter, after diffusion, must be at least one alignment tolerance larger than the contact in all directions. The size of the contact thus controls the emitter size, and therefore the base size and the size of the device. Thus the contact is usually fabricated at the smallest dimension the masking process can reproducibly create. Contact openings are also made to the base, the collector plug, and perhaps to the isolation. In general, these contacts do *not* have to lie wholly within the structure contacted. If the plug or base contact overlaps the high-resistivity epi, the current flow will still be through the lower-resistivity layer.

In fact, in our example we have made the base contact deliberately overlap into the epi, in order to create a *Schottky diode,* which will make our transistor *Schottky-clamped.* As we saw in Chapter 7, a metal-semiconductor junction can have diode properties, and a properly positioned diode can improve the speed of a bipolar transistor by preventing it from saturating. We will fabricate this diode simultaneously with the base contact, by allowing the aluminum base lead to form a Schottky with the epi. Figure 10-10 shows how this strategy works.

Metallization

After contact mask, metal leads are fabricated to allow electrical connection of devices. Aluminum is deposited over the wafer either by evaporation or sputtering.

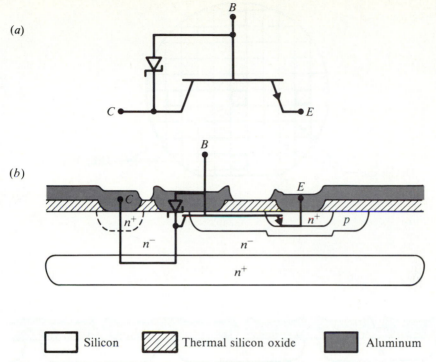

(a)

(b)

☐	Silicon	▨	Thermal silicon oxide	■	Aluminum

Figure 10-10 Formation of a Schottky diode concurrently with the base contact, by allowing the contact to overlap into the epi. (*a*) Schematic of the Schottky-clamped bipolar transistor. (*b*) Relationship of the physical configuration to the circuit schematic.

The thickness is determined by the current to be carried, and may be constrained by the etching process and by step coverage requirements of subsequent layers. A common thickness for single-layer metal is 1 μm; multilayer-metal processes strive for thinner first layers. Surface cleanliness, from the end of contact etching to the beginning of deposition, is vital, especially if a Schottky diode is to be fabricated.

After deposition, the thickness and film properties (reflectivity, resistivity) of the metal can be measured. Step coverage is also important, but is difficult to observe until after patterning.

Metal mask is the sixth mask in the sequence. The aluminum leads are patterned and etched. At a minimum, alignment must be good enough that the metal contact adequately overlaps the contact hole previously etched and does not overlap any other contact. If subsequent processes will bring an oxide or silicon etchant into the contact area, the metal will have to completely fill the contact hole to avoid damage to underlying layers. Linewidth of the leads is also important: narrow or notched leads cause excessive current densities and electromigration. Obviously, any bridging of metal will lead to a short in the circuit, so the area between the leads must be etched clear.

An *alloy* step follows metal masking. The substrates are heated to around 550°C to

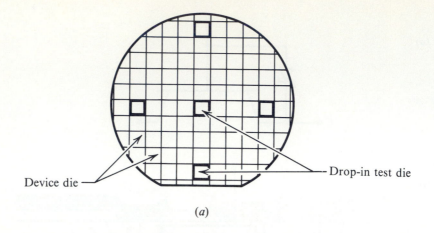

Device die —

Drop-in test die

(a)

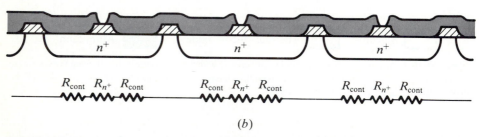

R_{cont} R_{n^+} R_{cont} $\quad$ R_{cont} R_{n^+} R_{cont} $\quad$ R_{cont} R_{n^+} R_{cont}

(b)

Figure 10-11 Tools for the electrical testing of integrated circuits. (a) Drop-in die, consisting solely of test structures, replaces a device die in the array. (b) A portion of a test structure on a drop-in, for testing contact resistance. The chain of contact in series multiplies the effect of any contact resistance.

form a aluminum-silicon alloy at the substrate surface, ensuring good electrical contact between the transistor structures and the leads. At this point, the device will undergo *electrical test*. Transistor electrical parameters will be measured. Many devices include *test patterns,* which may be either test structures within each individual integrated circuit (IC) or *drop-ins*. A drop-in is a special die pattern, consisting solely of test structures, which is substituted for some of the device die in the array. Drop-ins may include structures designed for sensitivity to particular fabrication weaknesses. For example, a long chain of metal-to-silicon contacts will detect variations in the alloying process before they are apparent in the behavior of single devices. Figure 10-11 shows the concept of the drop-in and of some test structures. Also at this step, a scanning electron micrograph (SEM) may be used to check the aluminum step coverage.

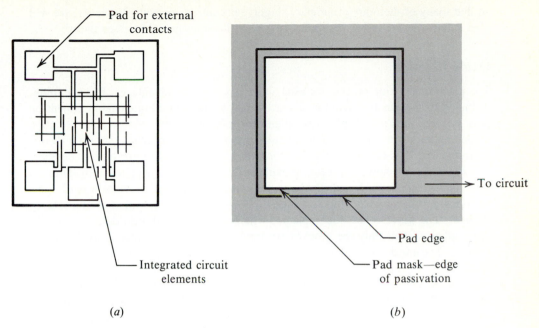

(a) Pad for external contacts

Integrated circuit elements

(b) Pad edge

Pad mask—edge of passivation

To circuit

Figure 10-12 Pad mask. (*a*) Arrangement of pads around the periphery of the device. (*b*) Close-up of pad showing the edge where the passivation is removed.

Passivation

Metallization is followed by *passivation,* in which an insulating, protective layer is deposited over the whole device. This protects it against mechanical and chemical damage in subsequent processing. Doped or undoped silicon oxide or silicon nitride, or some combination of them, are usually chosen for passivation layers. The layer is deposited by chemical vapor deposition at a temperature low enough not to harm the metallization. Passivation layers are fairly forgiving: thickness is not very critical. However, film chemistry must be controlled to prevent future corrosion, and the film must be nonporous and free from excessive stresses.

Passivation is followed by the seventh masking step, *pad mask.* It gets its name from the fact that openings are made only over the *pads,* which are the large metal structures to which the external leads will be attached during packaging. Pad mask is not visible in Fig. 10-6, because no patterning is done in the transistor itself. Figure 10-12 illustrates pad mask. Following this step, the fabrication process is completed. The device will now undergo functional testing to identify nonfunctional devices, followed by scribing and breaking, discard of defective devices, and assembly.

10-3 VARIATIONS ON A BIPOLAR THEME

The basic bipolar process we have just outlined is subject to a multitude of variations. It is, of course, the variations that increase the performance and versatility of devices, so that many of them are proprietary, highly specialized, or both. However, we will give a few widely known examples to give a flavor of how processes evolve.

Oxide Isolation

The junction isolation process we used above has some disadvantages. For one thing, diffusion of the junction requires a space equal to the minimum etchable dimension plus two epi thicknesses plus two alignment tolerances. For another, the isolation diodes interact with the transistors, resulting in *parasitic* properties that limit device performance. This has led to the development of *oxide isolation* for bipolar devices.

Figure 10-13 shows a process for oxide isolation.[4] Following epitaxial oxidation, a layer of silicon nitride is deposited and patterned. The nitride serves as a mask for a nonisotropic silicon etch process, using a HNO_3/HF solution, which produces deep grooves. The wafer is now ion implanted with boron. The oxide/nitride pattern masks the implant so that only the grooves are implanted. Now the wafer is oxidized in a furnace. The nitride now masks against the penetration of oxidant to the silicon, so that oxidation is selective and oxide grows only in the groove areas.

The oxidation process consumes silicon, so that at the end of the process, a layer of oxide fills the groove and extends to the buried layer. Boron is pushed ahead of the oxide layer, so that a p^+ region is formed at the bottom of the groove. This layer is called a *channel stop* and prevents any leakage at the epi-substrate interface. The nitride is now removed, and subsequent processing proceeds normally.

Oxide isolation has had some success in bipolar processing,[5,6] but has not eliminated junction isolation. However, as we shall see, it is standard in MOS processing.

Washed Emitters

Besides isolation, another significant contributor to the size of bipolar circuits is the emitter contact. To ensure the contact fits inside the emitter, the emitter must exceed the contact dimension by the emitter oxide thickness, plus two alignment tolerances. The minimum emitter size can be reduced to the minimum etchable dimension by a technique known as *washed emitters*.

Figure 10-14 shows the process, which makes use of the properties of phosphorus-doped oxides. The emitter is normally fabricated and diffused under oxidizing conditions. After diffusion, the emitter junction is outside the boundary of the original masked opening, and the opening is filled with a phosphorus-rich oxide. Such oxides etch much faster than normal in HF-containing etchants, and the difference can be exaggerated by choosing certain etch mixtures such as "P-etch."

Now the substrate goes to contact mask, and the base and collector contacts are fabricated normally. The emitter region is masked over and is not etched. After resist removal, the substrate is dipped or *washed* in a phosphorus-selective etch for a short period. The etch has little effect on the other contacts and oxide layers, but completely removes the fast-etching emitter oxide. The result is a set of contacts that are only a little smaller than the diffused emitter. Such methods, which use the properties of one

Initial oxidation
Buried layer mask
Buried layer diffusion
Epitaxy and oxidation
Silicon nitride deposition
Isolation mask

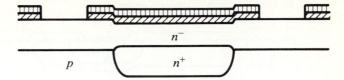

Silicon etch
Channel stop implant

Boron implant

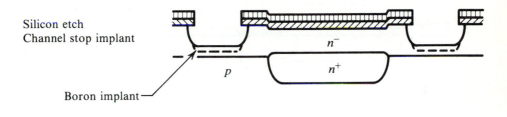

Oxidation
Nitride removal

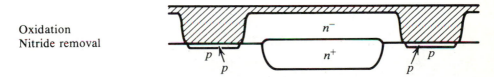

Figure 10-13 A process sequence for oxide isolation for bipolar circuits. (*After Ref. 4. This figure was originally presented at the Spring 1977 Meeting of The Electrochemical Society, Inc. held in Philadelphia, Pennsylvania.*)

circuit feature (the emitter) to define another one (the emitter contact) are called *self-aligning*. Since they do not require any tolerance for mask alignment error, they are powerful methods for reducing circuit dimensions.

Alternate Thin-Film Processes

Bipolar devices can be made with a variety of thin-film structures. One example, discussed in Chapter 7, is the use of a metal silicide and a barrier metal to form the Schottky diode. This process has become more standard than the aluminum Schottky and should be reviewed by the interested student.

Emitter diffusion (oxidizing ambient)

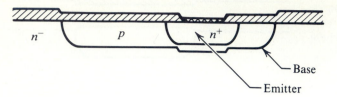

Contact mask

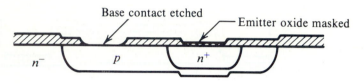

Emitter wash

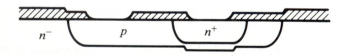

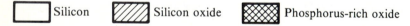

☐ Silicon   ▨ Silicon oxide   ▨ Phosphorus-rich oxide

Figure 10-14 Washed emitters. During contact mask the emitter contacts are protected from etching while base and collector contacts are being defined. The fast-etching emitter oxide is then washed away. After washing, the emitter contact is "self-aligning" to the emitter, reducing minimum emitter size.

Another example is the formation of *fusible links*, or simply *fuses*. The microelectronic fuse functions just as a macroscopic fuse does: it melts when used to carry an excessive current. Fuses can be used to make *programmable read-only memories* (*PROMs*). A read-only memory (ROM) is a memory device whose contents are permanently stored: they can be read from, but not written into. A programmable ROM is one in which the memory contents are entered electrically rather than using a masking step; it *can* be written into—but only once! This is done by electrically selecting, and blowing, an array of fusible links, one for each memory address.

Fusible links can be fabricated of metals or of polysilicon. The material must become nonconducting when heated by the programming current, and the change must be permanent. Metal links can be programmed when covered by passivation: presumably, the metal disappears into pores in the passivation. Polysilicon links must be exposed to oxygen during programming.

Multilayer Metallization

Our basic process used only one layer of metal to interconnect the individual transistors. For more complicated devices, it is highly desirable to have additional layers of interconnect; otherwise, elaborate detours of the metal leads are required to make all the necessary connections, and device size becomes limited by metal layout. At least two layers of metallization are quite common in bipolar applications.

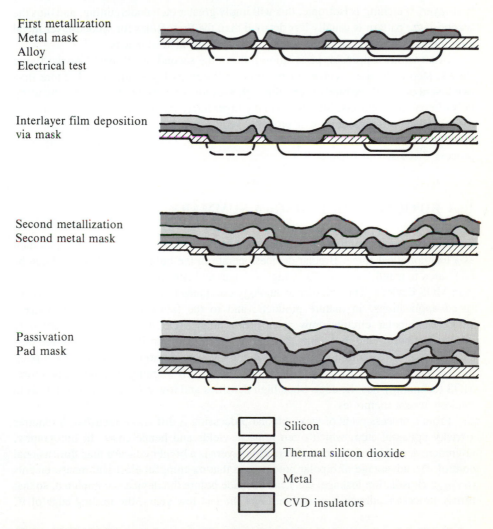

First metallization
Metal mask
Alloy
Electrical test

Interlayer film deposition
via mask

Second metallization
Second metal mask

Passivation
Pad mask

Silicon

Thermal silicon dioxide

Metal

CVD insulators

Figure 10-15 Multilayer metal process. This process begins after the first metal mask in Fig. 10-5.

Figure 10-15 and Table 10-2 show a two-layer metal process. The process takes up after the metal mask, alloy, and electrical test steps of our basic process. As before, an insulating thin film such as CVD phosphosilicate glass (SiO_2/P_2O_5) is deposited. However, now the thickness, conformality, and step coverage of the film become much more important. A rule of thumb for step coverage is that each layer be 1.5 times the thickness of the underlying layer.

Deposition is followed by *via mask*. This mask cuts holes, or *vias*, through which the second-layer metal can contact the first-layer metal. When a second-layer metal contact to silicon is desired, we have deposited a first-layer metal pad over the silicon to reduce the height of the step. Vias must be aligned to the underlying metal pattern so that good contact is made.

After via mask comes second-layer metal deposition and *second metal mask*. Following the 1.5 rule, second-layer metal will be two or more times the thickness of first-layer. If etching is isotropic, this will imply greater etch undercutting, and thus the second-layer pattern is usually less detailed than the first. Rules for second-layer metal alignment to vias are the same as for first-layer metal to contacts.

In a two-layer process, passivation follows the second metal mask. As the figure shows, step coverage becomes increasingly difficult as layers are added. More than two layers of metal are rarely seen, although additional layers can be obtained by using polysilicon interconnects. Successful many-layer interconnect, such as will be required for wafer-scale integration, will probably require use of planarizing processes that reduce rather than aggravate step coverage problems. One example is the polyimide process discussed in Chapter 8.

10-4 BIPOLAR PROCESSING: A SUMMARY

Bipolar integrated circuits were the first to come into production and remain important as a technological force. Bipolar devices operate in the bulk of the silicon and can be fabricated to handle comparatively high voltages and currents. They can be made faster than MOS devices. Thus bipolar technology continues to be the choice for very-high-speed applications, in mature products, and in the fabrication of analog devices. However, bipolar circuits draw more power than MOS devices. This is not only inherently undesirable, but limits the density with which transistors can be packed on a chip. Density is also limited by mask alignment considerations, because the bipolar process uses many masks and makes little use of self-aligning methods. Therefore, MOS technology has the edge when high densities and low costs are paramount, as in random-access memories.

From a process point of view, bipolar processing is diffusion-intensive. It requires a tricky epitaxial step, which often impacts yields and hence costs. In lithography, alignment accuracy of the many interacting layers is a bigger concern than dimensional control. An advantage of bipolar processes is that meaningful electrical measurements (BV_{CBO}, β, isolation leakage, etc.) can be made before the device is completed, so that timely process feedback is available. Over the last few years, the leading edge of IC

Table 10-2 Dual-layer metallization process for use in bipolar fabrication

Structure/step	Process	Purpose/considerations	In-process	Completed device
				Related parameters
First metallization				
First metallization	Evaporate or sputter aluminum	Metal must be continuous, cover steps well, and have sufficient thickness to carry required current	Thickness	Contact resistance, Schottkies
Metal mask	Lithography/etch (aluminum)	Leads must be aligned to contacts, be wide enough to carry required current	Pattern, alignment, dimension, step coverage	Shorts/opens
Alloy	Furnace heat treatment	Forms Al-Si alloy for good contact		Contact resistance, Schottkies
Electrical test	Parametric test	Identify process deviations		
Interlayer dielectric				
Interlayer film deposition	CVD (doped SiO_2)	Insulate first from second-layer metal. Must not distort first-layer metallization.	Thickness, composition	
Via mask	Lithography/etch (CVD SiO_2)	Allow second-layer metal to contact first layer and silicon. Vias must align to first layer metal.	Pattern, alignment	
Second metallization				
Second metallization	Evaporate or sputter aluminum	Metal must be continuous, cover steps well, and have sufficient thickness to carry required current	Thickness	
Second metal mask	Lithography/etch aluminum	Leads must align to vias, be wide enough to carry required current	Pattern, alignment, dimensions, step coverage	Shorts/opens
Passivation				
Passivation deposition	CVD (e.g., doped silicon dioxide)	Good integrity against moisture, alkalis, mechanical damage	Thickness, composition	
Pad mask	Lithography/etch (CVD layer)	Cut holes in passivation for attachment of package leads	Observe etch completeness	No contact

363

technology has been increasingly dominated by MOS technology. Nonetheless, bipolar technology will remain important in the semiconductor business.

PROBLEMS

10-1 Derive Eq. (10-3).

10-2 Derive Eq. (10-4). [You may wish to review the method used to derive Eq. (2-28) in Chapter 2.]

10-3 Observing Eq. (10-3) (for a uniform doping variation) and Eq. (10-4) (for a linear doping variation), predict the form of the equation giving the depletion zone width for a doping variation given by a quadratic equation.

10-4 Using the cross-section in Fig. 10-4b as a guide, find the percentage reduction that can be achieved in the linear dimension of a bipolar transistor by the following methods. (a) Replace junction isolation with oxide isolation. Allow 0.5 t_{epi} per side, for lateral etching and oxidation. (b) Use washed emitter contacts instead of patterned and etched contacts. (c) Implant base and emitter instead of diffusing them. Assume that transverse straggle during implant is equal to 50% of the junction depths.

For Problems 10-5–10-9, use the following process for an *npn* transistor fabricated in silicon:

Substrate resistivity: 15 Ω cm, *p*-type
Buried layer predeposition: arsenic, $Q = 1 \times 10^{16}$ cm^{-3}
Buried layer drive: 1250°C, 4 h
Epi resistivity: 0.20 Ω cm, *n*-type
Epi thickness: 3 μm
Base predeposition: boron, $Q = 1 \times 10^{15}$ cm^{-3}
Base drive: 1100°C, 1 h

Assume that the epi doping is uniform throughout and that the diffused layers have Gaussian distributions.

10-5 Use Fig. 2-6 to estimate the dopant concentrations in the substrate and epi layers.

10-6 Find the junction depth for the buried layer following buried-layer diffusion. You will need Eq. (3-38), Fig. 3-5, and the result of Problem 10-5. Estimate the resistivity of the buried layer using the "box" approximation, e.g., assume the effective dopant concentration in the buried layer is equal to the surface concentration. Then use Eqs. (3-80) and (3-82) to find the sheet resistivity and V/I for the buried layer.

10-7 Find the following for the base-collector junction, immediately following base diffusion. (a) x_j. (b) The concentration gradient c at x_j. (c) ϕ_{bi}. (d) The depletion region width w for no applied bias. (e) The resistivity, sheet resistivity, and V/I, using the box approximation. (f) BV_{CBO}. Let $\mathscr{E}_{crit} = 30$ V μm^{-1}. $\epsilon_s = 648q$ V^{-1} μm^{-1}.

10-8 Using the value of BV_{CBO} from Problem 10-7, find the depletion region width w at the voltage where breakdown begins. In the completed transistor, BV_{CBO} will be observed at the lesser of two voltages: the value for the isolated base-collector junction, or the value at which the junction depletion layer extends to touch the buried layer. The second condition is called *punchthrough*. Is this transistor likely to display punchthrough?

10-9 Emitter diffusion takes place at 950°C with a Q of 1×10^{19} cm^{-2}. Suggest a cycle time to bring the emitter junction depth very close to the base junction depth.

10-10 Using your answer to Problem 10-9 and some reasonable approximations, find the ϕ_{bi} for the emitter-base junction. When in operation, the emitter-to-base voltage will be fairly close to ϕ_{bi} regardless of base current. Why?

10-11 Comment on the effect of each of the following on the β parameter of a bipolar transistor. (a) Increased base resistivity. (b) Increased emitter resistivity. (c) Increased base drive time. (d) Increased emitter drive time. (e) Impurities decreasing the recombination lifetime in the base region.

10-12 Explain each of the following nonidealities in transistor β: (*a*) Early effect. (*b*) Transistor saturation at low voltage. (*c*) Breakdown at high voltage. (*d*) Falloff in β at low base currents. (*e*) Falloff in β at high currents. Suggest at least one process parameter that can be manipulated to reduce each of these effects.

10-13 Design a mask alignment scheme for a bipolar process. For each successive layer, specify which layer it must align with, whether the alignment tolerance may exceed the minimum achievable, and, if so, by how much.

REFERENCES

1. A. S. Grove, *Physics and Technology of Semiconductor Devices,* Wiley, New York, 1967, p. 165.
2. A. S. Grove, *ibid.,* p. 220.
3. J. M. Early, *Proc. IRE* **40**:1401 (1952).
4. T. Hirao, K. Kijima, and T. Nakano, "Characteristics of Schottky TTL Fabricated by Oxide Isolation Utilizing Ion Implantation," in *Semiconductor Silicon 1977,* H. Huff and E. Sirtl (eds.), The Electrochemical Society, Princeton, NJ, 1977, p. 1005.
5. P. C. T. Roberts, D. R. Lamb, R. Belt, D. Bostick, S. Pai, and D. Burbank, *IEEE Electron Device Lett.* **EDL-2**:28 (1981).
6. W. C. Ko, T. C. Gwo, P. H. Yeung, and S. J. Radigan, *IEEE Trans. Electron Devices,* **ED-30**:236 (1983).

ELEVEN

FABRICATION OF MOS DEVICES

In Chapter 10, we saw that transistors in integrated circuits can be made using either of two technologies: bipolar or metal-oxide-semiconductor (MOS). In this chapter, we explore the fabrication of circuits composed of MOS transistors. Since insulators other than oxide can be used in these transistors, this technology may also be called *metal-insulator-silicon (MIS)* technology. Additionally, because the MOS transistor operates by the effect of an electric field on the substrate, it may be called a *field effect transistor (FET)*. Various combinations of initials are in use, including *MOST, MOS-FET,* and *MISFET*. We will value simplicity over generality and use the terms MOS and MOS transistor.

MOS devices are based on the properties of the MOS capacitor, which we studied in Chapters 2 and 3. In the first section of this chapter, we review MOS capacitors and derive the properties of the MOS transistor. In Section 11-2, we present a typical process for fabrication of a silicon-gate, n-MOS device. In the third section, we will look at trends in MOS processing, including the increasing popularity of *complementary MOS (CMOS)*. In Section 11-4, we summarize key points about MOS processing.

11-1 THE MOS TRANSISTOR

Figure 11-1 presents the MOS transistor. Figure 11-1a shows the construction of the transistor, using p-type silicon. A layer of oxide has been grown on the silicon, and a metallic electrode rests on the oxide. We have seen this arrangement before in the

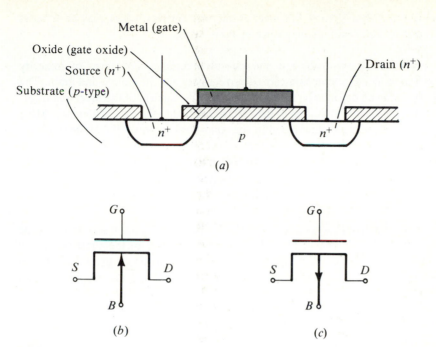

Figure 11-1 The MOS transistor. (*a*) Diagram of an MOS transistor, showing gate, gate oxide, source, and drain. (*b*) Symbol for an *n*-channel MOS transistor. (*c*) Symbol for a *p*-channel MOS transistor.

MOS capacitor. The electrode is referred to as the *gate* of the transistor, and the insulating oxide is the *gate oxide*. There are two additional structures in the transistor: identical n^+ regions on either side of the gate. In operation, both of them will be reverse biased, and the region with the lesser reverse bias is designated the *source*. The other is called the *drain*. Openings are made in the gate oxide to allow contact to the source and drain. This transistor, with n^+ source and drain in *p*-type silicon, is an *n-channel* device and is denoted by the symbol shown in Fig. 11-1*b*. Figure 11-1*c* shows the symbol for a *p-channel* device, in which the source and drain are p^+ and the substrate is *n*-type.

We introduced the *n*-channel MOS capacitor in Chapter 2, and looked further into properties of MOS gate oxides in Chapter 3. We found that MOS capacitors have the following characteristics:

Both the capacitance, and the way in which the capacitor stores charge, vary with the field applied to the gate electrode;
The effective field in the silicon depends not only on the actual gate voltage, but on properties of the oxide;
There are three regimes of charge storage: *accumulation* for negative effective gate voltages, *depletion* for small positive voltages, and *inversion* for sufficiently large positive voltages;

In inversion, the charge is stored in a layer of minority carriers, which may be either at the semiconductor surface or deeper in the silicon.

The MOS transistor operates in the inversion regime. The layer of minority carriers that is induced in inversion carries current between the source and drain. The resulting conductance depends on the number of charge carriers in the inversion layer, which in turn is a sensitive function of the voltage applied to the gate. Therefore, MOS transistor characteristics are determined by the conditions for inversion of the semiconductor surface.

We saw, from Eqs. (2-44) and (2-46) that inversion occurs when

$$V' \geq 2\phi_F + \frac{-qN_A}{C_o} x_d \tag{11-1}$$

where V' is the effective gate voltage, ϕ_F is the Fermi potential of the substrate, N_A the doping density, x_d the width of the depletion layer induced in the substrate, q the electron charge, and C_o the oxide capacitance. Oxide capacitance is given by the ratio of the oxide permittivity ϵ_{ox} to the oxide thickness x_o. Using Eq. (2-47), we can eliminate x_d and write

$$V' = 2\phi_F + \frac{1}{C_o} (4\epsilon_s q N_A \phi_F)^{1/2} \tag{11-2}$$

where ϵ_S is the permittivity of the semiconductor. Equations (3-66) and (3-67) described how oxide properties determine the effective gate voltage V' as a function of the actual voltage V. Since we shall be referring to a number of voltages, we replace the V used to signify the gate voltage in Chapter 3 with the symbol V_G, and write

$$V' = V_G - V_{fb} \tag{11-3}$$

where the flat-band voltage V_{fb} is given by

$$V_{fb} = \Phi_{MS} - \frac{1}{C_o} \left(Q_{ss} + \int_0^x \frac{x}{x_o} \rho(x) \, dx \right) \tag{11-4}$$

Q_{ss}, $\rho(x)$, and Φ_{MS} are properties of the materials and are defined in Chapter 3. Combining, we see that the condition for onset of inversion is

$$V_G - V_{fb} \geq 2\phi_F + \frac{1}{C_o} (4\epsilon_s q N_A \phi_F)^{1/2} \tag{11-5}$$

We will call this value of V_G, where inversion begins the *threshold* or *turn-on* voltage, V_T. Thus

$$V_T = V_{fb} + 2\phi_F + \frac{1}{C_o} (4\epsilon_s q N_A \phi_F)^{1/2} \tag{11-6}$$

Example A MOS transistor is constructed on a substrate with $N_A = 10^{15}$ cm^{-3} and $t_{ox} = 0.05$ μm. V_{fb} for the process is -0.5 V. Find V_T.

To determine V_T using Eq. (11-6), we need to obtain C_o and ϕ_F. From Eq. (2-16),

$$\phi_F = kT \ln \frac{N_A}{n_i}$$

$$= 0.0259 \text{ eV} \times \ln \frac{10^{15}}{1.45 \times 10^{10}}$$

$$= 0.289 \text{ V}$$

From Eq. (2-39),

$$C_o = \frac{\epsilon_{ox}}{x_o}$$

$$= \frac{3.9 \times 8.86 \times 10^{-14} \text{ F cm}^{-1}}{5.0 \times 10^{-6} \text{ cm}}$$

$$= 6.91 \times 10^{-8} \text{ F cm}^{-2}$$

As always, we have obtained the oxide permittivity ϵ_{ox} by multiplying the dielectric constant for oxide, 3.9, by the permittivity of free space ϵ_0.

We now compute the value of the expression under the square root in Eq. (11-5):

$$4\epsilon_s q N_A \phi_F = 4 \times 11.7 \times 8.86 \times 10^{-14} \text{ F cm}^{-1} \times 1.60 \times 10^{-19} \text{ C} \times 10^{15} \text{ cm}^{-3}$$
$$\times 0.289 \text{ V}$$

$$= 1.92 \times 10^{-16} \text{ C}^2 \text{ cm}^{-4}$$

We can now compute V_T as

$$V_T = -0.5 \text{ V} + 2(0.289 \text{ V}) +$$
$$[(1.92 \times 10^{-16} \text{ C}^2 \text{ cm}^{-4})^{1/2} (6.91 \times 10^{-8} \text{ F cm}^{-2})^{-1}]$$

$$= -0.5 \text{ V} + 0.578 \text{ V} + 0.2 \text{ V}$$

$$= 0.278 \text{ V}$$

We now want to determine the amount and location of the mobile charge in the inversion layer. We said in Chapter 2 that this charge was located at the silicon surface under "low frequency" conditions such that minority carriers could migrate to the surface in a time short compared to variations in the gate voltage. In the MOS transistor, the n-doped source and drain regions are copious suppliers of electrons, and minority carriers are always available. Therefore, the transistor displays the properties we called "low frequency" even at very high operating frequencies. The inversion charge appears right below the silicon surface and forms an inverted region called a *channel*. Since the channel is composed of electrons, the channel is an *n-channel* and the method is *n-channel MOS* or *n-MOS*.

Since the channel forms at the surface, the inversion capacitance is the oxide capacitance. Thus, from Eq. (2-49), the amount of charge stored as minority carriers is given by Q_{inv}, with

$$Q_{inv} = -C_o(V_G - V_{fb} - 2\phi_F) + (4\epsilon_s q N_A \phi_F)^{1/2} \tag{11-7}$$

From Eq. (11-6), we can write

$$Q_{inv} = -C_o(V_G - V_T) \tag{11-8}$$

We can now find the conductivity of the channel as a function of the gate voltage. Conductivity is the inverse of resistivity; hence, by analogy to Eq. (2.10), we can write

$$\frac{1}{\rho} = \frac{q}{x_j} \int_0^{x_j} \mu_n n(x) \, dx \tag{11-9}$$

where μ_n is the mobility of electrons, $n(x)$ the electron density, and x_j the depth of the channel. But the inversion charge is given by a similar integral:

$$Q_{inv} = -\frac{q}{x_j} \int_0^{x_j} n(x) \, dx \tag{11-10}$$

and so

$$\frac{1}{\rho} = |\mu_n Q_{inv}| \tag{11-11}$$

We can obtain conductance g_o from conductivity by multiplying by the breadth of the channel Z and dividing by its length L. In Fig. 11-1, L is the source-to-drain distance and Z is the extent of the transistor perpendicular to the plane of the paper. We then obtain

$$g_o = \frac{Z}{L} \mu_n C_o(V_G - V_T) \tag{11-12}$$

The channel conductance thus varies linearly with the gate voltage. If we apply a (negative) voltage V_D to the drain region, relative to the source, a current will flow from source to drain given by

$$I_D(V_G, V_D) = g_o V_D = \frac{Z}{L} \mu_n C_o(V_G - V_T)V_D \tag{11-13}$$

Figure 11-2 shows the current-voltage characteristics of the MOS device in this *linear* region. We can define a parameter called the *transconductance*, g_m, given by $g_m = \partial I_D/\partial V_G$. This is the variation in drain current with gate voltage and is analogous to the beta parameter for bipolar circuits. It gives the modulation in drain current resulting from a change in gate voltage. Its value in the linear region is

$$g_m = \frac{Z}{L} \mu_n C_o V_D \tag{11-14}$$

and so it is a function of V_D. Note that the transconductance increases with decreasing L and with increasing C_o. C_o in turn increases with decreasing gate oxide thickness x_o.

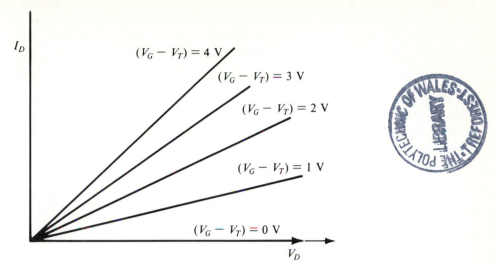

Figure 11-2 Current-voltage characteristics of an MOS transistor in the linear region.

For this and other reasons, MOS processing involves continuing efforts to reduce channel length and gate oxide thickness.

Effects of Additional Applied Voltages

This simple picture of MOS transistor characteristics becomes more complicated when additional applied voltages are considered. As Fig. 11-1 suggests, there are actually four voltages that can be applied to an MOS device: the gate voltage, the drain voltage, a source voltage V_S, and also a voltage applied to the bulk substrate, V_B. We will need to take one voltage as reference or ground; V_S is usually chosen. We have already considered the effect of gate voltage. We will now consider in turn V_B and V_D.

Applying a voltage V_B to the substrate has two effects. The first is to offset the effective gate voltage, which is the difference between gate and bulk potentials. The other is to change the width of the depletion layer induced in the substrate by the gate voltage. As a result, Eq. (11-6) becomes

$$V_T = V_B + V_{fb} + 2\phi_F + \frac{1}{C_o} [2\epsilon_{sq}N_A(V_B + 2\phi_F)]^{1/2} \qquad (11\text{-}15)$$

This effect, called the *body effect*, can be used to adjust the threshold voltage of MOS devices. Consideration of body effect considerably complicates the equations for conductance.

We have already considered drain voltage, but we implicitly assumed that it was sufficiently small that the properties of the channel were not effected. However, as drain voltage increases, it counteracts the effect of the gate voltage in the region near the drain. When the difference between gate and drain voltage equals the threshold voltage, a channel is no longer induced at the edge of the drain region. This condition, called *pinch-off*, is shown in Fig. 11-3. When pinch-off occurs, further increases in drain voltage no longer induce an increase in current, and the transistor is said to be in *saturation*. The voltage condition for saturation, V_{Dsat}, is simply

$$V_{Dsat} = V_G - V_T \qquad (11\text{-}16)$$

To obtain the saturation current, we can no longer use Eq. (11-13), because the effective gate voltage has now become a function of distance y along the channel. Instead, we consider the current-voltage relationship across each channel element of length dy. In Eq. (11-12), we replace the gate voltage V_G with $V(y)$, the gate-to-substrate voltage at y. In place of the channel length L we use dy. We now have

$$g_o(y) = \frac{Z}{dy} \mu_n C_o(V(y) - V_T) \qquad (11\text{-}17)$$

From the definition of conductance, it follows that

$$dV = \frac{I_D \, dy}{Z\mu_n C_o[V(y) - V_T]} \qquad (11\text{-}18)$$

along each element of the channel. Rearranging gives

$$[V(y) - V_T] \, dV = \frac{I_D \, dy}{(Z\mu_n C_o)} \qquad (11\text{-}19)$$

We now integrate voltage from zero to V_{Dsat} and length from zero to L, obtaining

$$I_{Dsat} = \frac{Z}{L} \mu_n C_o \frac{(V_G - V_T)^2}{2} \qquad (11\text{-}20)$$

Therefore, in saturation, current is proportional to the square of $(V_G - V_T)$. The transconductance in saturation, g_{msat}, is given by

$$g_{msat} = \frac{Z}{L} \mu_n C_o |(V_G - V_T)| \qquad (11\text{-}21)$$

Figure 11-3 shows the current-voltage characteristics of an MOS transistor over both the linear and saturation ranges, based on the fairly simple treatment we have given here.

Further Complications

A number of additional factors complicate the characteristics of real MOS devices. For one, we treated the body effect and saturation independently. When both V_B and V_D are considered, a number of additional terms appear in the equations for I_D, V_{Dsat} and I_{Dsat}. Interestingly, most of these terms are small if the gate oxide thickness is small relative to the depletion layer thickness. A full treatment is given by Grove.[1]

In addition, we have ignored some physical effects. One is the change in mobility

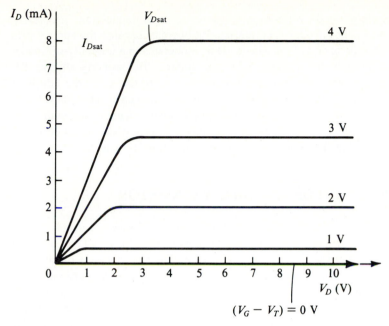

Figure 11-3 Current-voltage characteristics of an MOS transistor. A number of complicating effects have been omitted.

with increasing field. At sufficiently high gate voltages, this effect becomes significant, causing conductance to increase more slowly with V_G than expected. Another effect is *saturation resistance*. As drain voltage increases beyond V_{Dsat}, the pinch-off point moves farther up the channel away from the drain. This results in a shortening of effective channel length and therefore some increase in current with V_D even in saturation.

As designers strive for the shortest possible channel lengths, various *short-channel effects* become more significant. In short channels, electric fields induced by V_D become high. At a high enough field, the velocity of carriers in the silicon approaches a limit, a phenomenon known as *velocity saturation*. This tends to reduce V_{Dsat} and I_{Dsat} from expected values. If channel length is short enough that the source and drain depletion regions can interact, a decrease in V_T occurs. Due to these and other effects, the *I-V* characteristics of a short-channel MOS transistor can vary quite a bit from Fig. 11-3.

Finally, there are limits on the maximum reverse bias that can be applied to the source and drain regions of the transistor. At sufficiently high voltages, breakdown occurs at the *pn* junctions formed by the source and drain with the substrate, leading to large current flows into the substrate bulk. Alternatively, depending on the channel length, the source and drain depletion regions may touch, a condition known as *punchthrough*. In this case, large currents flow from source to drain.

Depletion Transistors

One important variation on the MOS transistor deserves separate mention. The device we described earlier is an *enhancement* transistor: it operates because the applied gate voltage *enhances* the inversion of the silicon. Thus it operates in a normally-off mode: the impedance is high in the absence of a gate voltage. By properly shifting the threshold voltage, *depletion transistors* can be fabricated. In this case, the threshold voltage is made negative, so that an inversion layer is present even at zero gate voltage. The transistor is now "normally on," and the applied voltage *depletes* the number of carriers in the inversion layer. This results in a change in origin and direction in the characteristic curves, but the general principles of operation remain the same.

11-2 CONSTRUCTION OF AN NMOS TRANSISTOR

Figure 11-4 shows a cross-section of an NMOS transistor. Compared to the bipolar transistor, there is less difference in configuration between the idealized and practical transistors. We note that the gate is made of polysilicon rather than a metal. This transistor uses the *silicon-gate* process, which is now standard. We also note the addition of a thick layer of silicon dioxide on either end of the transistor, and an n^+ doped region under the oxide. This oxide layer, called the *field oxide,* serves as the isolation between transistors. Oxide isolation is thus standard in MOS processing. The mechanism of isolation, however, is somewhat different from that of a bipolar transistor. In MOS devices, current flows only through the induced channel, and therefore transistors are inherently isolated as long as channels are not formed between them. However, in connecting the integrated circuit, stripes of metallization will be fabricated to run across the oxide. These metal stripes form the gate of a parasitic MOS transistor, with the isolation oxide forming the gate oxide. The threshold voltage of this

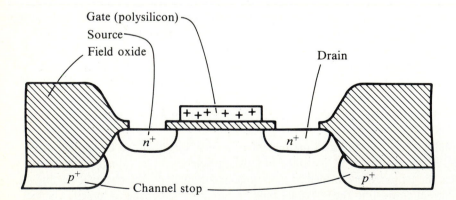

Figure 11-4 Cross-section of a practical NMOS transistor.

transistor must be kept higher than any possible operating voltage, to prevent unwanted transistor action. The thickness of the field oxide, together with the p^+ implant under the oxide, serves to elevate the V_T of this "field transistor" above the voltage where a conducting channel could be formed. The implant is therefore called a *channel stop*.

Isolation

Figure 11-5, and Table 11-1, describe a basic NMOS process. The first step is the fabrication of the isolation. First, a thin *initial oxidation* is carried out. This oxide serves mainly to protect the silicon surface during subsequent steps, and therefore the thickness is not too critical. Oxidation is followed by deposition of silicon nitride, using a CVD process. The nitride will serve as an oxidation mask for selective oxidation. It needs to possess sufficient thickness and integrity to block oxidant molecules effectively. Usually 0.1 μm or so of nitride suffices.

Nitride deposition is followed by the first lithography step, *field mask*. A resist pattern protects all of the areas where active devices will be formed, and the nitride is etched away in the areas to be isolated. Due to the difficulties of wet chemical nitride etching, plasma etching is usually used to remove the nitride. In fluorine-containing plasmas, the oxide will be attacked, at a fairly slow rate, after the nitride is removed. However, the slow-etching oxide serves to protect the underlying silicon, which would otherwise etch rapidly. After etching, the field mask photoresist is *not* removed. Instead, the resist is left in place as an implant mask for the next step, implantation of the channel stop. This is a p^+ implant into the areas where the nitride has been etched. By using the same masking step for both nitride etch and channel stop implant, the implant is automatically aligned with the nitride-free area.

The resist is now removed, and the field oxide grown. An extended steam oxidation cycle grows up to a micron of oxide in the areas where there is no masking nitride. Where nitride remains, oxidant does not reach the silicon surface, and oxidation is blocked. At the edge of the nitride, some oxidant diffuses laterally, generating the characteristic "bird's-beak" gradient in oxide thickness. Note that, due to the consumption of silicon during oxidation, the oxide extends about as far below the original surface as it does above. This helps to planarize the surface and simplifies step coverage during later processing steps.

Following oxidation, the silicon nitride layer is stripped off. In this application, a wet chemical etchant such as phosphoric acid may be used, since there is no need to maintain dimensions or otherwise tightly control the etching. Chemical methods offer good selectivity with respect to the underlying oxide, so considerable overetch can be used. During the oxidation, the channel stop implant will diffuse significantly. This helps to maintain a significant concentration of dopant below the advancing layer of oxide. However, it does lead to a concentration gradient in the source and drain areas. This effect, along with the gradual slope of the oxide "bird's-beak," increases the space that must be allotted to isolation. As device densities increase, this "overhead" becomes harder to support, leading to increased interest in less space-wasting isolation techniques.

Table 11-1 A process sequence for fabrication of an _n_-channel MOS integrated circuit

Structure/step	Process	Purpose/considerations	Related parameters	
			In-process	Completed device
Substrate	_p_-type		Resistivity	
Isolation				
Initial oxidation	Thermal oxidation	Noncritical. Provides silicon protection, etch stop for nitride, etc.	Thickness	
Nitride deposition	CVD	Mask material for selective field oxidation	Thickness, integrity	
Field mask	Lithography/etch (nitride). Resist not stripped.	Patterns nitride and provides resist masking for field implant		
Field implant	Ion implant, p^+	Implants area where field oxide will be grown to prevent channel formation.	Resistivity	V_T (field)
Strip resist	Strip resist	Removes field mask resist		
Field oxidation	Selective oxidation	Grows thick field oxide except where nitride protects the surface	Thickness	V_T (field)
Gate fabrication				
Gate oxidation	Thermal oxidation	Grow gate oxide	Thickness	V_T, ΔV_{fb}
Polysilicon deposition	CVD	Provide material for polysilicon gate		Gate resistance
Polysilicon doping	Diffusion (n^+) growth	Phosphorus-dope polysilicon to reduce resistivity	Resistivity, thickness	Gate resistance
Polysilicon gate mask	Lithography/etch (polysilicon)	Determines gate width and thus channel length	Dimension	L (V_T, I)

Process	Method	Description	Measurement	Defect/Issue
Source/Drain				
Source/drain implant	Ion implant (p^+)	Defines source/drain using gate and isolation as implant masks	Dose	
Source/drain anneal/diffusion	Diffusion or anneal	Anneal implant, diffuse dopant as desired to final location	L (effective)	L (V_T, I)
Glass/Contact				
Phosphosilicate glass deposition	CVD	Protect gate region from alkali metal contamination	Thickness, % phosphorus	V_T stability
Glass reflow	Furnace heat treatment	Smooths glass for improved step coverage		
Contact mask	Lithography/etch (glass/oxide)	Opens contacts in glass to gate, source, drain, etc.	Dimension, alignment	Shorts, opens
Metallization				
Aluminum deposition	Sputter	Deposits metal for conductors. Al/Si alloy used to reduce pitting.	Thickness	
Metal mask	Lithography/etch (aluminum)	Patterns conductors	Dimension, alignment	Shorts, opens
Alloy	Furnace heat treatment	Forms Al-Si alloy for good contact		R(contact)
Electrical test	Parametric test	Identify process deviations		
Passivation				
Passivation deposition	CVD (e.g., doped silicon dioxide)	Good integrity against moisture, alkalis, mechanical damage	Thickness, composition	
Pad mask	Lithography/etch (CVD layer)	Cut holes in passivation for attachment of package leads	Observe etch completeness	No contact

Initial oxidation
Silicon nitride deposition

Channel stop implant (p^+)

Selective field oxidation

Silicon nitride removal

Gate oxidation

Polysilicon deposition
Polysilicon doping/oxidation

Gate mask

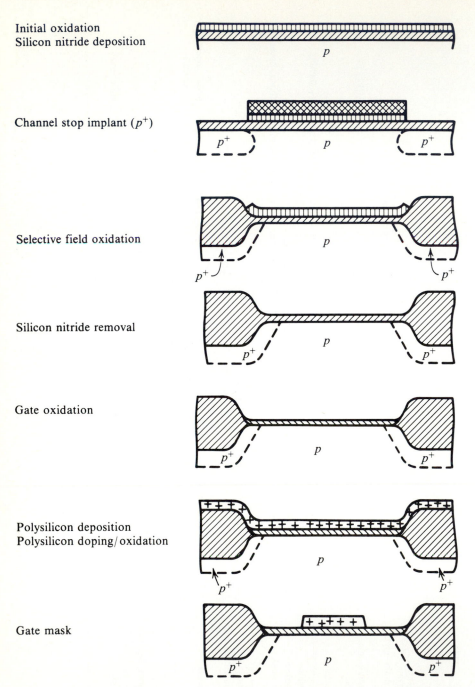

Figure 11-5 Steps in the fabrication on a NMOS transistor. Each panel depicts a cross-section, except for the final view which is a top view. In the top view, glass and passivation are omitted, although the shape of the contact opening through the glass is shown.

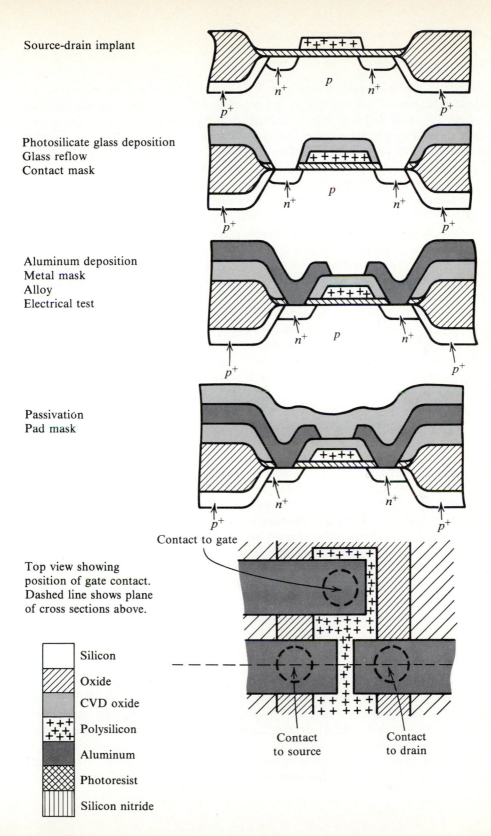

Source-drain implant

Photosilicate glass deposition
Glass reflow
Contact mask

Aluminum deposition
Metal mask
Alloy
Electrical test

Passivation
Pad mask

Contact to gate

Top view showing
position of gate contact.
Dashed line shows plane
of cross sections above.

Contact
to source

Contact
to drain

Silicon

Oxide

CVD oxide

Polysilicon

Aluminum

Photoresist

Silicon nitride

Gate Fabrication

Following nitride removal, the gate oxide is grown. This step determines both the oxide capacitance C_o, which appears in most of the MOS equations, and the flat-band voltage V_{fb}. Also, any holes in the oxide will short the gate to the substrate, destroying the device. However, we want the gate oxide as thin as possible: only a few hundred Ångstroms. The need to grow a defect-free, very thin, high-quality oxide without contamination makes gate oxidation a key step in MOS processing.

To get a high-quality oxide, the previous oxide is often stripped from the gate region, and a totally new oxide is grown. Here the thickness of the field oxide is again a benefit. Complete removal of previous oxides in the gate region requires considerable overetch, to compensate for any conceivable process variation. With an initial oxidation of tenths of micrometers, and a field isolation of roughly a micron thick, this overetching will not significantly compromise the field oxide thickness. After stripping the initial oxidation, the gate oxide is grown slowly and carefully, usually by dry oxidation in a chlorine ambient.

The gate material is now deposited over the gate oxide. Polysilicon is the standard gate material and is deposited using a CVD-type process. Deposition is followed by a phosphorus diffusion in a furnace, to dope the poly and thereby lower its resistivity. The polysilicon surface may next be oxidized, particularly if the gate will be patterned by wet etching. In this case, the oxide will be needed as an etch mask during poly etching.

Polysilicon gates are convenient because they allow high temperatures at later processing steps. However, even when highly doped, poly has a significant resistivity, which reduces the speed of the device. To obtain faster devices, much attention is being given to materials for refractory, low-resistivity gates, with the interest centering on various silicides of refractory metals. It is likely that some such material will soon replace polysilicon in the gates of the most advanced MOS circuits.

The gate structure is patterned in the second mask step, *gate mask*. Following lithography, either wet or plasma etching is used to form the gate. The width of the gate will determine the effective channel length L, which appears in the various MOS equations as a major determinant of circuit behavior. Gate dimension in turn depends on the dimension of the resist pattern and the undercutting during etching. Dimensional change in etching depends on the anisotropy of the etch process and (unless etching is totally anisotropic) the polysilicon layer thickness. Therefore, dimensional control of gate lithography and etching is critical, and gate masking is another key step in MOS fabrication.

Source-Drain Fabrication

With the gate in place, we are ready to fabricate the source and drain. This is done without an additional lithography step, by ion implantation using the gate and the field oxide as implant masks. An n^+ implant is used, at an energy insufficient to penetrate the gate or the field oxide. This *self-aligning* characteristic of the standard MOS process is a major advantage, allowing the basic transistor to be constructed using only two masks. It is worth noting that even these two need not be too carefully aligned. As long as there is reasonable space between gate and isolation on either side of the gate, some asymmetry of placement will not critically effect the device performance.

Following source-drain implant, various anneals or drive-in steps may be performed to position the dopant as desired. Source-drain processing is another factor affecting effective channel length. The channel length is the distance from the source *pn* junction to the drain junction, at the surface under the gate. If the dopant distribution around the windows approximates the classic diffusion distribution, the junction length will be decreased by $2x_j$ from the gate dimension.

While source-drain junction depth (or actually, width) is critical, doping density is less so. To a first approximation, device characteristics do not depend on the source and drain doping as long as the doping is reasonably heavy.

Glass, Contact, Metallization, and Passivation

After source-drain implant and any related diffusions, the MOS transistor is complete. The remaining process steps provide protection, metallic interconnect, and passivation for the circuit.

The next step is chemical vapor deposition of a phosphosilicate glass. The glass is phosphorus doped for two reasons. First, phosphorus is an excellent getter for alkali metals. We recall that contamination by alkalis such as sodium can cause instabilities in V_{fb} and thus in V_T: the doped glass prevents them from reaching the gate region. Second, doped glass flows at temperatures around 1000°C. This allows use of a *glass reflow* process following deposition. The glass is heated in a furnace to its flowing point, rounding the contours of the glass and smoothing out any sharp steps. This results in better metal step coverage over the lofty topography of the field oxide. Note that glass reflow is possible only with high-melting-point gate materials. It is not possible with most metallic gates.

Contact openings are now etched through the glass at the third masking step, *contact mask*. Contact mask can be a critical masking step, because contact size and alignment limits the minimum size of the device. The source and drain regions must be large enough to allow the contact to fit, with allowance for alignment tolerance. If the contact window touches the gate, the gate will be shorted to source or drain; if it touches the *p*-type substrate, the channel will be shorted. Similarly, the gate contact must fit within the gate. To keep the transistor as small as possible, the contact is usually printed at the minimum size achievable with the given process, and the alignment tolerance is made as tight as the equipment will allow. Being printed at the resolution limit, contacts, even if designed as squares, usually print as circles and are so shown in Fig. 11-5.

Contact mask is followed by metal deposition and by *metal mask*. The considerations here are similar to those in bipolar metal masking: metal leads must have sufficient width, thickness, and step coverage to keep current density below the electromigration limit, and adjacent leads must not touch, even with the greatest expected process variation. Alignment must be sufficiently good for the metal to cover the contact completely, thus sealing the source and drain against contamination and attack. In Fig. 11-5, we have shown only source and drain contacts and metallization in cross-section. To minimize the length of the transistor, the gate contact and the associated metal lead are not positioned in line with the source and drain. This arrangement is shown in the top view of the transistor presented at the end of Fig. 11-5.

Following metallization, the aluminum and silicon are alloyed to produce good contact. During this step, silicon dissolves into the aluminum, causing pits that can penetrate the source and drain regions if not controlled. As source and drain regions grow thinner, this pitting becomes increasingly intolerable. It is minimized by metallizing with an aluminum alloy containing a percent or two of silicon, which reduces the dissolution rate. Alloying is followed by preliminary electrical testing of devices, for the early detection of process variation.

Finally, the device receives a passivation layer and a *pad mask,* which exposes the pads for bonding during assembly. These steps are identical with those described for bipolar processing.

It is important to note the basic simplicity of MOS processing, relative to bipolar. The basic number of mask steps (exclusive of pad mask) is four rather than six. Alignment considerations are much simpler. The basic transistor structure is self-aligning, requiring only that the field oxide and gate are reasonably positioned relative to each other. Most of the doping steps are also forgiving: in the first approximation, no dopant density except that of the substrate appears in the device equations. Really, the only critical considerations are gate oxide properties, determined at gate oxidation, and channel length, which depends on gate mask dimension and source-drain junction depth.

The simplicity of MOS processing was apparent in the early years of microelectronics; successful implementation took somewhat longer. A major impediment was the sensitivity of transistors to contamination, particularly by alkali metals. We recall that metals such as sodium form mobile positive ions in the oxide, which can move about under normal operating conditions and shift the threshold voltage. Some contamination could be tolerated with PMOS, since the thresholds were driven up, but in NMOS the threshold was reduced, turning the device unexpectedly on. The contamination issue was therefore a major barrier to NMOS processing until it was controlled by the use of phosphosilicate glass. Another hindrance was that MOS transistors, unlike bipolar ones, cannot be meaningfully tested electrically until essentially complete. Cleanliness and control remain the absolute essentials of MOS processing.

In practice, modern NMOS processes involve several more masking steps than our example. Table 11-2 describes some of these additional steps. Many processes use a *buried contact* to polysilicon. In this method, contacts are opened to the substrate before polysilicon deposition. The poly thus makes contacts with the substrate at designated spots and serves as a second layer of interconnect. This reduces the size of the finished device. Additional implants may be used. One application is for threshold adjustment of some or all of the transistors. For example, if a design includes depletion transistors, those transistors will receive an implant to adjust the threshold voltage to the "normally on" level. During this implant, the enhancement transistors will be masked off.

Alternatively, multiple implants may be used to "fine-tune" the dopant distribution in the source and drain regions. Strong fields arise near the source and drain junctions,

Table 11-2 Additional process steps used in MOS processing

Step	Position in sequence	Purpose
Buried contact	Lithography/etch step prior to gate fabrication (polysilicon deposition).	Opens contacts in oxide to substrate, so that polysilicon contacts substrate. Allows poly to serve as another layer of interconnect.
Threshold adjust implant	After isolation, prior to gate fabrication	Implantation to adjust threshold voltage of some or all transistors. For example, allows formation of depletion as well as enhancement transistors.
Source-drain "shaping" implant	After principal source-drain implant, in combination with various lithography and etch steps	"Fine-tunes" the doping in the source-drain regions, to control fields and thus hot-electron effects, breakdowns, etc.

which can have effects on device performance and reliability. For example, *hot electron effects,* which we describe below, are sensitive to these fields. Varying the dopant density in this region by the use of additional source-drain implants can help control these effects. Such implants are becoming increasingly common in MOS processing.

MOS process complexity is also increasing due to the need for additional interconnect layers. MOS technology, using the poly layer both as gate and interconnect, managed to avoid dual-layer metal designs until fairly recently. However, as devices become more complex, the complications required to make all needed connections in one plane become insupportable. Having dealt with multilayer interconnect in Chapter 10, we will not examine it further here.

11-3 HOW TO MAKE A BETTER MOS TRAP

The history of MOS processing has been a quest for speed, density, and reduced power consumption. While each of these factors is important, their order of priority has varied with technology. Early on, speed was a principal desire, to offset the speed advantage of bipolar. As devices matured, high density and the resulting low cost became more important. Modern fast, high-density devices now generate so much power in a small area that chip cooling is becoming a major design constraint, thus increasing the importance of reducing power.

MOS performance in each of these areas depends in a systematic way on device dimensions, a relationship first worked out by Dennard et al. in 1974.[2] Beginning with an MOS transistor of some known properties, they considered the results of shrinking it in all dimensions by a scaling factor S. That is, we fabricate a new device with oxide thickness of Sx_o, gate length SL, etc. (In the case of shrinking, $S < 1$.) They also increased all dopant concentrations by a factor $1/S$. To keep electric fields constant in the shrunk device, all applied voltages are reduced by S. We now ask what effect this has on device currents and voltages.

The threshold voltage V_T scales roughly as the applied voltages, decreasing by S. Referring to Eq. (11-6), and assuming that V_{fb} is small relative to other factors, the threshold for the scaled device will change mainly due to the increase in C_o, which varies with the gate oxide thickness. There will also be a change due to increased dopant concentration N_A and a corresponding change in ϕ_F, but in general these effects are limited.[2] As a result, we can state that V_T scales as S.

Drain current is given by Eq. (11-13) in the linear region and by Eq. (11-20) in saturation. In both cases, current is proportional to the square of the voltages, to the oxide capacitance, to the carrier mobility μ, and to the ratio Z/L. The voltages scale as S, the capacitance as $1/S$ (varying inversely with thickness), and the Z/L ratio remains unchanged. So we can write

$$I \propto \mu C_o V^2 \propto \mu S \qquad (11\text{-}22)$$

The resistance of a transistor is given by the ratio of voltage to current. Thus

$$R \propto \frac{S}{\mu S} = \frac{1}{\mu} \qquad (11\text{-}23)$$

The time delay characteristic of a circuit element is proportional to the "RC constant" given by multiplying capacitance by resistance. C_o is capacitance per unit area; capacitance will thus scale as $C_o Z L$, which is proportional to S. It follows that time delay τ is given by

$$\tau \propto RC \propto \frac{S}{\mu} \qquad (11\text{-}24)$$

and "speed" (the inverse of time delay) increases linearly with increased mobility or reduced size. It also follows that power P is given by

$$P = VI \propto \mu S^2 \qquad (11\text{-}25)$$

and that the power-delay product, a frequently used figure of merit, scales as

$$P\tau \propto S^3 \qquad (11\text{-}26)$$

These expressions explain why improvement in MOS has always meant the pursuit of increased carrier mobility and, especially, smaller dimensions.

Scaling principles make the reduction of dimensions fairly straightforward. By scaling voltages, dimensions, and doping density as described, device characteristics remain (reasonably) invariant after scaling. Extensive redesign is not necessary, and scaling can proceed as soon as the processing challenges can be met. This principle is so taken for granted that engineers will describe an MOS process by one or two parameters, usually gate oxide thickness or channel length.

One problem in scaling has been that, in practice, voltages have not been scaled along with the other paramenters. This is because chips must use standard power supplies, and provide output signals to with standard logic. If power supplies generate 5 V, and TTL logic conventions require 5-V input, a 4.21-V chip will not be very marketable. However, there has been a general downward trend in voltages from 12 V to the current 5 V, with a 3-V convention widely anticipated in the near future.[3] "Stickiness" in the scaling of voltage leads to higher fields in the scaled transistor and the formation of "hot," high-energy electrons. Impact ionization by these electrons results in substrate currents that limit device performance. Hot electrons can also tunnel into the gate oxide, leading to fixed charge and eventual device failure. Since this second process has an activation energy around 3.1 V, a 3-V voltage convention should be effective in controlling it.[4]

The application of scaling has served MOS technology well. However, at channel lengths near 1 μm, factors not considered in the scaling equations are become limiting and the advantage of scaling is reduced.[3,5] Several neglected resistances are no longer negligible, including the resistance of the gate material, the metallic interconnect, the contacts, and the source and drain regions. Hot electron effects are another limiter. A third problem, mentioned previously, is the velocity saturation limit on carrier mobility. This limitation is particularly interesting because it reduces the traditional superiority of NMOS over PMOS. Although at low velocity electrons are more mobile in silicon than holes, the velocity-limited values are fairly similar. As a result, sufficiently small p-channel transistors again become competitive with n-channel.[6] As fewer advantages accrue from further scaling of conventional NMOS, the time is ripe for a quantum leap in MOS technology.

One approach is to change the semiconductor or the carrier type and obtain a substantive improvement in the carrier mobility. This happened when NMOS replaced PMOS as the principle MOS application, capitalizing on the higher mobility of electrons to achieve a large increment in speed. Similarly, the adoption of gallium arsenide as a "mainstream" MOS material promises a sixfold increase in electron mobility.

Another technology change currently under way is concerned more with power than speed. This is *complementary metal-oxide-silicon* technology, or *CMOS*. In CMOS, both n- and p-channel transistors are used in the same device. The result is a large reduction in power consumed, as shown in Fig. 11-6. Figure 11-6a shows a conventional NMOS *inverter,* a circuit element that produces a high output for a low input, and vice versa. Inverters form the core of many logic and memory circuits. When an "on" signal (5 V) is applied to the input, the transistor turns on, a current I flows through the resistor of resistance R to ground. The IR drop through the resistor pulls the output down to ground voltage, resulting in an "off" (0 V) signal. If the transistor is off, there is no drop across the resistor and the output is at power supply voltage V. The NMOS inverter consumes power when in the "on" condition at a rate of V^2/R. Power consumption can be reduced by increasing the resistance or by using a depletion transistor in place of the resistor, but cannot be eliminated.

Figure 11-6b shows a CMOS inverter. Two transistors are used, a p-type and an n-type. The threshold voltages are adjusted so that one transistor turns on and the other

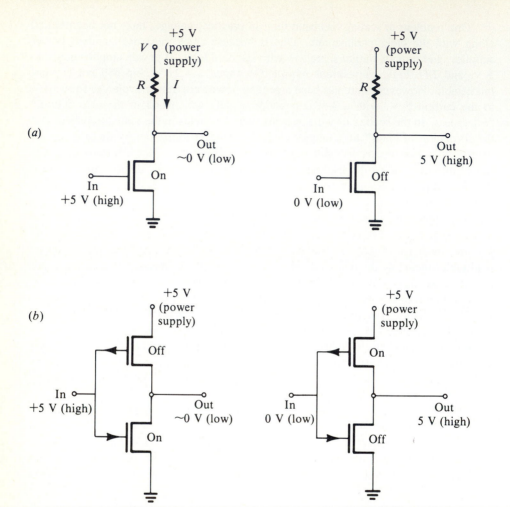

Figure 11-6 NMOS and CMOS inverters, showing why CMOS devices consume less power. (*a*) The NMOS inverter uses a resistor (or a depletion transistor) to "pull up" the output, and consumes power when in the "on" condition. (*b*) In the CMOS inverter, one transistor is always off except during switching, and no power is consumed except when switching from one state to the other.

off at about the same gate voltage. Application of the "on" input turns the top transistor off and the bottom one on, shorting the output to ground. An "off" signal turns off the bottom transistor and shorts the output to the power supply. Current is drawn, and power consumed, only by the load applied to the output. The CMOS inverter, therefore, consumes power only when being switched. The resulting power savings depend on the number of switching events expected, but are, in any case, substantial.

Power consumption is an important issue because of the high density of modern

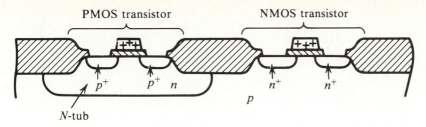

PMOS transistor   NMOS transistor

N-tub

Figure 11-7 Cross-section of a CMOS device, showing NMOS transistor fabricated in the substrate, and PMOS transistor fabricated in the n-type well or tub.

circuits. Power produces heat, and heat must be removed to keep the chip within operating temperature. Currently, many designs are more limited by heat removal than by other constraints. This fact has reawakened interest in CMOS. Simultaneously, velocity saturation of carrier mobility has reduced the performance penalty associated with the PMOS transistors that are required in CMOS. For these reasons, a general shift in MOS technology to CMOS is now under way.[7]

Figure 11-7 shows a cross-section of a CMOS device. (Metallization and passivation are not shown.) The construction is essentially similar to the NMOS transistor, except that one transistor—the PMOS one—has been constructed in a large n-type *well* or *tub* fabricated in the substrate. This CMOS device is thus built using the *n-well* or *n-tub* method. An alternative is to begin with n-type silicon, and construct the NMOS devices in a p-type well. Twin-tub approaches have also been tried. Each method has advantages; however, n-well, using a p-type substrate, is the most direct adaptation of the conventional NMOS process.

Table 11-3 and Fig. 11-8 give an outline of a CMOS process. Prior to isolation, the tub is fabricated using a *tub mask* as the first mask, followed by a tub implant or diffusion. The tub needs to be deep relative to the other transistor features, requiring a fairly long diffusion. Isolation and gate fabrication then proceed normally. The edge of the tub lies under the isolation, allowing for fairly loose alignment between the tub and field masks.

Following gate fabrication, both NMOS and PMOS source-drain structures must be fabricated. This requires at least two implants, one for each type, and two mask steps. However, the gate and isolation continue to self-align the implant. The photoresist is used only to block off one set of structures, and alignment and dimensional control are undemanding. After completion of source-drain implants, the remaining process follows the NMOS sequence.

CMOS requires several more masking steps than the basic NMOS process. At least one threshold adjustment by implant is also likely to be required. However, modern NMOS processes have also added masking steps, and the additional CMOS masks are, in general, undemanding. As a result, in practice, CMOS processing is not much more expensive than NMOS.

CMOS does suffer from one characteristic failure mode that needs careful control. This is *latch-up,* which results from the existence of parasitic bipolar transistors in the

Table 11-3 A process sequence for fabrication of a CMOS integrated circuit

Structure/step	Process	Purpose/considerations	Related parameters	
			In-process	Completed device
Substrate	p-type		Resistivity	
Tub fabrication				
Initial oxidation	Thermal oxidation	Noncritical; protects silicon surface, promotes resist adhesion, etc.	Thickness	
Tub mask	Lithography (resist not stripped)	Defines location of n-tub (or n-well)		
Tub implant	Ion implant (n)	Defines dopant dose for n-tub. After drive, the tub will be the "substrate" for the PMOS transistors and define their properties.	Resistivity	PMOS device properties
Tub drive	Furnace diffusion	With implant, defines tub doping and junction depth	Resistivity, junction depth	PMOS device properties
Isolation		Same as NMOS process		
Gate fabrication		Same as NMOS process		
Source-drain				
Mask tub area	Lithography (resist not stripped)	Noncritical mask protects tub area during NMOS source-drain implant		
NMOS source-drain implant	Ion implant (n^+)	Implants source-drain for transistors not masked off	Dose	
Source/drain anneal/diffusion	Diffusion or anneal	Anneal implant, diffuse dopant as desired to final location	L (effective)	L (V, I) for NMOS
Strip resist	Strip resist	Removes resist over tub		
Mask NMOS area	Lithography (resist not stripped)	Noncritical mask protects area of NMOS transistors during PMOS source-drain implant		
PMOS source-drain implant	Ion implant (p^+)	Implants source-drain for transistors not masked off	Dose	
Source/drain anneal/diffusion	Diffusion or anneal	Anneal implant, diffuse dopant as desired to final location	L (effective)	L (V, I) for PMOS
Strip resist	Strip resist	Removes resist over tub		
Rest of process		Identical to NMOS process		

Initial oxidation
Tub mask
Tub implant

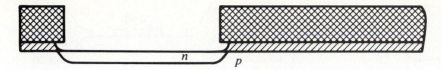

Resist strip
Tub drive

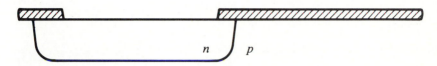

Isolation
Gate fabrication
Mask tub area
n-channel source-drain implant

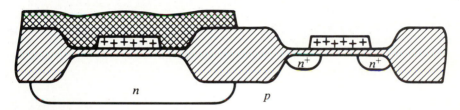

Strip photoresist
Mask n-channel area
p-channel source-drain implant

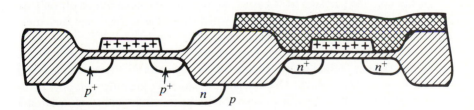

Strip photoresist
Glass deposition and reflow
Contact mask
Metallization and metal mask
Passivation and pad mask

Figure 11-8 Steps in the fabrication of a CMOS device. Tub fabrication precedes isolation and gate fabrication, which are identical to the NMOS process of Fig. 11-5. Two source drain implantations are required. Subsequent processing is identical to the NMOS process.

(a)

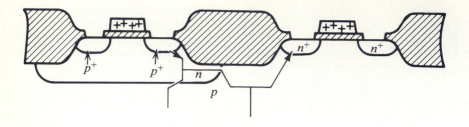

(b)

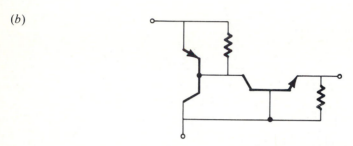

Figure 11-9 Latch-up sensitivity in CMOS devices. (a) The substrate, n-well, and p-channel source or drain form one parasitic bipolar transistor; the n-well, substrate, and n-channel source/drain form another. (b) The resulting device is a silicon-controlled rectifier or thyristor, capable of irreversibly switching large currents to flow through the substrate.

CMOS configuration.[8] Figure 11-9a shows a CMOS transistor, with the parasitic bipolar structures highlighted. Figure 11-9b abstracts these features to give a schematic of the resulting parasitic device. This device is a *silicon-controlled rectifier (SCR)* or *thyristor,* which is used to switch large currents in power supplies. When a CMOS device "latches up," the parasitic SCR has been turned on. It can be turned off only by removing all power from the transistor, by which time the device, and any units connected to it, may have been physically destroyed.

CMOS device and process design must ensure that latch-up is avoided in the finished device. This is facilitated by keeping the β of the parasitic transistors low and the bulk substrate resistance high. One method of achieving the latter, without compromising other device properties, is to deposit a lower-resistivity epitaxial layer on a high-resistivity substrate. As a result, the increasing popularity of CMOS may bring epitaxy, traditionally a bipolar method, into the MOS world as well.

11-4 MOS PROCESSING: A SUMMARY

In the competition between MOS and bipolar methods, MOS has had the advantages of simplicity and density; bipolar, of speed and flexibility. The MOS process is inherently simpler. Diffusion or implant parameters are fairly noncritical. There are fewer masking steps, and alignment tolerances are looser. Due to the self-aligning nature of the source-gate-drain structures, device size is limited principally by the gate dimension, and each increase in lithographic resolution could be promptly translated into more compact devices. As a result, MOS quickly dominated those markets where density and low cost were paramount. MOS has been particularly successful in memory applications, which require the inexpensive production of large arrays of identical elements. However, the sensitivity of MOS to contamination and the limitations on in-process electrical testing necessitate scrupulous attention to process reproducibility and cleanliness.

With time, both advantages and disadvantages are becoming less distinct. MOS evolution has erased much of the speed advantage of bipolar. Conversely, MOS processes have become more complex, and additional masking steps have been added for improved performance. CMOS will add additional steps. However, the bipolar nature of CMOS also offers the opportunity to build bipolar transistors into CMOS devices. These transistors may not give the optimum in performance, but they can serve effectively as interfaces with analog or higher-voltage peripherals.

Successful MOS processing requires tight reproducibility and control of key parameters, and scrupulous cleanliness to avoid alkali contamination. Purity in water, gases, and liquids is vital, as are the most contamination-free handling techniques. Frequent C-V plotting of furnaces must be conducted. Key process parameters such as gate oxide thickness and gate dimensions must be closely monitored, because in-process electrical checks are not useful with MOS devices. Thus, compared with bipolar processing, MOS involves fewer key parameters, but demands the utmost in control.

MOS is currently established as the mainstream VLSI technology, a trend that can be expected to continue for some time. In the next few years, we can expect to see a greater use of CMOS, along with modifications in gate materials, interconnect schemes, and isolation methods. Gallium arsenide processing is also receiving much attention, and a "standard" high-volume GaAs manufacturing methodology may soon emerge. Beyond this point, more radical extrapolations of the basic field-effect-transistor principle can be expected.

PROBLEMS

11-1 An NMOS integrated circuit is fabricated so that it has the following properties:

$$V_{fb} = -0.5$$

$$N_A = 3 \times 10^{14} \text{ cm}^{-3}$$

Gate oxide thickness $= 0.04 \ \mu m$

Find V_T for this transistor.

11-2 The transistor of Problem 11-1 has a field oxide thickness of 1 μm. (a) Assuming no other changes, what is the threshold of the NMOS transistor formed by metal leads, the field oxide, and the substrate? (b) A channel stop implant is to be used to raise this threshold to 8 V. What should the dopant density be in the field area to accomplish this?

11-3 Suggest three ways of raising the threshold of the transistor in Problem 11-1 to 5 V. Be specific about the quantitative changes required. Comment on the practicality of each alternative.

11-4 A transistor in the circuit described in Problem 11-1 has the following parameters:

$$L = 2 \ \mu m$$
$$Z = 3 \ \mu m$$

A voltage of 5 V is applied to the gate. Find: (a) V_{Dsat}. (b) I_{Dsat}. (c) Transconductance g_m in the linear region. (d) Transconductance g_{msat} at saturation.

11-5 The transistor of Problem 11-4 requires an effective channel length of 2 μm. The source-drain junction depth is 0.2 μm and the polysilicon thickness is 0.4 μm. The lateral diffusion of the source and drain is equal to the junction depth. The polysilicon etch process is isotropic, with 10% overetch. Specify the photoresist critical dimension at gate mask. Estimate the appropriate dimension for the mask plate if positive photoresist is used. Negative photoresist?

11-6 (a) Find the overall length required for the NMOS transistor shown in cross-section in Fig. 11-5 in terms of the minimum printable dimension D, the tolerance for alignment A, and any appropriate layer thicknesses, junction depths, etc. (b) If the source and drain were not self-aligning to the gate, but were fabricated by diffusion through a separate mask, what would the overall length be? (c) If $A = D = 2 \ \mu m$, and the field oxide thickness is 1 μm, what percentage of the length of the self-aligned device is attributable to isolation?

11-7 In our discussion of scaling in Eqs. (11-22)–(11-26), we assumed that voltage was scaled. In practice, this does not always happen. How will current, resistance, time delay, power, and power-delay product scale if voltage is kept constant while dimensions and doping density are scaled?

11-8 Suggest a way to eliminate a masking step from the CMOS process presented in Table 11-3 and Fig. 11-8.

11-9 A frequently cited condition for avoiding latch-up in CMOS devices is that the product of the β values of the two transistors shown in Fig. 11-9 be less than unity. Comment on factors that affect these β values and how reducing them might affect performance of the CMOS device.

11-10 You are the engineering manager for an MOS production operation. Make a list of which processing variables you will monitor, how they might be measured, and what your response would be to a deviation in any of them.

REFERENCES

1. A. S. Grove, *Physics and Technology of Semiconductor Devices*, Wiley, New York, 1967, p. 321.
2. Robert H. Dennard, Fritz H. Ganesslen, Hwa-Nien Yu, V. L. Rideout, Ernest Bassous, and Andre R. Leblanc, *IEEE J. Solid-State Circuits* **SC-9**:256 (1974).
3. Youssef El-Mansy, *IEEE Trans. Electron Devices* **ED-29**:567 (1982).
4. S. M. Sze, *Physics of Semiconductor Devices*, Wiley, New York, 1981.
5. VLSI Laboratory, *IEEE J. Solid-State Circuits* **SC-17**:442 (1982).
6. E. Takeda, Y. Nakagome, H. Kume, N. Suzuki, and S. Asai, *IEEE Trans. Electron Devices* **ED-30**:675 (1983).
7. Rick D. Davies, *IEEE Spectrum* **20**:27 (October 1983).
8. Genda J. Hu, *IEEE Trans. Electron Devices* **ED-31**:62 (1984).

TWELVE

FABRICATION OF NONSILICON DEVICES

Microelectronic processing techniques are applicable to the fabrication of many devices other than silicon integrated circuits. Microelectronic methods are the natural choice whenever a technology requires smallness of size and precision in control of materials, and as these methods become more widely understood, the list of possible applications continues to increase. In this chapter, we look at an assortment of microelectronic devices other than the silicon integrated circuit. In this survey, we will look at only a few possible applications, and those will not be presented in depth. Thus the purpose of this chapter is not to impart a working knowledge of any particular device, but to give some insight into how microelectronic methods can be applied to a variety of technical problems.

In Section 12-1, we deal with a close relative of the silicon MOS transistor, the gallium arsenide field effect transistor, which is used in extremely high-speed applications such as microwave circuits. In Section 12-2, we take up a second application of gallium arsenide, in optoelectronic devices such as the light-emitting diode and the laser diode. In Section 12-3, we turn to the fabrication of an electromagnetic device, the magnetic bubble memory. Another magnetic device, the thin-film recording head, is covered in Section 12-4.

12-1 THE GALLIUM ARSENIDE FIELD EFFECT TRANSISTOR

Gallium arsenide has superior electron mobility to silicon and is thus a natural choice for very high-speed applications. Field effect transistors made of gallium arsenide are in widespread use in microwave applications and are receiving increasing interest elsewhere. Devices based on gallium arsenide have been clocked at 10 GHz (Ref. 1). Despite its high speed, the development of gallium arsenide (GaAs) has been delayed by its significant processing disadvantages. There is no native oxide, and interfaces with deposited insulators tend to have high surface state densities. Therefore, fabrication of MOS transistors is not effective. Instead, GaAs transistors are usually of the *metal-semiconductor field effect transistor (MESFET)* variety. This is a cousin of the MOS transistor in which the field of a gate structure mediates the flow of current in a channel. Both the channel and gate, however, are formed differently, by methods that take advantage of the properties of GaAs.

In the GaAs MESFET, the channel is not induced by an applied field. Rather, it is fabricated as part of the device. Use is made of the *semi-insulating* properties of undoped GaAs. The intrinsic carrier density n_i is a factor of 5×10^{-3} less in GaAs than silicon, and the bulk resistivity is 1000 times higher. Therefore, undoped GaAs is a very poor conductor. Using undoped GaAs as an effectively insulating substrate, a conducting channel can be formed at the surface by *n*-doping the GaAs. Isolation is achieved by fabricating either physical or chemical interruptions in the conducting layer.

Current flow through this channel is modulated by the field of a *Schottky gate*. The metal of the gate structure is chosen to form a Schottky barrier when placed in contact with the semiconductor, and the depletion region of the Schottky constricts the channel. Therefore, the MESFET is "naturally" a depletion device; however, enhancement devices can also be constructed. Figure 12-1 presents a GaAs MESFET in cross-section. Note that the gate is not only in direct contact with the semiconductor, but recessed into it.

A number of process sequences are in use for the fabrication of MESFETs. The outline we give here is based primarily on a particular process recently published,[2] with commentary as appropriate on published variations.[3,4] Figure 12-2 presents the process sequence through formation of the gate.

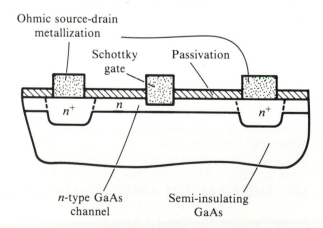

Figure 12-1 Cross-section of a gallium arsenide metal-semiconductor field effect transistor (GaAs MESFET), showing the semi-insulating substrate, the *n*-doped channel, source and drain ohmic metallization, and the metal Schottky gate.

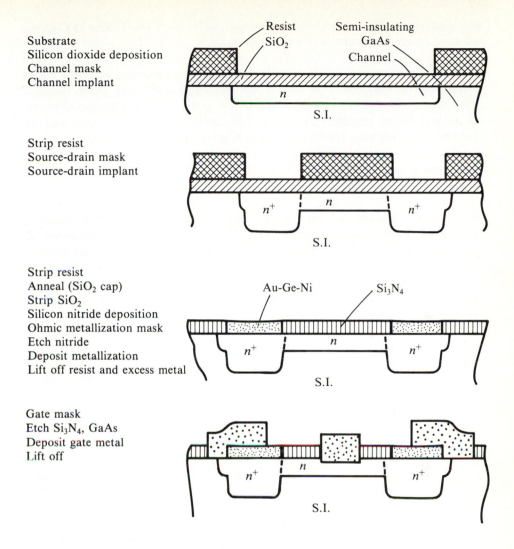

Substrate
Silicon dioxide deposition
Channel mask
Channel implant

Strip resist
Source-drain mask
Source-drain implant

Strip resist
Anneal (SiO$_2$ cap)
Strip SiO$_2$
Silicon nitride deposition
Ohmic metallization mask
Etch nitride
Deposit metallization
Lift off resist and excess metal

Gate mask
Etch Si$_3$N$_4$, GaAs
Deposit gate metal
Lift off

Figure 12-2 A process sequence for fabrication of a GaAs MESFET, based on Ref. 2. (*Reprinted with permission of* Solid State Technology, *published by Technical Publishing, a company of Dun & Bradstreet.*)

Substrate The substrate is semi-insulating GaAs grown by the *liquid encapsulated Czochralski (LEC)* method using boron nitride crucibles. This method, in which the GaAs is synthesized under a protective layer of boron oxide, in the same crucible used for crystal growth, produces a crystal free of silicon contamination. Earlier methods, in which a quartz crucible was used in contact with the GaAs, resulted in significant silicon contamination and thus *n*-type doping. Semi-insulating behavior could be obtained only by compensating for the silicon with a heavy doping of chromium, which adversely affected electron mobility and, as a result, speed. The LEC-grown substrate is semi-insulating without compensation.[5]

Channel formation and isolation In the semi-insulating substrate, we create a conducting channel by masking the surface and performing a selective *channel implant* using silicon ions. In our example, a layer of silicon oxide is first deposited to protect the surface during implantation. The silicon serves as an *n*-type dopant and the implanted area becomes conductive. An alternative method is to implant the entire surface with silicon, and then implant the areas to be isolated with protons, which make those areas insulating again.[6] Physical isolation, in which the conductive layer is etched through to the surface, can also be used. Figure 12-3 shows these alternatives.

Source-drain implant Another masking step is used to define source and drain regions, which are given a deep and heavy n^+ implant. The implant requires an anneal, but GaAs cannot be exposed to high temperatures unless precautions are taken to prevent dissociation and out-diffusion of arsenic. Therefore, a cap layer of silicon oxide, silicon nitride, aluminum nitride, or alumina (Al_2O_3) is deposited on the surface to seal it. After anneal, the layer may be stripped, or left in place as a lift-off layer for metallization.

Ohmic metallization Metallization of GaAs devices usually employs gold or gold alloys, and is patterned by the lift-off technique. Lift-off allows precise resolution without the chemical and physical complications of etching, and is facilitated by the use of a deposited film like the cap layer. In our process, we have stripped the cap layer, and thus deposit a new passivating layer of silicon nitride. This layer is now masked, exposing only the source and drain areas, where metallization is wanted. Etching the silicon nitride using the resist mask produces some undercutting, so that the resist overhangs the underlying film. This overhanging profile is excellent for lift-off. Several layers of metallization are now deposited: first a layer of germanium-gold alloy (perhaps 0.1 μm), then a nickel layer as a barrier metal, and then a thick layer of gold. The germanium-gold alloy makes good ohmic contact with the GaAs. However, when excess gold is freely available to the substrate, diffusion of the gold and decomposition of the GaAs is observed at temperatures as low as 260°C (Ref. 7). Therefore, the nickel layer is inserted to prevent diffusion of the gold metallization to the substrate.

After metal deposition, the resist is stripped off, lifting off the metallization everywhere except in the contact windows. The overhang created by etching the nitride ensures that step coverage is poor, and the excess metal breaks cleanly away from the metal contacts.

Gate metallization The Schottky gate is now fabricated, again using a lift-off technique. When the resist pattern for the gate mask is in place, the silicon nitride and even the GaAs can be etched in the gate region, opening a trench into which the gate can be recessed. Recessing produces a thin channel under the gate, which improves threshold voltage control while allowing the rest of the channel to be thick to minimize parasitic

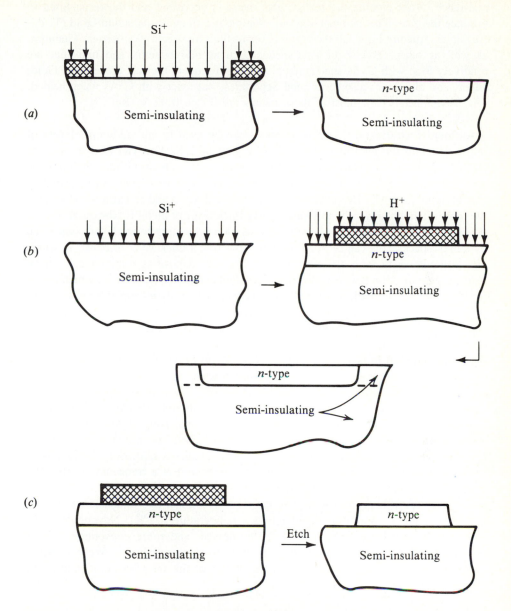

Figure 12-3 Alternatives for isolation of the GaAs MESFET. (*a*) Selective channel implantation. (*b*) Selective isolation by proton implantation. (*c*) Moat isolation by etching.

resistance. After etching the recess, gate metal is deposited and the resist lifted to produce the gate. The gate metal in our example is a titanium-palladium-gold (Ti-Pd-Au) layer structure in which the titanium forms the actual Schottky barrier. Titanium, as well as other choices for gate metals such as molybdenum and tungsten, are relatively inert with GaAs and stand up well to subsequent thermal processing. Gold, silver, and aluminum also form good Schottkies, but react with GaAs when heated. The barrier heights for most of these metals are 0.8 to 0.95 V (Ref. 8).

Additional metallization Various methods can be used to add additional layers of interconnect to the substrate. The Schottky gate metallization can also be used to fabricate one layer of interconnect. A passivation layer—silicon nitride, silicon oxide or perhaps polyimide—is used to protect the metal. A second layer can be added by conventional methods. However, many microwave devices add an extra metal layer in the form of an *air bridge*. In this case, a very thick layer of metal is fabricated so as to be suspended a few microns above the passivation protecting lower layers, separated by empty space. This can be done by first depositing a thick layer of resist, in which vias to the lower metal layers are patterned, and then depositing a thin layer of metal over it. Over this thin metal layer is then deposited another resist layer, in which the air bridge pattern is defined. Thick air bridge connectors are then formed by electroplating. The thin metal layer conducts the current required for plating, and metal is deposited wherever it is exposed to the plating solution by the pattern of the second resist layer. When the resist layers are removed and the thin "seed" layer etched away, the bridges are left in position. Figure 12-4 shows air bridge fabrication. This method can be used to fabricate inductor coils for microwave devices.[4]

The GaAs MESFET is an example of the use of III–V compound semiconductors (that is, semiconductors made by compounding elements from Periods III and V of the periodic table). This and similar devices are seeing increasing application, to reach speeds not attainable with silicon. For example, a MESFET using two layers of GaAs together with a layer of aluminum gallium arsenide, called a *selectively doped heterostructure transistor (SDHT),* has demonstrated operation at a frequency of 10 GHz. This device is somewhat less convenient than a conventional GaAs MESFET: it uses some quantum-mechanical trickery (in the form of a quasi-two-dimensional electron gas) and requires cooling to liquid nitrogen temperature (77°K). Nevertheless, the advances in speed are dramatic.[1] Both this device, and more conventional GaAs devices, are being integrated into circuits of increasing complexity. We can expect GaAs devices to be an increasingly important part of the semiconductor field in the future.

12-2 THE SEMICONDUCTOR LASER AND OTHER OPTOELECTRONIC DEVICES

Another important use of compound semiconductors is in *optoelectronic devices,* which interconvert electrical and light energy. As we discussed in Chapter 2, GaAs and similar materials are ideal for this purpose because the band gap is *direct;* that is, the

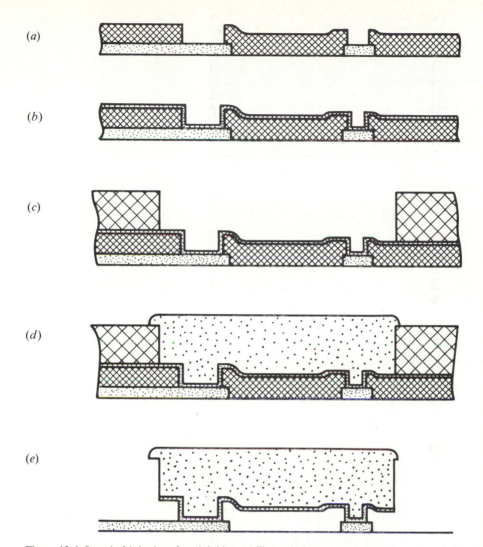

(a)

(b)

(c)

(d)

(e)

Figure 12-4 Steps in fabrication of an air bridge. (a) First resist layer deposited and vias patterned. (b) Thin metallic "seed" layer deposited. (c) Second resist layer deposited and air bridge pattern defined. (d) Plating of bridges. (e) Resist layers and seed layer removed.

conversion of hole-electron pairs to photons, and vice versa, occurs without a change in momentum. Hence there is no need for a third body to carry off momentum, greatly increasing the probability of the interconversion. As an example of an optoelectronic application, we will examine the fabrication of the semiconductor laser diode.

Laser operation ("lasing") can be visualized as occurring in three steps, as depicted in Fig. 12-5. We begin with the laser medium in a stable state, in which most of the electrons are at their lowest accessible energy. In the first stage of lasing, energy is put

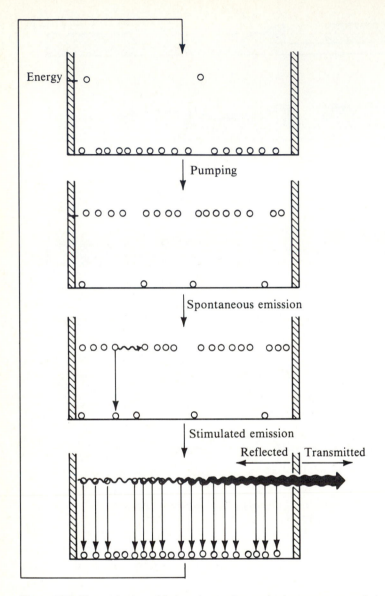

Figure 12-5 The mechanism of lasing. A pumping mechanism creates a population of excited electrons. Spontaneous emission then creates a few photons. Passing through the medium, these photons stimulate coherent emission by the rest of the population. This process can proceed sporadically or continuously.

into the medium to create a population of excited electrons, which can return to their stable states by emitting photons. This "pumping" step creates the potential for emission of light. The next step occurs when some electrons emit photons and return to their original step. This is *spontaneous emission,* and will produce low-intensity, incoherent light. In the third stage, the spontaneously emitted light interacts with other

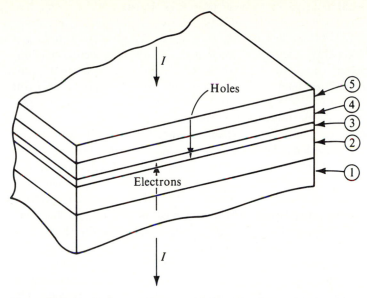

Figure 12-6 An aluminum gallium arsenide heterojunction semiconductor laser. Layer 1: *n*-type GaAs. Layer 2: *n*-type 36% AlGaAs. Layer 3: *p*-type 8% AlGaAs. Layer 4: *p*-type 36% AlGaAs. Layer 5: *p*-type GaAs. (*Reference 9. Reprinted with permission of* Solid State Technology, *published by Technical Publishing, a company of Dun & Bradstreet.*)

excited electrons, causing them to emit photons in synchrony through *stimulated emission*. If enough of the light can be contained in the laser cavity (for example, by mirrors at either end), most of the population can be stimulated to emit, creating the intense, coherent burst of light that characterizes the laser. A laser, then, requires a pumping mechanism by which an unstable population of electrons is created, and a means for containing some of the light in the laser cavity to stimulate sufficient emission.

In the semiconductor diode, a nonequilibrium electron population is created by injection of electrons and holes across a *pn* junction. (That is, conduction-band electrons are injected from the *n* side, while valence-band electrons are removed to the *p* side, creating a population of high-energy electrons and of low-energy available states.) Light confinement occurs by reflection at the crystal edges, and by refraction above and below the junction plane. The "mirrors" are the sharp cleaved edges of the crystal. Fabrication of the laser depends on manipulating both carrier concentration and refractive index by control of composition.

Figure 12-6 shows the layers of an *aluminum gallium arsenide heterojunction semiconductor laser.*[9] The term *heterojunction* refers to the fact that the various layers differ not only in doping, but in chemical composition. The variation in composition is achieved by the use of aluminum gallium arsenide (AlGaAs), an alloy of gallium arsenide in which some portion of the gallium atoms have been replaced with aluminum. Because aluminum can replace gallium in GaAs in any concentration without much change in the lattice parameter, we can grow epitaxial layers of AlGaAs on GaAs substrates, creating multiple layers of different properties. In our laser, there are five layers. The bottom layer is the substrate, of n-type GaAs. The second layer, 2 μm thick, is of n-type 36% AlGaAs; that is, $Al_{0.36}Ga_{0.64}As$. The next layer, 0.15 μm thick, is of p-type 8% GaAlAs. The pn junction thus lies between the second and third layers; dopant densities are chosen so that most of the recombination occurs in the third layer. The next layer is a 2-μm fourth layer of p-type 36% AlGaAs capped with 1 μm of p-type GaAs.

Each layer has a specific function. The first and fifth layers are electrical contacts, and the substrate, of course, contributes mechanical support as well. The second and fourth layers serve to confine both photons and charge carriers within the active third layer. Layers with higher aluminum content have a lower index of refraction, so that photons leaving the active layer are refracted back into it. They also have a wider band gap, discouraging charge carriers from "overshooting" the active region after injection.

The middle layer is the active layer in which laser emission occurs. The aluminum content of this layer determines the band gap and thus the wavelength of the light. By a proper choice of alloys, III–V compounds can span a significant range of frequencies. In addition to gallium, aluminum, and arsenic, the Group III element indium and the Group V elements phosphorus and tin can be used. Operating frequencies can be extended from the optical red, at 0.63 μm (Ref. 10) out to several microns. The wavelength emitted by 8% AlGaAs is 820 nm, in the infrared region of the spectrum.

Figure 12-7 illustrates the fabrication of the heterojunction laser. On a wafer of silicon-doped GaAs, the four layers are grown by liquid-phase epitaxy. The wafer is placed in a slider, which travels under a carbon boat containing wells for each layer to be grown. In the wells are placed the proper compositions of reactants to form each layer. The boat is then heated in a furnace to melt the reactants. The slider is brought in turn under each well, the temperature is lowered slightly, and the appropriate epitaxial layer is formed by crystallization from the melt. By this method are grown the tellurium-doped n-type AlGaAs layer, the germanium-doped p-type third and fourth layers, and the heavily germanium-doped contact layer.

To convert the layered wafer into a functioning laser, three additional things are needed. To localize the emission, parts of the surface need to be made inactive. This can be done by masking the surface over the active region, either photolithographically or mechanically, and implanting protons into the remaining area. As shown in the figure, this results in a channel-like active region. Second, electrical contact must be made to either side of the junction. The germanium doping of the surface layer facilitates good contact. Usually, a metallic sandwich similar to those used in GaAs MESFETs is deposited onto the wafer, and gold bonding pads are fabricated over them by plating. Finally, the wafer is carefully cleaved into chips. Since the edges of the chips are the mirror surfaces of the laser, this operation must be done with great care.

The semiconductor laser is one example of an optoelectronic device. There are a number of other devices that interconvert light and electrical energy, including the *light-emitting diode (LED)*, a nonlaser light-emitting device; *optical detectors* that convert light energy to an electrical signal; and solar cells designed to produce useful amounts of electricity from sunlight.

The light-emitting diode is, like the semiconductor laser, a layered structure in

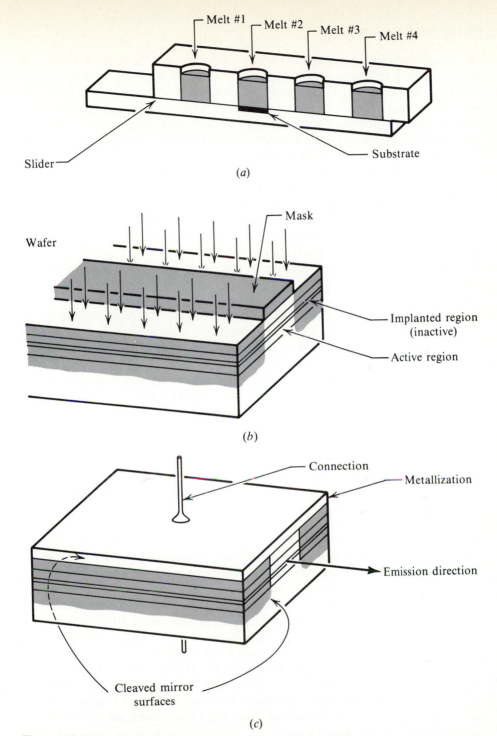

Slider

Melt #1   Melt #2   Melt #3   Melt #4

Substrate

(a)

Mask

Wafer

Implanted region
(inactive)

Active region

(b)

Connection

Metallization

Emission direction

Cleaved mirror
surfaces

(c)

Figure 12-7 Fabrication of a heterojunction laser. (*a*) Liquid-phase epitaxy to produce the heterojunction structure. (*b*) Proton implantation to restrict the active region. (*c*) Metallization, bonding pad fabrication, and cleaving.

which injection across a *pn* junction leads to light emission from an active region. A similar sequence of layers is used. Since the *p* contact is separated from the active region only by a thin epitaxial layer, most recombination will take place in an area right below the contact. Therefore, the active area for LED emission can be shaped by properly patterning the *p*-contact. LEDs can be made either edge-emitting, like the laser mentioned earlier or surface emitting, in which emission is perpendicular to the active plane. The emission frequency again depends on the semiconductor band gap, and a range of frequencies is attainable by the use of various III–V compounds. For light communication systems not requiring lasers, various indium-gallium-arsenic-phosphorus alloys (InGaAsP) are convenient because they emit at wavelengths around 1.3 μm, low enough for good transmission through silica optical fibers. GaAlAs emits around 0.8 μm and suffers significant attenuation. Where laser light is not necessary, the simplicity of the LED is an advantage.[11]

Light-emitting devices can function in optical communications systems only if combined with optical detectors. Such detectors are functionally the reverse of emitting diodes: light energy generates hole-electron pairs in the area of a reverse-biased *pn* junction, resulting in current flow and an electrical signal. Heterojunctions of III–V materials are frequently used. Composition of layers is again determined in part by wavelength. The active layer must be absorbing at the wavelength to be detected, but layers between the light source and the active layer must be transparent, requiring a wider band gap. With proper reverse bias, the initial number of generated charge carriers can be multiplied by avalanche impact ionization within the space-charge region, resulting in self-amplification in the detector. Additional layers may be used to increase amplification or to reduce noise, dark current, and other interferences.[12]

Although it is of macroscopic size and therefore not strictly microelectronic, the solar cell is clearly an optoelectronic device. Again, light creates electron-hole pairs that are swept apart by the reverse bias on a *pn* junction. Silicon remains the leading material for this application, and such microelectronic methods as ion implantation, thermal oxidation, metal evaporation and even photolithography (to define front-side contacts) are used in their fabrication.[13]

The discrete semiconductor laser and most other optoelectronic devices require the fabrication of heterostructures consisting of multiple epitaxial layers of substantially different chemical composition. The primary emphasis is thus on compositional control of materials, with less demand placed on photolithography and etching. In this way, optoelectronic fabrication resembles the early work with discrete transistors. However, since both optoelectronic and electronic devices can be fabricated on GaAs and similar materials, integrated optoelectronic circuits can be made. For example, if a laser is to be modulated, as in a communications system, it must be driven by a transistor driver, which can be integrated onto the same substrate. If a detector and amplifier are integrated with the laser, we have an optical repeater that will amplify and retransmit a signal. A number of such devices have been reported.[14] It seems logical to expect the integration process to proceed, resulting in more complicated devices with a need for more intricate patterning.

The manufacture of optoelectronic devices must be considered still in its infancy.

The potential, in optical communications and other applications, appear large, yet processing techniques remain unwieldy. For example, liquid-phase epitaxy is a sensitive process that does not lend itself to large quantities. In addition, thickness control is hard to maintain. Since gas-phase epitaxy is not appropriate to GaAs, there is considerable interest in molecular-beam epitaxy (MBE) or other epitaxial methods in this application.[10] The cleavage of laser crystals to create mirror surfaces is another delicate operation. As the market for semiconductor lasers becomes wider, we can expect substantial progress in these and other areas.

12-3 THE MAGNETIC BUBBLE MEMORY

Microelectronic methods are not limited to use with semiconductors. We now turn to a device that depends on the magnetic, rather than electrical, properties of the substrate: the magnetic bubble memory.

A memory device can be constructed in any medium that allows the storage of ordered sequences of "on" and "off" signals, and a number of media have been exploited for this purpose. In the magnetic bubble memory, information is stored in a magnetized garnet substrate by the presence or absence of a "bubble" of oppositely magnetized material. Circuitry is fabricated on the substrate to generate and detect bubbles, to store them in a well-defined sequence, and to move them about. These elements are created by microelectronic techniques.

The bubble memory occupies a niche between semiconductor memories and rotating memory devices such as disks. The semiconductor memory is very fast and reliable, but relatively expensive. It is randomly accessible, which means that any memory address can be inspected at any time. Disks, tapes, and similar storage media are very cheap per bit, but are bulky, slow, and are more or less sequential: information is recovered by first finding the beginning of a sequence and then reading to the desired address. The bubble memory has many advantages of semiconductor memory: it is faster and more compact than disk or tape. As compared with semiconductor memory, the bubble memory is simpler to fabricate, capable of high density, and is cheaper per bit. Bubble memories can be fabricated in multimegabit capacities, while the 1-Mbit semiconductor memory is only now (1985) approaching production. However, because bits travel along loops within the memory, the bubble memory is a sequentially accessible device, which results in long recovery times for the "last bit."

Figure 12-8 shows some of the concepts used in a magnetic bubble memory. Figure 12-8a shows the overall organization of the memory. Bubbles are stored in loops. In our example, a generator that creates bubbles is tied to the upper horizontal loop; the lower loop contains the detector, which reads bubbles. Bubbles are stored in the vertical loops. Transfer gates (not shown) at the points where the loops nearly touch allow the transfer of bubbles between the top, bottom, and storage loops.

Figure 12-8b shows in more detail how the bubbles are kept on the loops. Each loop consists of a serrated structure that has the property of attracting bubbles. Bubbles tend to settle into stable locations at the cusps of the structure. To move bubbles along

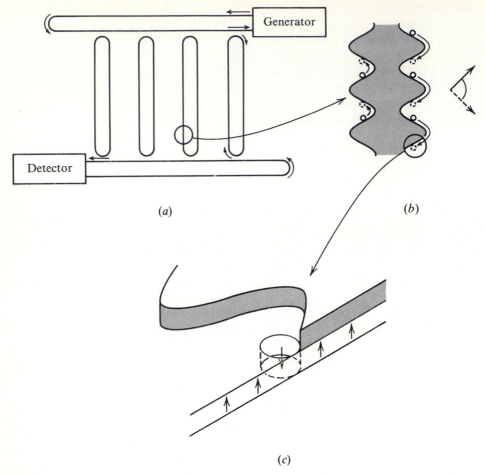

(a)

(b)

(c)

Figure 12-8 Organization of a magnetic bubble memory. (*a*) Overall organization. (*b*) Detail of bubble movement along track under rotating field. (*c*) Close-up view of bubble in garnet layer below the track structure.

the loops, a rotating magnetic field is applied to the device in a direction parallel to the substrate surface. The field sweeps the bubbles from cusp to cusp. The figure shows the bubble movement during a quarter-cycle of the rotation.

Figure 12-8*c* gives a closer look at the nature of the bubble, as it hugs the wall of the propagation structure. The bubble is a cylindrical magnetic domain, in the substrate material, with a polarization opposite that of the rest of the substrate. A constant magnetic field, perpendicular to the substrate surface, is applied to keep the bulk of the material polarized. The bubble is stable under a range of values of this field: excessive field will collapse the bubbles, and insufficient field allows them to expand and lose their shape. Bubbles are generated in the substrate by the field of a hairpin-shaped collector. It is important to note that the moving bubble is simply a localized magnetic condition. There is no chemical or physical change or movement in the substrate material. However, bubbles can be visually observed under polarized light.

In the figure, the propagation structure is shown as a separate layer deposited on top of the garnet substrate. In the original type of bubble memory, the propagation structure consisted of discrete chevrons of *Permalloy,* a iron-nickel alloy with very high magnetic permeability (flux-carrying ability) and low coercivity (resistance to changes in magnetization). These chevrons were assembled into herringbone structures roughly resembling the sawtooth pattern we have shown. More recently, selective ion implantation has been used to create propagation structures by creating a stress gradient in the substrate. In this case, the actual structure is *in* the substrate: the overlying layer would be the implantation mask.

Fabrication of a bubble memory requires at least three structures: the supporting magnetic film; the propagation structure; and conductor patterns comprising the generator, the detector or sensor, and transfer gates. The detector, because of its criticality, may require an additional separate operation, and insulating layers with electrical vias may also be required. Several alternate sequences can be used to build these structures.[15] Therefore, we will give a somewhat generic treatment of processing methods; it should be understood that the order of steps can vary.

Substrate The substrate is a wafer-shaped crystal of garnet, which is a compound of the general composition (rare earths)$_3$Fe$_5$O$_{12}$. An active magnetic layer is grown on the nonmagnetic bulk crystal by liquid phase epitaxy and is of complex composition, e.g., Y$_{1.4}$Lu$_{0.3}$Sm$_{0.3}$Ca$_{1.0}$Fe$_{4.0}$O$_{12}$. Such materials have a threefold symmetry of magnetization, which must be considered in planning the rectangular (fourfold) symmetry of the loop design. This material is capable of supporting a complex variety of bubbles, most of which are not well-behaved. Ion implantation of the substrate introduces additional stress and shifts the direction of easy magnetization, suppressing these unwanted "hard" bubbles. Helium, neon, or hydrogen ions are usually used.[16,17]

Propagation patterns After the magnetic film is prepared, a propagation pattern must be formed to contain and control the bubbles. This can be done either by fabricating Permalloy chevrons on the surface or by selectively implanting a pattern into the substrate. In the implantation method, the serrated pattern shown in Fig. 12-8b is created in a masking material, and the substrate implanted with an ion such as helium. This creates a discontinuous change both in stress and in the easy direction of magnetization, at the boundary of the masked area. The stress gradient attracts bubbles to the discontinuity, forming a "rail effect," which allows them to be guided along the loops. Furthermore, when the applied magnetic field interacts with the discontinuity, forces are created that sweep the bubbles along the loops. Once the implantation is complete, the masking material may be removed.

In the chevron method, propagation patterns are formed by depositing, masking, and etching or ion milling Permalloy. To function properly, the discrete chevron patterns need to be more complex than the implanted shape. This is a disadvantage, since the minimum chevron size is determined by the ability of the imaging process to resolve the smallest detail required. Implanted patterns have less fine detail and can thus be of smaller overall size, supporting smaller bubbles for a given photolithographic resolution.

Conductors Electrical conductors are used to manipulate the bubbles in the memory. A hairpin-shaped conductor can be used as a generator. Another conductor loop is used as part of the sensor, to stretch the bubble out for better detection. Since the sensor is magnetoresistive, additional conductors are required to supply a current to it. Interloop transfer also requires conductors, either patterned into gates, or to apply transfer current pulses to the loop patterns. A number of different gate configurations are available. Conductors are fabricated by conventional deposition, photolithography, and etching or ion milling. Usually an additional dielectric layer is required to insulate conductors from other structures, and vias must be etched in the insulator. Silicon dioxide is often chosen for the dielectric.[18] Polyimide has been used as a dielectric with Permalloy-chevron memories.[19]

When an ion-implanted propagation pattern is used, the propagation pattern and the conductor layer can be made self-aligning by using the conductor layer as the implant mask. In the propagation areas, the "rails" at the edge of each implanted pattern run along the edge of a conductor pattern. Transfer between loops can be effected by pulsing the conductors appropriately. Generation loops are also surrounded by implanted regions. While reducing the process sequence by one mask, this approach requires certain compromises in the device circuitry.[15]

Sensor The sensor is usually a strip of Permalloy to which a voltage is applied. As bubbles pass under the Permalloy, the flux in the sensor changes and the resistivity varies due to the magnetoresistive effect, leading to a change in the current transmitted. For best detection, the bubble is simultaneously stretched out by a conductor. The sensor and the stretching conductor are usually fabricated one atop the other, separated by the dielectric layer. The dielectric is perforated by vias to allow electrical connections to the sensor. The magnetic domain structure in the Permalloy, and thus the sensor efficiency, can be affected by the sharpness of the edges of the sensor pattern. To obtain maximum sharpness, the sensor is usually patterned by ion milling.

Figure 12-9 presents a possible fabrication sequence for a bubble memory. After ion implantation for hard bubble suppression, the propagation pattern is masked and the wafer implanted. We show a cross-section through a part of the loop; bubbles will propagate perpendicular to the plane of the paper. In our example, Permalloy is next deposited, masked, and ion milled to form the detector. Bubbles will be brought to the detector by another propagation pattern that is not shown in our cross-section.

Next, an insulated layer of SiO_2 is deposited, and vias are patterned through it to allow contact to the detector. Following this, the conductor material (often aluminum or gold) is deposited, masked, and etched or ion milled. We show the conductor connection to the sensor through the via, and another area of metallization overlying the propagation area. The conductor portion of the sensor, lying over the Permalloy strip and separated by the dielectric, is out of the plane of the cross-section shown.

After the bubble memory chip is completed, tested, and scribed, it must be assembled into a package that will provide both the fixed perpendicular and the rotating in-plane magnetic fields. Because the device can function only under the influence of

Substrate
Ion implant for hard
  bubble suppression

Propagation pattern masks
Ion implant

Strip resist
Permalloy deposition
Detector mask
Ion mill
Strip resist

Deposit SiO$_2$ via
  mask/etch

Conductor deposition
Conductor mask/etch

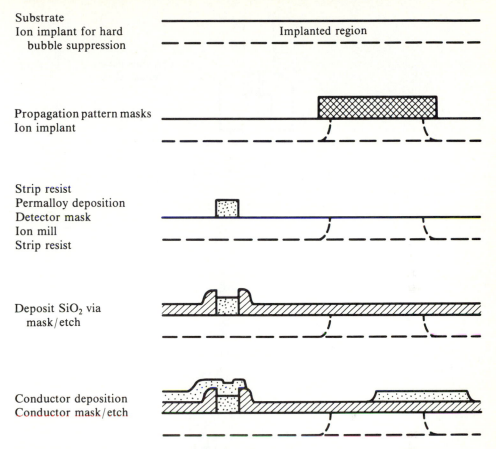

Figure 12-9 Fabrication of a magnetic bubble memory. The order of operations can differ.

these fields, the bubble memory package is an active part of the device rather than a passive container. Usually a permanent magnet supplies the fixed field, while sinusoidal signals are applied to two coils to produce the rotating field. The package also shields against stray external fields.[16]

The bubble memory, like the transistor, depends both on significant modification of substrate properties and on aggressive photolithography. The next device we will examine is also magnetic but differs in that the substrate plays no active part in device operation.

12-4 THIN-FILM RECORDING HEADS

The thin-film recording head is an application of microelectronic techniques to the miniaturization of the conventional ferrite recording head. Figure 12-10a shows how heads are used in disk drives to read information of a magnetic disk. The head consists

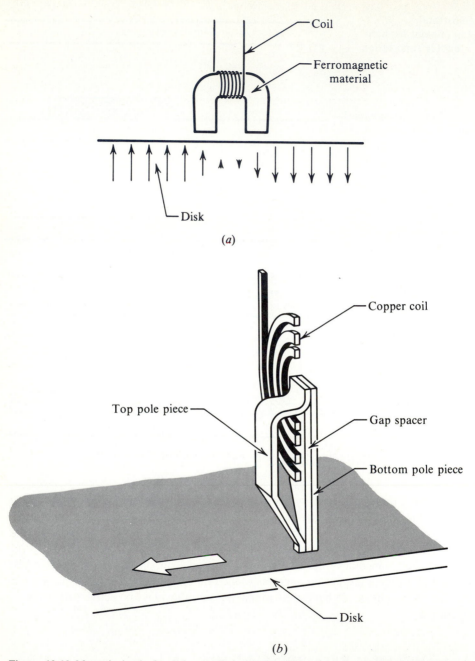

Figure 12-10 Magnetic heads for disk recording. (*a*) Operation of a conventional head: Change in magnetization of the moving disk results in a change in flux through the magnetic core and induces a current in the coil. (*b*) Thin-film realization of a recording head, showing pole pieces comprising the magnetic core and copper coils.

of a ferrite core surrounded by a conducting coil. The head is suspended over the rotating disk and mechanically positioned over the appropriate "track" of the disk. Information is stored on the disk as strips of magnetization of differing polarities. When the interface between polarities passes under the head, there is a change in magnetic flux through the ferrite, creating a current in the coil. The magnitude of the current depends on the rate of change of magnetic flux and the number of turns in the coil. The head can also be used to write information into the disk by passing a current through the coil to induce a magnetic field in the core. In this case, the field depends on the current and the number of turns.

Figure 12-10*b* shows the configuration adopted in a thin-film recording head. The "U" shape of the magnetic core is obtained by fabricating two pole pieces. The flat "bottom" pole piece rests on the substrate (not shown). The top pole is of a more complex shape. It joins the bottom pole at the "back" of the device to complete the magnetic circuit. At the "front" end, near the disk, the poles approach, but do not touch. They are maintained at a precise separation by a gap spacer. Conducting copper coils are fabricated to pass between the two pole pieces. Current is induced in the coils if the flux flowing through the pole pieces changes.

Critical parameters in head performance include the number of coil turns, the magnetic properties of the pole material, the pole tip and gap dimensions, and the shape of the pole pieces. Read current depends on the number of turns and the rate of change in flux, so many turns and a fast-moving disk both increase the signal. It is desirable to keep all of the induced flux within the pole pieces rather than letting it "short-circuit" some of the coil turns. This is encouraged by high permeability in the pole material, and by keeping top and bottom poles widely separated until they join at the "back" of the head. This requires an extremely short "throat"—the region near the disk where the poles are close and parallel. Short distances between the pole tips and the back closure are also desirable, but this constricts the space available for extra coils, requiring a trade-off. Designers give attention to the thickness of the poles along their length, their angle of approach, and prevent magnetic saturation in areas behind the pole tips.[20]

Additional trade-offs are required in determining the thickness and width of the poles and the gap spacer. The thinner the poles and gap, the smaller the "spot" on the disk that can be detected and the higher the resolution in reading. Similarly, the pole dimension perpendicular to the disk should be short in order to allow many tracks to fit on the disk. However, in writing, too small a pole cross-section may not allow generation of the field strength required to change the disk magnetization. Thus the pole and gap dimensions are a compromise between the need for high resolution in reading and sufficient field strength for writing.[21]

Figure 12-11 shows a processing sequence for the fabrication of a thin-film head. The substrate is a ceramic material, chosen for its mechanical rather than electrical or magnetic properties. Unlike substrates we have previously studied, this material serves the device only as a mechanical support. The substrate may be coated with an *undercoat* film prior to depositing the first pole material.

The poles are fabricated of a material with high magnetic permeability and very low coercivity, so they serve as efficient carriers of magnetic flux and carry no residual magnetization. Permalloy (a nickel-iron alloy) or some similar material is used. It may be deposited by plating, evaporation, or by sputtering. Materials like Permalloy have a preferred easy direction of magnetization and form a magnetic domain structure determined by this direction and the shape of the pole pattern. Domain structure effects the ease and speed of switching the poles. Therefore, deposition conditions are chosen to control not only the permeability and coercivity, but the easy direction of magnetization. This can be done by deposition in a suitably oriented magnetic field. Pole

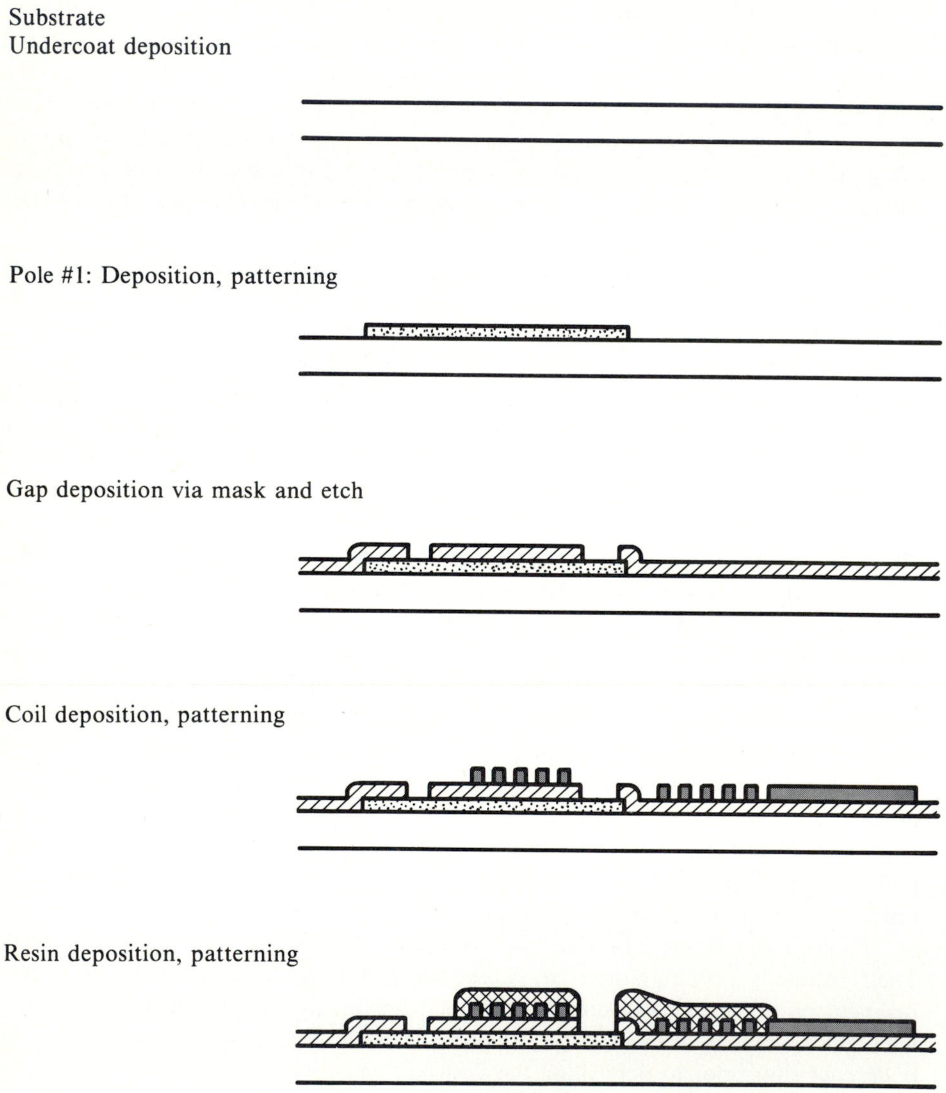

Substrate
Undercoat deposition

Pole #1: Deposition, patterning

Gap deposition via mask and etch

Coil deposition, patterning

Resin deposition, patterning

Figure 12-11 Process sequence for manufacture of a thin-film recording head.

Pole #2: Deposition, patterning

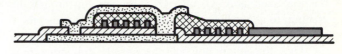

Overcoat deposition pad mask

Testing
Mechanical lapping

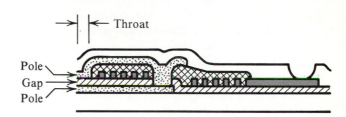

Slider machining

Final assembly

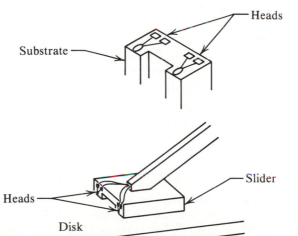

patterning is done by photolithography and etching. Ion milling is frequently chosen in preference to wet or plasma etching, because of the sharp edge profiles that can be obtained.

Over the bottom pole, the gap material is deposited. This is a durable insulating material such as Al_2O_3 or SiO_2. Vias are cut through this material by conventional etching, to allow closure of the pole and the back and the front. The back closure will remain in the finished device; the front closure serves another purpose.

Over the bottom pole and the gap, the coil structure is now fabricated. Copper is the usual choice for the coil structure: it may be evaporated, sputtered, or plated. It is photolithographically patterned into a coil, with bonding pads for attachment of leads. The coils must now be embedded in a medium that will support the top pole, separating it from the bottom pole. Planarization of the coil topography is also desirable, to avoid undulations in the top pole structure that can effect performance. Organic resins are ideal for this purpose. One choice has been photoresist, baked at elevated temperatures to increase its stability. Polyimide films can also be used.[22] The resin encapsulating the coil is deposited by spinning, and is patterned and cured. Photoresist has the advantage that it can be self-patterned; polyimide films require photolithography and etching. A via is formed in the resin to allow closure of the poles.

The top pole is now deposited and patterned by a technique identical to that used for the bottom pole. The shape of the top pole and the angle at which it descends at the "front" of the device depend on the contours of the resin layer. The poles meet at the front and back closures, so a loop of magnetic material surrounding the coil has now been formed. A passivation layer or *overcoat* is used to cover the poles, and a pad mask allows attachment of connections to the coils. This completes the microelectronic phase of thin-film head construction.

Machining techniques are now used to finish the head. The front closure of the poles makes a closed magnetic circuit that can be tested by passing a current through the coil. After testing, the substrate is diced into chips, and each chip is mechanically lapped to grind away the front closure. The length of the "throat," an important parameter, is determined by the length of the grinding process.

The substrate now assumes another function. To operate properly, the head must be suspended a few microns above the disk surface. Too great a separation between disk and head results in a signal that is undetectably low, but physical contact can lead to a "crash" with destruction of the disk drive. Therefore, the head is designed to "fly" over the disk. This is accomplished by machining the substrate chip (or *slider*) into an aerodynamic shape that will be suspended over the moving disk. The last panels of Fig. 12-11 show the machined shape of the substrate and the substrate in its flying position over the disk. Note that, in operation, the slider is rotated by 90° so that the head is positioned at the back of the slider, on the trailing edge. In final assembly, the slider is attached to a carriage that scans it radially over the disk.

The thin-film head is an example of a device requiring a combination of microelectronic and conventional fabrication techniques. As microelectronic methods become more widely understood, it is logical that they will be applied to the miniaturization of other devices previously made solely by mechanical means. In these applications, we can expect to find a similar mixture of methods.

PROBLEMS

12-1 As a successful engineering manager of a silicon MOS integrated circuit operation, you have been asked to oversee your company's first commercial operation for the manufacture of gallium arsenide MESFETs. Make a list of the fabrication steps, and for each step note the type of equipment you will require, and the controls you intend to have. Note any steps that you will control particularly tightly and any areas in which you need to seek more information before making decisions.

12-2 After finishing your doctorate in a laboratory concerned with VLSI fabrication methods for silicon, you have accepted a postdoctoral position in a group studying optical integrated circuits. You will be asked to produce optical ICs in which a semiconductor laser is integrated with various other circuit elements. List the processing operations you will want to be able to carry out, the general types of equipment required, the measurements you will want to take after each processing step, and any areas requiring further information.

12-3 A desire to escape the housing costs of Silicon Valley leads you to arrange a transfer from a silicon IC operation to your company's magnetic bubble memory plant. As engineering manager, which processing steps will you expect to be performing and what controls would you like to see for each process? Which steps will receive your particular attention? In what areas will you want additional information from your staff?

12-4 A start-up company producing thin-film heads has made you an offer of founder's stock that you cannot refuse. As an engineering professional with extensive silicon integrated circuit experience, what processing steps do you intend to carry out, what types of equipment must you buy, and what controls will you implement at each step? In what areas do you feel an immediate need to hire additional expertise?

REFERENCES

1. Raymond Dingle, *IEEE Trans. Electron Devices,* **ED-31**:1662 (1984).
2. Ajit G. Rode and J. Gordon Roper, *Solid State Technol.* **28**:209 (February 1985).
3. Rory L. Van Tuyl, Virender Kumar, Donald C. D'Avanzo, Thomas W. Taylor, Val E. Peterson, Derry P. Hornbuckle, Robert A. Fisher, and Donald B. Estreich, *IEEE Trans. Electron Devices* **ED-29**:1031 (1982).
4. T. Andrade, *Solid State Technol.* **28**:199 (February 1985).
5. Harvey Winston, *Solid State Technol.* **26**:145 (January 1983).
6. Donald C. D'Avanzo, *IEEE Trans. Electron Devices* **ED-29**:1051 (1982).
7. D. C. Miller, *J. Electrochem. Soc.* **127**:467 (1980).
8. Sorab K. Ghandi, *VLSI Fabrication Principles: Silicon and Gallium Arsenide,* Wiley, New York, 1983, pp. 467 ff.
9. William Sponsler, *Solid State Technol.* **24**:93 (March 1981).
10. Izuo Hayashi, *IEEE Trans. Electron Devices* **ED-31**:1630 (1984).
11. R. H. Saul, *IEEE Trans. Electron Devices* **ED-30**:285 (1983).
12. Gregory E. Stillman, Virginia M. Robbins, and Nadar Tabatabaie, *IEEE Trans. Electron Devices* **ED-31**:1643 (1984).
13. Mark B. Spitzer, Stephen P. Tobin, and Christopher J. Keavney, *IEEE Trans. Electron Devices* **ED-31**:546 (1984).
14. Amnon Yariv, *IEEE Trans. Electron Devices* **ED-31**:1656 (1984).
15. Yeong S. Oin and Ian L. Sander, *IEEE Spectrum,* **18**:30 (February 1981).
16. J. C. North, R. Wolfe, and T. J. Nelson, *J. Vac. Sci. Technol.* **15**:1675 (1978).
17. T. J. Nelson, R. Wolfe, S. L. Blank, P. I. Bonyhard, W. A. Johnson, B. J. Roman, and G. P. Vella-Coliero, *Bell Syst. Tech. J.* **59**:229 (1980).
18. T. J. Nelson, P. I. Bonyhard, J. E. Geusic, F. B. Hagedorn, W. A. Johnson, and W. D. P. Wagner, *IEEE Trans. Magn.* **MAG-17**:1134 (1981).

19. Hiroshi Umezaki, Hideki Nishida, Norikazu Tsumita, Naoki Koyama, Hisao Nozawa, and Yutaka Sugita, *IEEE Trans. Magn.* **MAG-18**:753 (1982).
20. G. V. Kelly and E. P. Valstyn, *IEEE Trans. Magn.* **MAG-16**:788 (1980).
21. H. Aoi, M. Saitoh, T. Tamura, M. Ohura, H. Tsuchiya, and M. Hayashi, *IEEE Trans. Magn.* **MAG-18**:1137 (1982).
22. M. Hanazono, S. Narishige, and K. Kawakami, *J. Appl. Phys.* **53**:2608 (1982).

AFTERWORD

In the course of this book, we have studied a number of microelectronic techniques. Starting with a review of semiconductor properties, we have presented the basics of diffusion, patterning, and thin-film deposition methods. In the last few chapters, we have seen how these methods can be applied to the fabrication of a variety of innovative devices.

Microelectronics is still a young field, and continuing change is to be expected. The drive for smaller dimensions and finer control seems likely to continue for at least a decade or two, and during this period we can expect dramatic changes in equipment, in applications, and in the relative importance of various techniques.

On the other hand, microelectronics is past its infancy, and the list of basic techniques is probably close to complete. Diffusion may become less important relative to implantation; photolithography may be performed with more exotic forms of radiation and more complicated equipment; gas-phase epitaxy of silicon may become unusual and liquid phase epitaxy of III–V compounds may become common. A few entirely new methods may be introduced. Nevertheless, we can expect that the microelectronic devices of the year 2000 will be fabricated using the principles and general methods presented in this book.

INDEX